Praise for Thomas Lockley

"With fast-paced, action-packed writing, Lockley and Girard offer a new and important biography and an incredibly moving study of medieval Japan and solid perspective on its unification. Highly recommended."

—*Library Journal* (**starred review**)

"This fact-checked portrait of an often-mythologized warrior with manga and anime variations is an exciting and illuminating tale of action and intrigue. Fans of manga and anime will enjoy reading this excellent work of Japanese history."

—*Booklist* (**starred review**)

"A readable, compassionate account of an extraordinary life."

—*Washington Post*

"Eminently readable…. The solid scholarship on and imaginative treatment of Yasuke's life make this both a worthwhile and entertaining work."

—*Publishers Weekly*

"A rich portrait of a brutal age."

—*Kirkus Reviews*

"The time has come for history to embrace the amazing story of Yasuke. In *African Samurai* words flex their muscles and pay tribute to a man of physical strength and combat skills. The writing is seductive and the reader sees the world through Yasuke's eyes. There is much to learn about the wonder of his life, and his story is a sharp blade cutting into invisibility."

—**E. Ethelbert Miller**

"This book is not only the best account in English of Yasuke, the famous African samurai. It's also a delightful introduction to the vibrant and multicultural world of Asian maritime history. Written novelistically, with a light scholarly touch… Exciting and informative!"

—**Tonio Andrade, author of *The Gunpowder Age***

"Rarely do I read a book that challenges my worldview of history, but *African Samurai* certainly alters my understanding of African and Japanese history. *African Samurai* gripped me from the opening sentence—a unique story of a unique man, and yet someone with whom we can all identify."

—**Jack Weatherford,**
author of *Genghis Khan and the Making of the Modern World*

"*African Samurai* sounds like a novel, a freaking amazing novel. But Yasuke is real, and Lockley and Girard bring him and his world to life with incredible research and style. Yasuke may have lived in the 1500s, but he is a hero for our modern world. Seriously...when is the movie?"

—**Bret Witter, #1 *New York Times* bestselling**
coauthor of *The Monuments Men*

AFRICAN SAMURAI

The True Story of Yasuke, a Legendary
Black Warrior in Feudal Japan

THOMAS LOCKLEY
and
GEOFFREY GIRARD

HANOVER
SQUARE
PRESS

**HANOVER
SQUARE
PRESS™**

Recycling programs
for this product may
not exist in your area.

ISBN-13: 978-1-335-04498-3

African Samurai

First published in 2019. This edition published in 2021.

Copyright © 2019 by Thomas Lockley

Published by arrangement with Tuttle-Mori Agency, Inc., Tokyo, in association with Foundry Literary + Media, New York.

This edition published by arrangement with Harlequin Books S.A.

Hanover Square Press
22 Adelaide St. West, 40th Floor
Toronto, Ontario M5H 4E3, Canada
HanoverSqPress.com
BookClubbish.com

Printed in U.S.A.

For my mother, Ruth,
who gave me a lifelong love of books,
and David, her husband and stalwart

—T.L.

TABLE OF CONTENTS

Prelude: *Yasuke de gozaru*   13

PART ONE: Warrior
C1: A Welcome to Japan   19
C2: Only the Grace of God   38
C3: The Ghosts of Africa   50
C4: Seminary Life   60
C5: The Terms of Employment   75
C6: The Witch of Bungo   89
C7: Pirates and Choir Boys   99
C8: A Riot on Monday   115
C9: *Tenka Fuba*   132
C10: Feats of Strength   147

PART TWO: Samurai
C11: Guest of Honor   159
C12: Treasures Old and New   172
C13: The Way of Warriors   194

C14: His Lord's Whim   204
C15: Oda at War   219
C16: The Dead are Rising   230
C17: Collecting Heads   242
C18: *Fuji-san*   260
C19: Battle Cry   272
C20: The Honnō-ji Incident   289

PART THREE: Legend
C21: Japan, Tomorrow   307
C22: The Guns of Okitanawate   319
C23: Possible Paths   331
C24: Yasuke Through the Ages   358

Afterword   381
Author Note   395
Notes   403
Index   457

An orphaned blossom
returning to its bough, somehow?
No, a solitary butterfly.

—*Arakida Moritake*

When a lion runs and looks back,
it's not that he is afraid.
Rather, he is trying to see
the distance he has covered.

—*African proverb*

PRELUDE
Yasuke de gozaru

June 21, 1582

Before daybreak, the Honnō-ji Temple already glowed brightly. Flames engulfed its roof and walls in climbing waves of gold and crimson. Scattered around the main temple, another half-dozen smaller structures crackled and sparked like festival bonfires as thick smoke spread over Kyoto.

Deep within the growing fires, Lord Nobunaga and his small entourage had clustered together to fight it out. They only delayed the inevitable. They were outnumbered a hundred to one, surrounded by multiple lines of gunmen and archers. Their only defenses burning. The gunfire had paused and the traitor Akechi now ordered the advance of his veteran samurai from all sides into the smoke, wielding swords and spears. The vengeful lord would not let the fire complete his retribution.

Yasuke emerged from the inferno to face them. He'd managed to escape out the side of the burning temple. Lost within the confusion of the flames and smoke, he now faced only three men in the tightening circle of hundreds. With the blaze raging behind him, he'd hoped to cut through them quickly. To somehow escape before another three, or *thirty*, blocked his escape.

The Japanese proverb "gossip about a man and his shadow will appear" was proving far too literal for the traitorous Akechi soldiers frozen before the foreign warrior. They knew Yasuke only from camp rumors. Nobunaga's "black man." The African samurai.

In person, they had never before seen a shadow so tall, a man so dark. Nobunaga's bodyguard stood above them like an adult over children, their helmets barely reaching his chest. And his half-concealed face was more than dark-skinned—it was freshly smeared with ash and blood from the battle to appear more terrifying. He also, perhaps most daunting of all, clutched a samurai's sword, its blade already lacquered in blood.

The three warriors had not expected this. They'd imagined only vanquished foes, a few mortally wounded survivors ready for the final blow, or perhaps a cowering maiden fleeing the blaze. This was no terrified servant girl.

Yasuke loomed over them focused, undaunted. Wrathful.

One of the soldiers glanced at the sword in his own trembling hand and his look revealed all: *it was not weapon enough to fell such a man.*

Yasuke smiled grimly. Fear was a much-needed ally this night. This would be the last mission for his lord. The cloth bundle cinched at his hip sat heavily against Yasuke's upper thigh as if urging him onward. Moments before, he had vowed to carry Nobunaga's mortal remains to his lord's heir, and he'd not journeyed halfway across the world to break such vows.

The three soldiers remained spellbound, unable to move. Even words failed them.

"Yasuke de gozaru," the African samurai challenged, stepping forward into attack position.

I am Yasuke.

PART ONE
Warrior

CHAPTER ONE
A Welcome to Japan

Yasuke arrived in late July. Even at sea, the air was warm and heavy and a steady, hot offshore wind drifted off the bordering coastline. He traveled on a *nao*, a top-of-the-line Portuguese merchant ship carrying lustrous Chinese silk, Chinese scrolls of learning, Chinese medicines and Portuguese guns.

Crates of the world's latest weaponry, and all the lead and saltpeter that went with it, belonged to the man Yasuke had been employed to protect: the Jesuit missionary from Europe, Alessandro Valignano. With the guns, Valignano and his fellow Jesuits also brought a fine stash of Christian artifacts and, they believed, literal salvation. But even they understood it was the guns Japan waited for. And a few crates of firearms to save millions of souls was, in the end, a fair bargain.

The year was 1579. Elsewhere, Sir Francis Drake had just

landed in California and claimed it for Queen Elizabeth. Ivan the Terrible, of Serbian-Turkic heritage, now firmly ruled Russia. French Catholics and Protestants were ending a brutal thirty-year religious war largely orchestrated by their former queen, Catherine de Medici of Italy. In Peru, the very last remnant of the Incan Empire had just been defeated by Spain. And, the bubonic plague, which began in China, was now killing thousands in Venice. The world had grown much smaller.

In this same spirit, Japan had appeared again off the Portuguese ship's port bow earlier in the morning. Stipples of rock and green surfacing from the restrained waves, uninhabited islets framed by the smaller specks of fishing boats. The mainland now ran low on the horizon, its scattered islands finally becoming a solid belt of green trapped beneath summer haze and infinite white clouds. It would be only another half day into port. The crew and passengers were comforted at finally reaching the end of their long journey, cheering and clapping backs in shared relief. Yasuke had every reason to celebrate with them.

He'd spent the last two years working as a bodyguard and attendant for Valignano, drifting ever eastward from Goa (in India, where he'd been first employed by the Jesuits) to Melaka (in modern-day Malaysia) and onto Macao (in southern China), Valignano settling affairs in the Jesuit outposts in those places before finally continuing to their farthest destination: Japan.

It was the most successful of all the Asian missions for the Jesuits and a source of considerable pride and promise for the Church in Rome. Here, Valignano planned to remain for several years and, for the foreseeable future, Yasuke would no longer have to journey by sea.

Travel in any century teaches patience and tolerance, and the African warrior had certainly exercised his share of both. From standing guard in stifling heat awaiting countless transports, slogging across muddy trails, safeguarding the Jesuit baggage—barrels and boxes filled with candles, devotional works of art,

The Portuguese Nao or "Black Ship" from Macao arrives in Nagasaki. Late sixteenth century, Kano Naizen.

relics, clothing, wine, food supplies and gold—along crowded docks, to suffering seasickness and worse as the ships heaved and rolled over rough waters.

Their current passage had proven little different than the other Portuguese ships he'd traveled on in the last two years: uncomfortable, filthy, dangerous and disease-ridden. Valignano often wrote home of his loathing of travel by water and it was easy to appreciate why. The ships were garbage dumps with sails, crew and passengers alike living in their own filth for weeks. Rats thrived and multiplied between decks, and only strengthened as passengers grew weaker during the later stretches of any long voyage; fresh cadavers or those too frail to resist became an extra source of ready food for the vermin. (In turn, rat meat was a main course several weeks into most journeys for the common mariner.) And, all too often, deadly diseases also signed up for the voyage. Syphilis, typhus, malaria, hepatitis, bubonic plague, smallpox, meningitis, rabies. Combined with the atrocious diet, horrendous work conditions, strict punishment and even conventional murder, life at sea always took its toll. Such voyages often killed half of those aboard. To truly learn to pray, one needed only to go to sea.

Fortunate survivors stumbled off the ships when reaching their final destination. If they were very lucky, the end point would have a missionary hospital; most of the beds in such hospitals were taken by recovering mariners. Otherwise, they recovered in flea-ridden flophouses. However unpleasant such lodgings were, even these were grand improvements over the ships. Thus, as they neared land again, hope and the inimitable accomplishment of survival once more filled the ship's crew and passengers.

In the face of such dangers and discomforts at sea, the voyage had not been all bad for Yasuke. He'd gotten on well with the sailors and would miss their company. It was a mixed crew: Portuguese, Indians, Chinese and even several other Africans. Portuguese was the default language of the maritime world and Yasuke, having spent two years traveling mostly by sea, now spoke enough to share in their gripes and jests. As a likable fellow with an easy smile, he would have garnered the camaraderie of all the working men from the first moment he'd stepped onboard with the Jesuits.

Yasuke had also eaten better than most aboard. The crew ate mainly hardtack biscuits (also called "molar breakers" or "worm castles"), dried meat or fish and drank mealy water; rations were given out *monthly* if not gone bad, and if the officers were honest enough not to have sold them for personal profit. Avoiding such dangers, the Jesuits had brought onboard their own supplies which Yasuke had defended, cooked, served and eaten. Livestock (and their feed) as well as fine food such as figs, honey, salt, olive oil, flour and even wine had all been available to the Jesuits *and* their entourage of attendants employed in India and China. This was not entirely done from charity; Valignano recognized a man of Yasuke's size and profession needed to eat well.

Standing at the rails, Yasuke glanced back at the man he now protected. Valignano conferred with his half-dozen sun-browned European colleagues and their Chinese and Indian fellow Jesuits. This mission was no ordinary group of proselytizing Catholics. Their leader, and Yasuke's direct charge and concern, was the most important Catholic in all of Asia.

While Alessandro Valignano's name means almost nothing today, in the late 1500s, it moved armies, assembled fleets and razed cities. He'd been given the official position "Visitor to the Indies" by Pope Gregory XIII and was then dispatched to inspect and develop the flourishing Catholic footholds in India, China and, finally, to the easternmost of the missions, Japan. An enigmatic place the Chinese referred to as the Land of the Rising Sun but largely ignored and avoided, a land of pirates and disorder. Valignano's first stop there was to be Nagasaki, a deep anchorage in a Jesuit-friendly province. From there, he and Rome hoped the rest of Japan awaited.

Yasuke grinned at the idea of Japan and looked back toward its land. The prospect of fresh food, fresh air, physical freedom and solid earth for the first time in three weeks was quite appealing. As was the prospect of a new place. Many enemies awaited here too, yes, but he was still curious to see what this mysterious country offered.

As they approached, however, their ship was met at sea by another smaller craft. Yasuke moved casually down the deck to stand closer to Valignano.

There were several anxious moments as the crew prepared for possible violence, knowing pirates from China or Japan occasionally were bold, and ravenous, enough to strike a Portuguese vessel. A wild flurry briefly overtook the ship: the decks were doused with water to prevent fire and then strewn with sand for footing; the many cooking fires aboard were snuffed, half a dozen small cannon were loaded and run out on the top deck; the powder and shot rammed; swords were distributed and several muskets primed. Throughout, Yasuke remained calm and silent, revealing no emotion, keeping any concerns to himself. Valignano, in turn, ignored the commotion, coolly watching the stretch of land running beside them. Having spent two years together in equally dangerous settings, the two men fed off each other's composure and tacitly agreed panic and fear aided no one.

As if to demonstrate this very point, the ship's concern proved for naught. Drawing closer, Yasuke and the others could now clearly see a Jesuit banner—a black pointed Order of Christ cross (evenly formed like a plus sign) atop the letters IHS (the Greek-to-Latin monogram for the name of Jesus Christ) upon a white background—run high on the approaching unarmed coastal craft and fluttering in the wind. The new ship rounded up into the wind and then lay wallowing, hove to. Moments after, a small Japanese-style sampan boat, which the Jesuit ship had towed behind, rowed jerkily across the dancing waters between the two ships.

Up the ladder to the deck came the "robust" Brother Ambrosius Fernandes, the Jesuit ship's captain. Brother Ambrosius looked and moved more like a battle-hardened soldier than any scroll-clutching religious official, and for good reason. He'd first come to Asia as a mercenary chasing the riches that came with such skills, but had offered his life to God in exchange for survival when a ship he'd been sailing on was caught in a tempest off Macao and several comrades were swept into the sea and lost. Fernandes saw the next dawn and joined the Jesuits shortly after returning to shore. After proper respect was paid to Valignano, Brother Ambrosius delivered a request from the local mission superior, asking the grand Visitor to change plans and make for another port: Kuchinotsu, instead of Nagasaki.

The request was part of a political maneuver by the Jesuits to punish a Japanese lord—Ōmura Sumitada, in whose territory Nagasaki was—for not being accommodating enough to their mission, by denying him the massive taxes, weaponry, revenues and honor he would have earned through hosting Valignano and the heavily laden Portuguese vessel. Instead, all that was now to anchor in the smaller port of Kuchinotsu, in a rival local domain. There, the mission superior, and a more "cooperative" Japanese lord—one desperate for the arms, and ready to agree to just about anything for them—would greet the Visitor and his team.

Yasuke and the other Jesuit attendants exchanged knowing looks as Valignano considered, then agreed. The switch meant another day of travel.

They were all getting another lesson in patience.

Yasuke was in his early twenties, no more than twenty-three.

He'd been a soldier for half his life, visited a dozen sultanates, kingdoms and empires. A young warrior who knew himself and more of the world than most ever do.

He was a very tall man, six-two or more—a giant for his time, comparable to meeting a seven-footer by today's standards. He was also muscular by the standards of *any* day, thanks to relentless military drill and a childhood, and lineage, built on a diet of abundant meat and dairy.

Arriving in Japan, Yasuke's eclectic attire revealed his familiarity with a much wider world. He was primarily dressed, quite smartly, in Portuguese clothing—baggy pantaloons to stop mosquito bites, a cotton shirt with a wide flat collar, a stylish doublet of dark velvet. But he carried a tall spear from India, its blade crafted into an unusual wavy shape with two "blood grooves" cut into the steel to make the blade lighter while still sturdy. He also carried a short curved Arab dagger at his side; both weapons shone like mirrors with constant attention from Yasuke's whetstone. His dark head was wrapped in a stark white turban-like cloth to protect it from the sun.

This was not, clearly from his garb alone, Yasuke's first arrival somewhere new, someplace utterly foreign. He was, rather, an experienced and well-traveled man in an ever-shrinking world.

He'd been on the move since he was a boy. From the swamps and plains of his birthplace on the banks of the Nile, to the mountains and deserts of northeast Africa, the fertile coasts of the Arabs, dusty Sind and the green of Gujarat. He'd likely fought alongside, and against, Hindus, Muslims, Africans, Turks, Persians and Europeans, and escaped death as a teenage soldier countless times before being employed by Valignano in Goa. The abducted child soldier was now simply the soldier. Well trained in weapons, strategy and security. Even, thanks to time spent beside leaders from several cultures, conversant in diplomacy. His experience and skills were of a caliber sought across the whole world, highly in demand among the rich and powerful.

This unexplored *Japão* (as the Portuguese called it) was merely the next place he was to be for some time as he put those same skills to work and did the job of protecting an employer and staying alive. The tide, he understood, must be taken when it comes.

The ship reached its new anchorage at midday.

Dark birds of prey and white seabirds swooped together overhead as the crew trimmed the sails. Yasuke and the Jesuits, free from any such responsibilities, lined the rails together to stare out at this nation of green forest and grey rock. The smells of land—rotting fish, moist dirt, sweating forests, drying seaweed and damp leaves—filled the stifling air for the first time in nearly a month. It was a fine substitute for the familiar sour stench of unwashed men. (Neither the multinational crew nor the Jesuits ever bathed; once a year if absolutely necessary was quite enough,

View of Kuchinotsu Harbor in the late nineteenth century.

for bathing—as *everyone* in civilized Europe knew—encouraged disease and licentiousness.)

The seaboard of Kuchinotsu was remarkably deceptive, presenting the appearance of a flat and narrow landscape. Mostly low craggy shores and thick forest. It was not until the Portuguese ship swung to port round the slight bend of land and finally turned into the narrow harbor that Yasuke saw the real Japan.

The bay opening before them was massive, and immense mountains loomed behind it like crouching gods. Beyond the peaks, initially appearing to be low-hanging clouds, were more towering ridges. The deeper the ship sailed into harbor, the more the Japanese backdrop expanded in the distance, slow slopes rising from the beaches and merging into those faraway mountains, seemingly stretching north without end. Anything could be awaiting them there. Anything at all.

The Portuguese ship was, in 1579, one of the largest in the

entire world. At around five hundred tons, she was built to hold vast cargoes and not for speed; looking neither sleek nor deadly, but showcasing unequalled bulk. The kind of craft people journeyed for miles to gaze at, that rulers dreamed of taking for their own. It was easily the biggest ship in the harbor by several times, the second biggest being a Chinese junk, of not much more than one hundred tons, loading sulfur. Now having dropped the longboats, the Portuguese ship was being towed toward land along glassy indigo water, the sailors straining at the oars. The hard work done, a hush fell over the crew remaining aboard and only the rigging's harping and the gentle swash beneath the stern could be heard. Dozens of Japanese fishing boats and smaller coastal merchant ships scattered aside for the approaching Portuguese vessel, then fell in line behind to follow in its wake like ducklings trailing after their mother. Except for the tired weather-greyed sail, the European craft was entirely black—a consequence of the dark and hard local wood used by the shipyards in India where it had been constructed. She, and her many sisters, would forever after be known in Japan as "The Black Ships."

The channel was shoaling fast and the Portuguese captain, Leonel de Brito, shouted and cursed into the strange new quiet to confirm the line of deepest water. De Brito came from an old and influential Portuguese family, and was making a fortune as the ship's captain, an appointment from the King of Portugal himself. He shot a livid look at Yasuke as he passed, blaming everyone in the Jesuit party for today's outrage. The captain was still furious with the change in plans as Nagasaki was a known and trusted port, while Kuchinotsu hadn't been visited for more than a decade, and the Chinese pilot employed to direct the ship was clearly unsure about the ideal spot for mooring. The pilot was suspicious of how shallow the harbor might be and continually had his assistant at work with the plumb line. Despite his blue Portuguese blood balking at taking orders, Captain de

Brito knew the best thing was to follow the Chinese pilot's advice, having learned to respect the skill of the Chinese mariners since arriving in the East. They finally dropped anchor amid fresh shouts of crew, pilot and captain alike as the ship swung round to the wind and tide and the longboats disengaged from towing duty and were prepared for disembarking.

Yasuke scanned the landing area. Valignano and the Jesuits were debarking in a trusted and safe port, yes, but their ally and host, Arima Harunobu, was surrounded by enemies. The Japanese lords to the north, under the leadership of Ryūzōji Takanobu, were at war with those like Arima who'd sought, and gained, an alliance with the Europeans by accepting their strange new religion.

A typical small port—declined some recently thanks to Nagasaki's emergence as the foremost anchorage of the eastern region—Kuchinotsu looked like a hundred other such refuges dotted along the seaboards of Japan. In mountainous and war-ridden islands like these, most goods and people traveled by coastal boats, not by land. The Kuchinotsu beach led up from the sea to the town and then to the greenery and mountains looming above. Fresh sea breezes nodded the tall grasses edging the port. The town comprised maybe sixty wooden buildings with thatch or tiled roofs; none were made of stone. Cove paths snaked between wooden houses, bamboo huts and larger wooden temples, decorated for the arrival. They'd anticipated Valignano's acceptance of the change of port.

A large welcome party was gathered ashore. A combination of missionaries, the local lord, his entourage, villagers, eager merchants who'd gotten wind of the change in destination, and the curious. As many as six or seven hundred, it appeared, amid large tents pitched for the occasion and adorned with flapping banners sporting the symbol of the Arima clan. (A blooming flower nested between five larger petals; a design markedly similar to the crest Yasuke himself would one day be sworn under.)

The Jesuit missionaries and their attendants fumbled slowly down the side of the ship into a single longboat, their long black robes hoisted and tied at the waist to enable better movement. Yasuke next passed down his spear and climbed in after the missionaries, each gentle swell lifting the boat against the ship as men fought to fend off the larger vessel. Valignano was the last to join the party, lowered down in a chair by two burly sailors, then taking a seat in the front of the boat facing shoreward ahead of Yasuke, who stood guard behind him.

In India, Yasuke had worked for men who favored their security *hidden*, defenders blending into the surroundings so the enemy never knew where the true danger and defense lay. Not Valignano. He wanted his security seen by all would-be thieves or assassins. It was clearly part of why he'd selected Yasuke from all the other potential guards; with his dark skin color and giant frame, potential enemies would see him coming and stay clear. Lions normally went for the easiest prey and often tracked their kill for days, carefully selecting the weakest member of the herd. With Yasuke always beside him, the most important Catholic in Asia would never present such weakness.

Thus, Yasuke stood balanced near the front of the rowboat directly behind Valignano, holding himself as gracefully as he could manage, wielding his tall spear. Behind him, two Jesuits struggled as the bracing sea wind threatened to tear their devotional banners from strong grasps. Japanese merchant and fishing boats bobbed all around them, their work delayed to scrutinize the newcomers. In contrast to the local boats in the bay (propelled by men standing and facing forward), the Portuguese longboats were being rowed "backward" with the rowers' backs to their destination.

Many of the Japanese fishermen and sailors in the surrounding boats sported simple wooden crucifixes and bowed their heads deeply as Valignano and the longboat passed. While a port town, few were accustomed to seeing such a group as this:

a half-dozen European men seated in long black robes despite the heat, several holding up tall cross-topped standards, the foreign symbols IHS prominent on the banners; and then the largest man among them, the giant Yasuke.

An odd sound drifted out to Yasuke and the others on the warm wind. As they drew nearer, it became music. A choir of the converted Japanese singing onshore. And closer still, a specific hymn revealed itself to the approaching longboat: *"Te Deum Laudamus,"* one of the Jesuits determined in surprise. The same hymn chanted so jubilantly when Joan of Arc and the French army liberated Orléans from English rule. The Latin words were now muffled by tongues unfamiliar with the language but distinct enough to bring, as hoped by the Jesuits waiting ashore, recognition and a brief satisfied remark from Valignano.

The first Jesuit missionaries had reached Japan almost thirty years before to the day, and more had continued arriving ever since. To build churches, hospitals and orphanages and to befriend as many local lords as possible as they worked to convert tens of thousands of Japanese to Christianity. The Kuchinotsu area had been Catholic territory for years, and dozens of the keenest faithful waded into the ocean up to their chests, still singing, to welcome their holy visitor. The Pope's, and hence God's, direct representative on earth had come to grace them with his presence and bring them closer to the divine. Thus, Valignano himself held demi-godlike status in their eyes.

The choir's voices grew quicker, both anxious and joyful, as other men took hold of the nearly beached longboat and, ignoring the confused oarsmen, hauled the boat up so the Jesuits could step ashore without getting wet.

Valignano gave blessings in Latin as the boat lurched to a stop, its back end still lifting gently on a swell as he stood. His arrival in Japan was the end of a six-year journey from Rome, via Portugal, Mozambique, India, Melaka and Macao. Throughout, there'd been stops and starts, successes and failures, and, above all, numerous prayers and blessings—in many languages—to worship, ward off evil and even preserve life.

Meanwhile, Yasuke again scanned the crowd of ragged peasants, stylish traders and weatherworn fishermen gathered around the tented pavilion. Looking for choke points, avenues of escape, prepared and ready for anything that might happen next. The approaching welcoming committee consisted mostly of village notables and merchants, who'd been awaiting the holy ship for days, everyone outfitted in their best garb. Senior warriors in light summer kimonos of cotton and silk, and topknots, with two swords thrust through their belts; and the Japanese, Chinese, European and Indian Jesuits sweating in their long dark robes.

The Japanese, he noted, *were* smaller than the Chinese he'd recently spent months with, and he'd many times heard Chinese refer to their island neighbors as dwarves. (The original character the Chinese used for "Japanese" was often translated as *short person* or *dwarf*.) Their average height was just under five feet.

Still, size notwithstanding, a line of formidable-looking warriors stood in differing brightly colored garb, their muskets, spears and other fearsome-looking pole weapons grounded. The first Jesuits to arrive in Japan had commented they'd never before seen "people who rely so much on their arms" and the few Japanese men Yasuke had encountered in Macao were almost always armed with two swords, one long and one short, and walked with a self-possessed and unique swagger. Now, he was on an island filled with millions of such armed men, where to be a warrior was an honor and an aspiration. He knew the highest-ranked of the armed warriors here were called *samurai*, elite and revered battle-hardened killers, a station or more above the common soldier. Another security concern they'd not had in China, where soldiers were looked down upon as ruffians and troublemakers and normally only combatants on a battlefield carried weapons. In Japan, warriors were the pinnacles of society and virtually all men bore some kind of weapon in daily life.

Armed or not, all here today had officially come to see this magnificent ship, get their hands on her trade cargo *and* to pay respects to Valignano. If a more important person had ever visited this rural backwater port, no one could remember who it had been. Beside his grand title, "Visitor to the Indies," Valignano had been born into power as an aristocrat and held himself like one. He had a worldwide repute of "magnificence." The Visitor was also a head taller than everyone, aside from Yasuke, and commanded easily with only stern looks that conveyed both brilliance and confidence. He was, in all ways, the kind of man people journeyed for miles to look at.

From a security standpoint, Valignano and this entire mission was a challenge. Japan was a country embroiled in civil war and division, hence the weapons, and the region they now entered was half controlled by enemies of the Church. Only the previous year, the Jesuits had backed the wrong side in a battle in a nearby domain and barely escaped with their lives. Valignano was a prime assassination target for an enemy power wanting an easy and potently symbolic victory.

While those territories further inland and north were largely unfocused on matters involving Catholics and their new local allies—the rival powers on the coasts of southern Japan knew perfectly well when and where the Pope's Visitor was arriving. Also, the geography offered few avenues of escape, and Valignano would meet many anti-Catholics during his lengthy stay. Still, despite the risks, he planned to remain for many years.

Lord Arima Harunobu—the boy ruler of Arima where Kuchinotsu was located—waited at the very front of the welcoming crowd. Several of the mission's Jesuits stood to his right, the Japanese lord's own attendants to the youth's left. Arima had just turned twelve, about the same age as Yasuke when he had likely first been enslaved and forced into slave-soldiery. If Arima were seven or a hundred though, the only concern now was whether the Japanese lord could successfully help protect the Jesuits.

Arima was—fortunately, to offset his youth—tall for his age, but slender with a thin mustache and a chiseled face that was feminine or, perhaps, even angelic. This youthful, ambitious, but lesser-ranked lord had barely brought enough guards and attendants from his castle seven miles away to make a proper showing along his own seashore that day. Even so, it was a massive risk. His castle was under siege and they'd had to leave quietly by night, their oars muffled as they glided quietly past the dozing besieging forces. A gamble worth taking to secure the goodwill of Valignano and the prizes attendant upon it. To see Lord Arima typically meant a one-day journey by boat around the coast, but for Valignano, the young warlord had been the one to make the trip. This arrival was perhaps the biggest event in his domain's history and the young lord's growing alliance with the Catholics was key to ensuring his survival and any greater aspirations for the future.

Yasuke vaulted over the side of the boat and the crowd of well-wishers who'd pulled the boat ashore melted back. The purpose of his fearsome spear, and the blade at his side, wedged into and held by his belt, was clear for all to see. The mercenary welcomed their stares and evident awe. Valignano was again getting his money's worth. Yasuke knew Valignano had employed him mostly for his size, a built-in intimidation factor beyond the mere fighting skills the African warrior had acquired in a lifetime of bloodshed.

Several of the faces gawked back at Yasuke in wonder, a look he'd seen before in China that went beyond admiration of his size and obvious martial skills. It was his skin color. Those who'd never seen anybody like him; those openly wondering if—from his color alone—he were truly human or some form of god or devil. Many locals had already made their judgment: from his skin color and pure white turban, they assumed erroneously Yasuke must surely come from the land of the black gods: *Tenjiku*. India. How wrong they were.

Yasuke made sure to make eye contact with possible threats on the beach, to promise them he was watching. Yet, after the bustle of previous landfalls in Melaka and Macao in southern China, this appeared to be nothing but a quiet fishing village. He breathed deeply. It felt good to be back on dry land so soon. This trip had been only three weeks long, but there was no denying the comfort that came from again standing on solid earth.

Lord Arima, accompanied by two European missionaries—Francisco Cabral, the local mission superior, and Father Fróis—advanced to help Valignano ashore.

Valignano remained in the prow of the boat as Lord Arima approached, and the Jesuit lifted his hands and gave another blessing to the awed crowd, interrupted only by the murmur of low surf and dragging shingle. Arima helped Valignano down, and the tableau of a petite twelve-year-old aiding the unusually tall missionary added further to the spectacle and strangeness of this historic meeting. All the actions were carefully staged to preserve each side's honor and show the highest degree of respect, friendship and fealty. Courtesy is the ornament of Japan, and this day it was on full display.

The singing had resumed, and the mission superior, Cabral, knelt in the sand to kiss Valignano's ring. Yasuke had repositioned himself so he was directly behind Valignano, the ocean to their backs, seeing everything from Valignano's viewpoint; possible holes in the local security, potential escape paths. Throughout the proceedings, Valignano spoke Portuguese and Father Fróis rendered his words into Japanese. Fróis was a Portuguese man who'd, in sixteen years in Japan, mastered the local language, despite enjoying little support from his superior, Cabral, who held that Europeans should not, and *could not*, learn the outlandish Japanese tongue.

Cabral proved to be as Yasuke had heard. A fifty-year-old Portuguese man and mission superior since 1570, he was a red-faced and notoriously short-tempered man; his brow and cheeks a little redder, perhaps, as Father Fróis managed the greetings between Valignano and the Japanese lord while Cabral stood mute. As the interpreting continued, Yasuke stifled a knowing smile; that Valignano had come to remove Cabral from his duties was news only to Cabral. He would be informed later when it became most opportune for the Visitor to do so.

The entire Jesuit party moved from the shoreline to the nearby tents for further arranged niceties. As they began walking, however, the gathered crowd of villagers shifted forward, wanting a closer look, perhaps to even touch a banner or one of the missionaries' cloaks.

Yasuke stepped closer to Valignano, eyeing the Japanese villagers for anything more than curiosity as Arima's soldiers advanced on the crowd, shouting in warning. Yasuke did not yet know Japanese and prepared for the worst, assuming treachery from the guards themselves, gently placing his hand close to Valignano's back to shift him one way or the other depending on what happened next. They pressed on to the waiting tents.

There, after another round of greetings and an exchange of gifts, the day's agenda and security matters were addressed. The immediate area was safe for now, and Arima would provide a skeleton guard for the priests along with a number of extra servants, reflecting the increased status of the mission personnel. More than that, however, the hosting Japanese lord could not offer. He was at constant war with the neighboring Ryūzōji clan (who were anti-Catholic and surrounded him now on all sides) and had a siege to attend to; but if at all possible, he assured them gravely in his high-pitched boy's voice, he would do more in the future.

The next stop followed a short walk to the mission building, where the entire Jesuit party would be served a modest banquet of European-style food. Arima would not join them this time, but would host Valignano in his castle at a later date when it was safe to do so. As soon as possible, he assured, as matters of great significance were to be discussed: plans and promises, both men believed, to last centuries.

A stirring of disquiet ran through the crowd while villagers at the fringes of the celebrations pressed closer again, struggling for a better look before the exotic foreigners moved on despite fresh warnings from the Arima guards.

The arrival of Valignano and the Jesuits was the most important thing to happen to the region in at least a century. But, as the Jesuit party prepared to leave the beach, it became clear—whispers in the crowd, pointing, open gawking even—that the gathering's attention had refocused.

For as tall and powerful and legendary as Valignano was, this *other* man—the priest's bodyguard—was taller than the Jesuit by a head or more, and, if that were not enough, physically built like something out of myth. (These common people had been raised on fish, roots, seaweed and rough grains.) But, it was not only the warrior's size that captured their attention. Or his unusually dark skin, for most around this port town *had* seen African and Indian sailors before. It was something else. Something neither they, nor he, could yet explain.

Some had come to see the colossal foreign ship.

Most to see the almightiest Catholic in all of Asia.

But everyone was looking at Yasuke.

CHAPTER TWO
Only the Grace of God

For Valignano, as Yasuke had learned during the last two years, divine grace was never slow. They began working their very first day in Japan. The Jesuit was a champion of the notion that God makes, but only man shapes. Having barely eaten, and still drained from the voyage, Valignano had set to the business of saving Japanese souls.

Following the warm audience with Arima, they'd proceeded quickly to the mission building, a small converted temple set back from the road and enclosed by a wide grey ceramic tile–topped wall. The temple was wooden, blackened with age, and roofed with slim wood shingles. Plum trees grew in the adjoining courtyard, where a ceramic incense burner sat empty. All other Buddhist decor had been removed years before. Now, a large wooden cross formed the backdrop to the main hall. Before it stood the altar on which rested a lone, empty candelabra. (The church had run out of European wax candles months before.) Curtained alcoves lined the wall behind, each containing a devotional picture to be revealed during holy days, special festivals, and as a reward for those ready to progress to the higher mysteries.

Mission business took place in a simple chamber adjoining this place of worship. There, Yasuke collected his land legs and stood guard several feet away from Valignano while the Jesuits ate and talked and Japanese servants bustled about them silently like late-afternoon shadows.

The Jesuits enjoyed a modest feast made from locally assembled ingredients: wild fowl, cooked fish, vegetables, some rough bread. They paired it with one imported essential from further west—European wine. Ordinarily, such spirits were conserved for communion or given as gifts to local dignitaries and the Jesuits would drink the local rice wine, sake. But, today was a special occasion: the arrival of the Pope's personal representative at the most distant outpost of Catholic Christianity.

The "Visitor to the Indies," Alessandro Valignano, was a nobleman born in the city of Chieti, in the Kingdom of Naples. (A Spanish territory at the time, but wholly "Italian" still in taste and temperament.) Having received his doctorate in law from the University of Padua at only eighteen, he'd then entered the Church, rising quickly until a personal scandal—he slashed a woman's face with a sword for reasons which were hushed up—cost him more than a year in prison. His time in jail apparently gave him some opportunity for reflection. Shortly after his release, in March 1566, Valignano joined the Jesuits, a new and rather militant—as many of the brethren were former soldiers, and not shy about using force when necessary—Catholic missionary order.

The Jesuit's founder, Saint Ignatius of Loyola—who'd died

Seventeenth century engraving of Alessandro Valignano.

twenty years before Valignano's voyage to Japan—had been a Spanish soldier-turned-mystic and perhaps the only saint with a notarized police record: for nighttime brawling with intent to cause bodily harm. (Like Valignano, this came *before* his conversion to missionary Catholicism.) Over the course of the next two centuries, Ignatius's new order swept across the world, spreading Catholic Christianity in whatever way necessary—on occasion even attempting to create entirely new state-like entities or colonies under its rule, especially in Brazil and Paraguay where they often organized the native people against European colonists and slave dealers. But first came the gifts, schools, charity work and the peaceful teachings and assurances of Christ's deliverance. One Jesuit forefather described their typical game plan: "We came in like lambs and will rule like wolves."

Consequently, with the conversions, over time the Jesuits were also *expelled* from at least eighty countries and cities, for engaging in political intrigue, infiltration and subversive plots of insurrection against host governments. Two centuries after Valignano and Yasuke, Napoleon would conclude, "The Jesuits are a military organization, not a religious order."

Valignano's criminal record evidently did not hinder his career with the Jesuits, or perhaps his brilliant mind and energetic Catholicism redeemed him in the eyes of his new superiors. He had a flair for dramatic spectacle (as evidenced by the hiring of Yasuke), and did not mind sticking his neck out or taking a chance to get big ideas done. He'd ruffled feathers in India and Macao by insisting European missionaries learn local languages, and—ignoring derision and protests from his Jesuit brethren— laid the foundations for the first Western attempts to academically study Chinese languages and culture.

By the age of thirty, in 1570, he was ordained. And, a mere three years later, after a stint as the rector of the College of Macerata, Valignano received one of the highest appointments a Jesuit could receive: "Visitor to the Indies."

The "Indies" meant all the territory between East Africa and Japan; one third of the Earth. As Visitor, Valignano now oversaw all Jesuit finances, business, trade, church law, mission policy and diplomacy in Asia. Wherever Valignano sat was the head of the table. He had the authority to admit and dismiss Jesuits, to appoint and discharge local superiors (the priest charged with leading a local mission), and to send any member of the order wherever he pleased. In theory, only the Jesuit Superior General or the Pope, both in Rome, could overrule his decisions. And, since any communication with such authority figures took several *years* to be delivered and answered, it meant he could do as he pleased. A "hindrance" that satisfied both sides, allowing Valignano to get things done with full deniability still possible for the Vatican.

Valignano's faith was deep, and his self-confidence in his own abilities to spread the influence of Rome in Asia was unshakable. As an aristocratic legal scholar with a deep understanding of theology, he ran his operations with authority and decisiveness. He had absolutely no doubt about the rightness of what he was doing, entrusting that nothing—no matter how unsavory during the process—ever ends ill which began in God's name. He was exactly the kind of man the Jesuits sought.

Still, his judgments about other people could be scathing, and his attitude toward the "lower" classes was dismissive at best. (One of the most prolific writers of his age—thousands of letters, books and logs—Valignano does not mention Yasuke, or any other attendant, once in his writings.) The commoners he introduced to Christ were not really his concern beyond being soul fodder, quantitative proof he was succeeding in his job. He was a patrician and expected to move among similarly exalted people wherever he was in the world. If the rulers and upper classes could be persuaded of the correctness of the Catholic God, then any other riffraff would soon follow their example. He went about this proactively and was often willing to turn a blind eye to, or even support, activities such as gunrunning, slave trading and other questionable practices, if it would ultimately help him save more souls.

Having inspected the Jesuit missions in Mozambique, India, Melaka and Macao in China, this single-minded, class-conscious, innovative, highly ambitious and genuinely devout man had finally brought his vocation to Japan.

During their celebratory meal, the Mission Superior Cabral's briefing was mixed. Reports on individual Jesuits maintained that most were doing their jobs well. Numbers of converts had reached one hundred thousand, often through mass baptisms of regional lords and thousands of their vassals, rather than individuals personally accepting Jesus as their savior. Thus, a vast majority of new "converts" had no true idea what they'd been sworn into or had even a basic understanding of Christian doctrine or faith. Often, unfortunately, even the priests and brothers themselves could not read Latin well, and shared scripture and ceremony from memory only.

There was also disquiet among the Japanese lay brethren who were pushing to be admitted as full members of the Jesuits rather than having their low rank decided by race alone. Cabral remained wholly opposed to the idea. It was bad enough, he argued, to have mixed-heritage Indian/Portuguese men, and "new Christians" (former Jews or Muslims) from Europe, without now allowing Japanese locals to get "above their station" too; a racist outlook quite common in this age.

The state of the missionary infrastructure, however, was improving, as reasonably safe bases had been established in Nagasaki and several other secure locations under Japanese patronage. There were, though, no defenses and transport was largely contracted out to locals. Thus, the Jesuits were almost completely at the mercy of the local Japanese lords who supported them.

While the Jesuits' discussion provided insights into the overall political landscape and helped Yasuke better understand the security situation, he had other, more immediate, concerns. He did not know the layout of the entire building yet, or the remaining areas of the port town. He preferred to have time to check doors, to learn the ins and outs of all the surrounding buildings, to inspect the kitchen and utensils. Worries of poison and secret passages that could be exploited by would-be assassins were very real. What perplexed him most was the sheer flimsiness of the buildings. They *looked* beautiful, but the wood was wafer thin and the doors and shutters were made of white paper on a simple wooden frame. Paper! Ideal to let light into the rooms, yet not translucent and far too easy for an enemy to simply walk through. For a man trained to identify and resolve weak points of entry, this was ludicrous. Further, Yasuke did not yet know the language, setting or supporting cast. All he could bring to the table until he did was his size, skill and vigilance. He hoped it would be enough.

Yasuke's concerns about the buildings around them were quickly eclipsed by the next turn of Valignano's conversation: plans to again see the boy lord, Arima—but this time in Arima's own castle. A stronghold currently under siege by his deadly enemy, Lord Ryūzōji, who specifically despised the Catholics. It was folly to even discuss. Yet, discuss it Valignano did.

If Arima had sneaked though the siege to welcome his distant visitor and get home again, surely Valignano could, likewise, reach Arima. A couple of local guides and soldiers would spirit them inland, over the river and through the mountain trails which formed the backdrop to the Japanese lord's castle. What choice did they have? After all, the Jesuits could not convert the rest of Japan while hiding in some remote fishing village. They needed to press ahead regardless of the danger and, God willing, live to proselytize the Catholic faith. Death was a constant companion in this age, whether at the hands of an assassin's blade, disease or from everyday threats such as an infected cut. It was not an age for the timid.

Days later, Yasuke was on another damned boat. They journeyed up the coast in the middle of the night, navigating toward Arima and the young lord's Hinoe Castle. After their landing, they'd head into territory surrounded on all sides by troops who were expressly, even fanatically, anti-Catholic. If captured, they would face torture and probably death. It was only their first week in Japan.

They traveled the seven miles in a tiny sampan boat, no more than a fishing craft, timing their arrival for the predawn hours hoping to avoid enemy discovery ashore or being intercepted at sea. The oarsman stood at the rear of the boat, handling the one huge oar. Valignano and a Japanese acolyte who'd accompa-

A sampan boat similar to the type used for coastal travel. Photographed and touched up with paint in the nineteenth century.

nied him to interpret huddled under a primitive shelter. Yasuke stood guard. While he'd do *his* best to make sure everything went smoothly, too much of the day was out of his hands. All he could do now was safeguard against a pitch-black, barely visible shoreline. While he watched, whispered conversation slipped from beneath the small boat's awning. Valignano was taking the opportunity alone with a Portuguese-speaking Japanese person who he could communicate easily with, trying to gauge the mood and attitude of the local man who so dearly wanted to enter into the Catholic order.

After the short boat ride, they landed on a pebbly beach miles from any port. There, several of Arima's soldiers waited and then sneaked Valignano, with Yasuke and the interpreter, silently up into the damp, fragrant, semitropical forests of southern Japan. Along steep back trails into the mountains, then down to a river which they forded, and then up a mountain again on the other side. The darkness was filled with the universal sounds of unfamiliar birds, scratching rodents and unseen creatures of good size moving deep in the nighttime. Whether they were wolves, monkeys, bears or boar, Yasuke could not tell, but he much preferred such sounds to the slosh of water and creaking wood. He was on his guard. Enemy soldiers probably scouted the surrounding woods, and though he was prepared to fight to the death, he knew that only the abilities of Arima's warriors— men they did not yet know or entirely trust—would save them all from capture and a grisly end. Or would they fade into the forest at the slightest sign of danger?

Following the narrow hidden mountain tracks, they finally reached the rear gate of Arima's castle just before dawn as planned, the locals delivering without a hitch. It was a promising, and comforting, start.

Hinoe Castle was, in reality, little more than a minor fort, but at least it was sturdy and defendable. The walls were stone, the gates of thick hard wood opened onto a path leading down the mountain to the river and, beyond, the sea. The stronghold perched atop steep slopes and cliffs, almost inaccessible except from the mountainous rear approach, which was passable for a few dedicated souls such as Yasuke and Valignano. No army would ever venture that way. Arima had recently erected a large wooden cross in the enclosed courtyard and Valignano uttered a brief comment of approval. As the sun rose above the seascape spread out beneath them, Yasuke took in the striking view. The waters they'd traversed in the dead of night were revealed as speckled with verdant sand-framed islands, the sea a deep vivid blue. On the horizon was a huge hazy, smoking volcano, Mount Aso.

They were welcomed less jubilantly than on their first day in Japan, for Hinoe Castle was still under seige. An enemy army camped at the bottom of the hill and it was difficult to bring in supplies; access was available only through the mountain passes. Thus, Arima's staff was suitably fawning, but clearly reduced in size and quite ragged from the time spent under the partial blockade. His soldiers were equally weary and strained. But, Yasuke noted, war hardened. Men who would—despite their current state—still fight well, and to the death if necessary.

Yasuke was already getting a good sense of the existing Japanese hierarchy. Farmers, fishermen, artisans, merchants and priests were protected by citizen-soldiers and lorded over by professional warriors—of various degrees of influence—*samurai*. Lord, *tono*, was a loosely assigned title of respect (like calling someone *sir* today) and normally reserved for men of landed wealth or military authority, including some samurai and nobles in the imperial court at Kyoto. Lord Arima was a moderately powerful lord, but with boyish dreams and aspirations of becoming a major force throughout the region.

Valignano and his small team were fed. Although highly irregular for the Jesuit, Valignano also asked if he might rest for a few hours before beginning serious discussions. After the hardships of their journey through the night, even Valignano needed respite, and he was given a room with a supple, sweet-smelling floor made entirely of tightly woven rice straw—tatami.

At midday, Valignano rose and they were fed again while silent, but attentive, servants brought them the kind of food which Arima and his steward presumed Valignano *wanted* to eat: meat (a rarity in Japan) with rice and some broth made from white-fish and vegetables. The meat, alas, had clearly been prepared by someone who'd little experience cooking meat. It was tough and very overdone. Mountain boar, most likely. Of course, in a castle under continual siege, they were lucky to get anything at all.

Following their second meal, Valignano was ready again for business and Yasuke stood guard while the interpreters struggled through the complicated and delicate exchange between the Catholic Visitor and the Japanese lord.

Young Lord Arima was in dire military straits—lacking munitions, under siege and ironically facing disunity in his domain due to his earlier persecution *of* Christians. (He'd previously been fiercely anti-Catholic; leading a youthful rebellion against his now-deceased Catholic father and forcing the apostasy of fifteen thousand of his father's subjects.) Now, several years later, as he was threatened each day with being overrun by the Ryūzōji clan, Arima welcomed Valignano—and God—with an open heart and empty arsenal.

Like so many others, Arima had been won over by the guns and trade Jesuits always seemed to carry alongside their hymnals and rosaries. The lone Portuguese black ship a few miles away in his harbor transported a small fortune in lead, saltpeter and guns—more than enough to break the siege and give the Arima clan some breathing space. Valignano made assurances that more munitions were on the way.

One particularly juicy piece of ordnance Valignano suggested for the future was a pair of cannon which could be specially made to order in the Portuguese foundries of Goa in India. These, Arima was assured, would seal a final victory against Ryūzōji and ensure dominance of the region for decades; no one else in Japan would be able to match their power. It went without saying, however, that such weapons would only be available to those Japanese lords best allied with the Church's ministry.

Arima requested the holy rite of baptism.

And, to make the depth of his conversion perfectly clear to Valignano, the young lord also offered to host Japan's first Jesuit seminary and to destroy all Buddhist temples and shrines within the territory which he still controlled. "Just as well." Arima grinned. "I need materials to improve this castle. Look at it! It is nothing but a hovel. Imagine what it would become with the help of your God. I will make it a shrine to him and use the destroyed heathens' places of worship in its construction." (Archeologists have recently determined the steps leading up to Arima's remodeled castle were, in fact, made from the tombstones of Buddhist graveyards.)

Before any of that, there was one more matter to attend to: the boy lord's not-so-secret mistress. Common knowledge to all, the girl was hidden away for the visit in the ladies' quarters. Valignano made his expectations perfectly clear. There would be no baptism, seminary *or* cannon unless the matter of this woman was settled. Monogamy and Catholicism were inseparable and Japanese customs of insouciant licentiousness and infidelity would not be tolerated by Mother Church.

Arima, all denials of any such wrongdoing spent, was heartbroken but understood this foreign custom was one that he would have to pay more than lip service to, and consented. He would marry the Jesuit's preferred wife in a Christian ceremony—they had a nearby lord's daughter in mind—which would, hopefully in the longer run, further his aims and forge a stronger Catholic alliance.

Yasuke hid his amusement at Arima's telling combination of desperation and Catholic exuberance. His charge, Valignano, was certainly off to a good start. Accordingly, the Jesuit Visitor was in fine spirits on their harried return several days later back through the woods and night to the beach to meet their boat again. Only the waning moon trailed their escape, and they reached the Kuchinotsu port again without incident.

Yasuke's confidence in the local guard increased another notch.

Valignano ordered the promised cannon later that same week, the letter to sail away from Japan on the same ship which had brought them.

Not a bad start for their first week in Japan.

CHAPTER THREE
The Ghosts of Africa

Thirty miles north, near Nagasaki, another minor warlord sat in his castle being taught a lesson.

Lord Ōmura Sumitada had been the first high-level Catholic convert in all of Japan and was, by all accounts, sincere in his Christian faith; or, at least very *very* good at pretending to be. For years, he'd accepted the Jesuits' power—and large numbers of imported guns and munitions—and used both to help defend himself against rival family members (including the young lord, Arima) and far more powerful regional enemies like the Ryūzōji. As with Arima, Jesuit military supplies, and a share of the Portuguese/Chinese trade profits, were Ōmura's hope and lifeline. Valignano and his Portuguese ship were a true godsend in every possible way.

But the ship, and the most powerful European in Asia, were

now moored a day's travel away with that child, Arima. And how, exactly, had Ōmura wronged the Jesuits that summer? His daughter had refused an arranged marriage—the one formulated and urged by the Jesuits—*to* Arima, and Valignano's potential alliance of "Christian lords," schemed, theoretically at least, years before he even reached Japan, crumbled as Ōmura stood by his daughter's decision.

The setback was, to Ōmura anyway, sadly predictable. His whole life had been one long stream of small triumphs countered by larger new obstacles. Only his alliance with the Jesuits had safeguarded his survival and enough breathing room to make it through another year, but now even *that* was threatened to be snatched from him by his young rival. (Who, in this world of confusing alliances and warlord intermarriage, also happened to be his nephew. Ōmura's daughter, if the Jesuits had their way, would be marrying her cousin.) Unless he was to physically force his daughter, something he was unwilling to do, Ōmura would need to up the stakes to make amends with the Jesuits. He was *already* baptized, years before, along with every single one of his sixty thousand subjects. Something somehow bigger had to be done. Ōmura was angry and panicked, but in a proactive and innovative state of mind, so he boarded a ship for Kuchinotsu to pay his respects to Valignano in person. If his daughter's hand could not be offered to the Church, he had to proffer the next best thing: Nagasaki.

All of it. The whole port city, and the land around it besides. One of the true jewels in his territory's crown. Until 1571, Nagasaki had been nothing but a few fishing huts, occasionally host to Chinese pirate bands, an insignificant part of Ōmura's domain. But, with a constant eye on the advantages he could win, Ōmura had given the Catholic foreigners permission to reside there in 1571. And it had grown as Christian merchants and those who were persecuted for their outlandish faith elsewhere in Japan, settled in large numbers while Ōmura retained legal control. It was now one of the best harbors in southern Japan, a new and thriving Christian-only settlement.

The gift of a Jesuit colony in perpetuity! That would, he believed, surely fix things between them. A substantial gesture to regain Valignano's favor and deny any other lord the rich taxes he'd lose if Kuchinotsu or another destination became the new port of call for the annual arrival of the Portuguese black ships. If the Jesuits agreed—and how could they not?—Ōmura would retain the benefits of the trade and still have Nagasaki as a safe retreat, should his enemies ever overwhelm him.

Lord Ōmura and Valignano met in the mission building in Kuchinotsu. There, Valignano listened—Yasuke standing guard a few feet away—without comment as a second Japanese lord now made long-winded promises that were supposed to last generations. Ōmura's offer of Nagasaki was an enormous step toward that greater purpose.

Still, Valignano did not accept Nagasaki straightaway, but instead sent Ōmura off with empty hands and an emptier war chest, while the Visitor feigned to consider the matter. Valignano preferred to keep Ōmura on tenterhooks and thus he "considered" for weeks, which then became months.

Nor did he grant Arima's request for baptism until seven months later.

For now, the Jesuits could afford, simply due to their ready access to weapons and ability to gift the black ship's docking right, to keep both Japanese lords disappointed and nervous. All just a small part of the great game being played out across the entire country. And only a tiny taste of the drama that was sixteenth-century Japan.

Japan in the sixteenth century was a patchwork of ministates, mountain domains and pirate-infested islands. It was united only by a distant memory of having once been part of a Kyoto-dominated realm under the nominal rule of a shogun, a military ruler who still recognized, and *permitted*, the spiritual and cultural authority of an emperor (or *dairi*) in Kyoto to perform symbolic rituals as a traditional figurehead.

But that balanced and peaceful realm had long since ceased to exist in any meaningful way. There hadn't been a shogun in nearly a decade and the current emperor now lived in the decrepit remains of a crumbling palace, powerless beyond tradition and etiquette. Regional power brokers, *daimyō* (warlords)—or men who styled themselves as such—now vied for control of parts of what had once been a vast island empire. Their power came from family clans and geographic or religious alliances. These "lords" sometimes controlled several domains, or like Arima and Ōmura, were left with only slivers of their birthright lands as others conquered, pillaged and dispossessed them. The common people under weak lords suffered death, starvation and enslavement. Those under stronger rulers prospered, but their men were regularly conscripted into their liege's war machines.

Adding to the complicated map of rival lords' domains, there were also vast areas under the control of fanatical Buddhist warrior monks who rejected all other masters, cities controlled by oligarchic merchant elites, *and* peripheral areas of mountains and island-dotted seas which were controlled by pirates or self-governing peasants.

The Japanese islands had now been at brutal civil war for more than one hundred years and were, despite the efforts of several major new and rising warlords, divided into warring factions, both big and small. The politics and bloody machinations—ruthless backstabbing, politics and massacres—found here, caused reverberations that have lasted until the present day. This period of Japanese history is called, quite simply, The Age of the Country at War.

But by the time Yasuke arrived in 1579, there were only a few major players left in these war games.

Lord Oda Nobunaga—the warlord who would soon change

Yasuke's fortunes—stood atop them all. He'd set up his home base in Azuchi, close to Kyoto, de facto capital of the country by virtue of being the traditional residence of the emperor. Nobunaga's rise, however, had not yet completely stopped the surviving minor warlords, bands of warrior monks and mercenary peasants from struggling for control of every remaining tiny fief and castle across the country. A perfect setting for the intentions, and deft diplomacy, of Valignano.

Months after its arrival in Japan, the Portuguese black ship prepared to depart at last, sailing back to Macao on the fall winds. A forlorn scene had presented itself as Yasuke accompanied Valignano to the docks to bid the vessel farewell when the last goods were loaded onto the departing *nao*.

Human goods. Children.

Most less than ten years old, being herded aboard by the sailors. There was no resistance, their young faces bewildered or terrified. Yasuke wondered what would happen to them. The comments of other Jesuits around him made it all too clear. These were "lucky" ones, those who would have their souls saved. These children, orphaned or abandoned, the *suteko* (the "thrown away" kids), would grow to adults in good Christian households in Christian-ruled regions of China, India, Manila, and some would be transported as far as Europe and Spanish America. The fee those good Christian households would pay, and the labor the children would perform, including sex slavery, would become the compensation for their raising and delivery from paganism. The children didn't yet understand any of this, of course, as they held hands for succor and the sailors herded them aboard the small craft to be transported to the waiting galleon at anchor in the harbor. There was no telling how far that ship would take them.

The scene would have brought back familiar memories for Yasuke.

★ ★ ★

The images from his past were faded, but his own village had been of perhaps ten large round shelters, with distinct conical roofs of thatched river reeds, and with a granary on stilts at the back. The lodging of the *Nhomgol*, or headman, took center stage in the village and the homes of his numerous wives and their children surrounded it. The entire village was encircled by a rough palisade of woven reeds to keep wild animals out, but was otherwise undefended. The cow herds, the center of his people's culture, wandered freely outside the main palisade. Yasuke often slept with them, as it had been the boys' job to guard the family livestock. It was peaceful work. But the memories which stirred now were traumatic. Anything but peaceful.

The attackers began predawn, encircling the settlement so none could escape. The rough palisade did little to keep them out.

Yasuke had just woken, puzzled by a strange unnatural silence. The cows' lowing, small animal noises and scratching which normally pervaded the night and got louder as the sun rose, were strangely absent this day. As if all the animals were holding their breath, waiting for what must surely come next. The silence had not lasted long. Horns, whistles and shouted whoops cut through the night fog which came off the Nile, as the raiders burst into the middle of the sleeping village, painted white with river mud, emerging from the mist beyond the rough village stockade. Torches now lit the whole night. The men seemed ghoulish, otherworldly, ghosts or spirits from hell. There was no resistance as the ultimate surprise of the attack had rendered the whole village powerless; there was only screaming and fire. The houses—made of reeds, wood, dung and dried mud—burned as easily as the torches which lit them.

As Yasuke's village fled their burning dwellings, they were taken easily; clubbed, speared, beaten and rounded up. Corralled. Any who attempted escape from the human pen were run through with spears or chased down and clubbed. One desperate woman, huge with child, had attempted to escape across the river and now lay impaled beside the shore. Many shook uncontrollably, others were frozen in shock, brutalized by the sudden new, horrific reality. They were shackled and tied together by the ghost-like raiders until more than half the village had been captured. Tales were told of slave raiders, but never in living memory had such men reached these people of the *Jaang*, the Dinka, so far inland, hundreds of miles away from the sea.

When the killing started, the raiders first pulled aside any surviving male with the distinct *Jaang* scar marks of adulthood on his face. These were the warriors, warriors who'd failed in their most basic task, the defense of their home. Roped together, unable to move, each was run through, decapitated or clubbed to death in the mud. Now there was no one left to even *attempt* resistance. After the men, the slavers decided which of the remaining captives were of value. The babies were brained against the hard earth. As they were ripped from their mothers' grasp, their wails pierced the early morning air, only to be cut short instants later. Next, the small children who also clung to their mothers screamed as they were dispatched. And, if by chance the long steel blade should pass through a child to pierce the mother, what of it?

It was only the older children the ghoulish slavers wanted. All boys and girls between five and twelve, and some of the younger women, were roped separately and taken away from the killing field, out of the village, and down to the riverside. There, they were made to squat, waiting helplessly—the enemy guards standing tall over them like giants in their distinctive one-legged pose, weapons ground into the reed-stripped bank. Yasuke and the others shivered in the early morning cold and the mist that came off the river. The reek of smoke and fresh death filled the destroyed village and shrouded them all.

A 1771 French map of Northeast Africa, by Rigobert Bonne. The port from where Yasuke left Africa is marked as Suaken. His birthplace was probably somewhere in the uncharted territory left of the Nile.

Soon, the unmistakable sound of their mothers screaming as they were raped, then slaughtered, reached the crouching children. The guards at the riverside went one by one to take their turn in the village. They left at a run, whooping and jumping in delight. Although Yasuke could not understand the language of his captors, he could feel their jokes as they talked, see the crude gestures. These men were delighted at their morning's work and comparing notes, no doubt joking and boasting at their prowess in the kill and in the brutal sexual assaults which now followed. When at last something resembling silence—at least there were no more screams, only the whimpering of the children around Yasuke—descended again, the victorious warriors who'd been in the village rejoined their brethren, carrying the grain from the granary, and gathered up their lances, clubs and spears. The whole band was in a great mood, chattering and laughing, even singing with seemingly genuine joy.

Waiting boats appeared and the young captives were herded aboard. Only thirty of the hundred villagers were left alive. Yasuke boarded as if in a dream. He'd wondered about his cattle, the cows that were his job and his family's whole life. *Why did these horrid men not take them?* Normally, raiders took only cattle; people were of little value compared to livestock. You could not give a child as dowry to your bride's father, nor would a child provide life-giving milk for the entire village. As they pushed off, he could hear the cows lowing in distress at the strange atmosphere and the lateness of their morning milking, one which would never come again.

Late the next day, the boats stopped and then they walked. And walked. Those who fell behind were run through. Many of the young women were pulled aside each night by the warriors for rape; the prettiest, however, were left alone to fetch a better price as virgins in a rich man's harem. While the land near the river was green and verdant, the farther they got from the water the more sandy and desolate it became. Eventually it transformed into hard desert, rocky outcrops peeking through an endless sea of sand.

Days after, their captors sold them to another band of lighter-skinned warriors. These were men garbed in long robes with cloths wrapped around their long black hair. The first people Yasuke had ever seen who did not have his ebony-dark skin color. And he was frightened. The raiders who'd destroyed his village had *painted* themselves white with Nile mud, but these new men had no need of paint. Were they ghosts, taking Yasuke and his kinsmen to be eaten? They rode huge horses and carried curved swords and ornate knives rather than clubs, yet still had spears to prod the captives along their way. They rode on horseback, and dwarfed even the giant men of the slave-raiding party. They also spoke in another language. The tones sounded like poetry or song, but belied their facial expressions. Hard and cruel. By the time they'd reached the sea, half of the original children and young women had perished. Only fifteen were left alive, and in a terrible state. Starving, dirty and covered in their own filth.

Suakin, in modern-day Sudan, was a shining city built of coral on an island in a lagoon surrounded by the azure waters of the Red Sea. Its walls and palaces were like nothing Yasuke had ever seen. It rose out of the flat desert like a dream as they approached. To the stunned, exhausted and hungry Yasuke it seemed a magnificent and ephemeral mirage.

Yasuke and his comrades did not enter a palace though, but found themselves in a dark crowded chamber beneath the slave market. The odor of human despair and fear in those cellars was a sharp contrast to the wealth and glory of the city outside, more so when contrasted with the sweet perfumes that the rich slave merchants used to cover the unpleasant stench their wealth was built on.

Yasuke shook himself back to the present.

That trace, that city of death and despair, was a wavering image from long ago. Fifteen years. Ten thousand miles.

He was a grown man now. A warrior.

If anything came out of the mists in Japan, he would be ready.

CHAPTER FOUR
Seminary Life

The port of Kuchinotsu resumed its slow and measured pace. The Portuguese ship had long ago departed for Macao, and the hundreds of Japanese merchants who'd gathered from every region of the country to buy its cargo had returned to their homes. All that was left now were the fishermen, some farmers and a dozen local merchants.

Still, Yasuke went through his security routines each day. Those he'd learned from the bodyguards while in India. Every day, every meal, every service. *How would I attack Valignano?* And then: *So, how would I defend him?* With needed help and translation from Father Fróis and the Japanese acolytes who were beginning to be able to speak decent Portuguese, Yasuke worked with the local soldiers Arima had left behind to develop defense strategies. He'd developed contingency and emergency escape plans too. He checked and rechecked the guards and lookouts at the village perimeter throughout the day and night. One of Valignano's later innovations was to recruit and arm a highly effective and dedicated Catholic citizen militia in Nagasaki.

In 1579, however, there was only Yasuke.

The Jesuits had no true power or military force beyond the promise of their access to finance and modern weapons, and the protection granted them by local lords. And Valignano, the head Catholic in Asia, was in a war zone. Only miles from a violently anti-Catholic enemy poised to overrun the fief of their host at any time.

Valignano had little concern about becoming a martyr—a glory often sought by Catholic missionaries—but the position of the Church in Japan was at stake. Yasuke stood as one of the few armed men around him with no local ties or opposing obligations. The African warrior was the last line of defense against any political or religious assassin, a very real and common threat in Japan, as well as against any petty criminal who might have wanted to take advantage. There were usually a few local guards assigned by Arima to help out, but these men were as much spies as protection. In the last resort, they'd obey the whims of the twelve-year-old boy. Aside from Yasuke, Valignano could truly count only on the grace of God.

In letters back to Europe, Valignano admitted to few of the genuine concerns he faced, besides being very tired and seasick. Military and security matters were essentially left out, especially when it involved anything that could make trouble in Rome. The legality of much of what he did—gunrunning, political maneuvering and tacit slaving—and much of what others did in the name *of* the mission, was only considered after the act, if at all. Hence a lot of what was done can only be ascertained from vague hints, aspersions and slips in the carefully edited letters which arrived in Rome, or from Japanese sources.

It was indeed far easier to obtain forgiveness than permission. The mission, in short, could get more done without the Vatican getting in the way and insisting on things being done "as in Rome."

For example, the Portuguese king had forbidden his subjects, which included the Jesuits who were under his patronage, from engaging in Japanese slavery in 1570. But such slaving (occurring long *before* the Jesuits) carried on regardless and the Jesuits were now prime participants in this human trafficking. So far from Europe, the king's writ held little weight. Jesuit authority legalized slave exports and the order clearly took a cut from the proceeds. Their mission house in Nagasaki was, in fact, recorded as the center of the business. More than one thousand Japanese slaves a year were exported at this time and then found in bonded labor in places as distant from home as Africa, India, Spain and even Mexico.

Valignano's accommodation to non-European languages and non-European norms (such as forms of dress and general adherence to the Japanese diet) was also highly unorthodox, but missionaries argued—as they still do in places like China today—that it was necessitated by local factors, to gain the respect of the people whose souls they strived to save. Valignano essentially established the first European scholarship on China when he ordered that four scholars be assigned in Macao solely to the study of China and its literature, since without the language it was impossible to attempt the so longed-for conversion of China. These scholars "should have their own teachers, a house apart and all the facilities they needed," he ordered. In Japan, he advocated imitating the ways of Buddhist priests and his leadership led eventually to the first Portuguese-Japanese dictionary, Japanese translations of European texts such as *Aesop's Fables*, and Portuguese translations of Japanese tales such as *The Tale of Genji*, a popular and often racy eleventh-century drama with hundreds of characters, and probably the first-ever widely published work of fiction written by a woman anywhere in the world. Nothing like this would have happened without the loose reins of Rome.

Yasuke had learned a modicum of Japanese in conversation with the locals and began to feel at home with this generally quiet, but friendly people. His appearance still caused a stir, and he'd surely become something of a local celebrity. Valignano had inspected several of the smaller missions within a day's travel of Kuchinotsu, and he and Yasuke became more familiar with the immediate area. The people of Kuchinotsu reminded Yasuke a great deal of other minor seaport communities he'd resided in during the past five years. It was in the lesser inland villages they visited outside of Kuchinotsu where Yasuke could not help but think again of his own home on the banks of the vast mother river, the Nile.

For the next nine months, Yasuke and Valignano mainly split time between two locations—the Kuchinotsu seaport where they'd first arrived and Kazusa, a nearby fishing village where Lord Arima had granted the Jesuits some buildings for a seminary. From a security standpoint, Kazusa was the safest of these two, specifically chosen for the unlikelihood of attack. It was several miles farther removed from the anti-Catholic Ryūzōji forces and Arima still had firm control over this location. Kazusa was home to only a few hundred souls, but there was an old Buddhist temple facing out onto a beautiful beach in the lee of a heavily wooded mountain promontory. A peaceful brook bubbled behind it.

Instead of being destroyed like many ancient places of worship (as Arima had promised), Valignano converted it into the first Catholic seminary in all of Japan. Later, when it was safer to do so, the seminary could move to a location nearer Arima's Hinoe Castle.

Valignano wrote the curriculum himself, from scratch. Initially the student body was made up of twenty well-born young men who were under the supervision of one priest and one brother, tasked with preparing the boys for priesthood. There were also guest appearances from visiting clergy.

The Jesuit belief in education was second only to their faith, and they extended it wherever, and, however, they could. Seminaries and schools were built to train priests who could refute Protestant arguments for reformation, promote Catholic thought and act as missionary beachheads. The Jesuits had opened their first school in Europe in 1548 and founded more than thirty there within the next decade. Valignano had similar plans for Japan.

Constancy proved an immediate ally in all that progress. Mission life has always been a regulated affair, scored by the ringing of hourly bells, gongs or human cry and on a cyclical, immutable schedule. In this way, the school at Kazusa was no different. The routine made it easier to integrate new acolytes and students into the ever-growing program. When not on the move to neighboring missions with Valignano, Yasuke found himself also constrained to this strict unremitting timetable.

He rose before Valignano, an hour or more before daybreak, to be ready to start his work when the Visitor awoke. He was already walking the seminary grounds when the mission's first cocks squawked. Valignano's first act was to prepare for prayers and mass at dawn while Yasuke completed another security round. There were two schoolwide meals: breakfast at 9:00 a.m. and dinner at 5:00 p.m. At each, Yasuke normally supervised the preparation to ensure poison was not introduced.

The central portions of Valignano's day—and, thus, Yasuke's— were taken up in meetings, writing and study sessions, interspersed by periods of prayer and reflection. Yasuke stood guard or attended to the external security throughout. Most days, Valignano met with one of the native Jesuits, Paulo Yohoken. An elderly scholar of seventy who'd retired from secular life as a medical doctor to follow Christ and join the Jesuits as an acolyte. Highly educated, he'd quickly become the Europeans' prime source of information on Japan and wider Confucian and Buddhist thought.

Usually, Father Fróis (the interpreter from the beach), would translate between Valignano and Yohoken, as Valignano took notes from the venerable Japanese doctor and asked seeking questions. Fróis had been in Japan since 1565 and was considered the mission's most talented linguist and scholar of things Japanese.

In this manner, the three men thrashed out the system by which the Jesuits would convert all of Japan. Their key strategy was to logically refute the native Shinto gods and Buddhist ways while adapting European religious customs to local tastes and norms. They believed that, where possible, converts—at least the socially important ones—should be persuaded through reasoned debate and sincere religious enlightenment. Promises of guns, silk and lead were fine for getting a foot in the door, but these couldn't be the whole motivation if Christianity were to thrive in Japan. They needed the ruling Japanese to truly accept Christ.

The evenings that followed were short. After dinner was litany at 8:00 p.m. and then sleep. It was wasteful to burn candles, though Valignano (notorious for his all-nighters) often made an exception for himself so that he could complete some pressing piece of writing or reflection. And the next day, the entire seminary did it all again.

While Valignano was known as a harsh taskmaster, and sometimes resented for it, he was also generally fair with his subordinates. He understood his bodyguard needed to keep in shape *and* have some releases. When Valignano felt he could dispense with his services, Yasuke was given time off.

In these moments of freedom, and depending on where they happened to be, Yasuke no doubt enjoyed an occasional swim in the river with the seminary students or, when they were in the bigger villages—such as Kuchinotsu where there were small entertainment quarters to cater to visiting sailors—he likely enjoyed a drink with the other soldiers or servants in town. It would have been a good chance to practice his Japanese and to catch up on news and gossip. When he was sure Valignano would not find out, he also most likely pursued brief dalliances with the local women.

Foreigners were rather surprised, some scandalized, regarding the willingness of both sexes in Japan to engage conjugally outside of marriage. To the Japanese, having sex was often something people merely did for enjoyment. Yasuke arrived in Japan at a turning point for the country and its culture, including sexual culture. Despite long exposure to Chinese ideas, Japanese society was still in the process of absorbing Confucian ethics of human relations, where women take a decidedly inferior role to males. Such ideas had not yet taken root among the less educated lower and rural classes, where women could still pursue their own enjoyment as equally as the men.

At a minimum, Yasuke likely stole an hour a day for genuine physical and martial training. He had to keep up his fitness and readiness. The seminary students spent their days in study, but there was also ample time for recreation and exercise. Valignano believed in staying physically active. Yasuke's exercise would have included sparring with a variety of weapons, swimming, running and wrestling. The Japanese samurai guards were equally faithful in taking time to train, and Yasuke often joined them. For years, he'd made sure to keep up with his physical strength and combat skills. Running, climbing, stretching. He'd learned as a teen from other African slave mercenaries and from the best soldiers in India: the only easy day of training was yesterday. Now, for the first time, he was studying the movements of the Japanese sword—so different from any other weapon he'd tried before. He discovered the power of the first strike, intended to kill an enemy instantly. Then there were new hand-to-hand methods, new wrestling grips, the beautiful power of the *naginata*, essentially a deadly razor-sharp sword on a long pole that could mow down foes as a sickle cuts back hay. He was learning how he would need to fight if he ever came up against Japanese blades.

In an era when literacy was limited to the privileged few and the scriptures used by the Roman Catholic Church, along with church services, were in Latin, imparting the Word of God was not always an easy task. In the majority of Western Christendom, where the Catholic religion was essentially inherited without question until the great cleaving ruptures of the sixteenth century, missionary work was not a huge challenge; the Church had a captive audience. But elsewhere in the world, where the stories of Christ's sacrifice were unknown, such tales often made little sense, especially as even the educated local leaders could not read or understand Latin.

The answer was to bring carefully selected bible stories to life through pictures and dramas. In Yasuke's Japan, the Jesuit mission was always crying out for more devotional pictures which had to come all the way from Europe. They would never have enough until a printing press arrived in 1590 and started to mass-produce Christian artwork and texts. The "divinity play," a drama enacted from a bible story, took on an enormous significance. The Japanese Christians learned about their new religion, its stories, its morals and its meaning, through both acting and viewing these live dramas. In the words of Goethe, who witnessed Jesuit dramas in the eighteenth century, "This public performance has again convinced me of the cleverness of the Jesuits. They despised nothing which could in any way be effective... There are some also who devote themselves with knowledge and inclination to the theatre and in the same manner in which they distinguish their churches by a pleasing magnificence, these intelligent men here have made themselves masters of the worldly senses by means of a theatre worthy of respect."

And so it was, at Christmas 1579, that Yasuke found himself

Adoration of the Magi, *Andrea Mantegna 1495–1505. Balthazar is clearly depicted as an African.*

in the grounds of the newly formed Kazusa seminary acting the part of Balthazar, the black-skinned magus who, by Catholic tradition, gave the baby Jesus his present of myrrh. The grounds of the former temple were well lit, the stone *toro* lanterns blushing with flickering flames in the faint sea breeze and the trees hung with red-and-white paper lanterns. Behind, the waves broke gently on the golden beach. People had walked, sailed or rowed for days to be there and Yasuke estimated the crowd, sitting and standing around the stage area, the former temple entrance, at nearly a thousand. Children adorned the trees, clamoring for a better view. These Christmas spectacles and huge gatherings had been a tradition ever since the first Christmas mass was celebrated in Japan nearly thirty years before.

Despite the winter chill, the mass of humanity and the flames warmed the atmosphere as the spectacle of the Christ Child's birth unfolded before them. The other actors were local boys, both seminary students and village boys who attended the catechism classes; they struggled with the Latin words, but that was unimportant. It was the spectacle the audience had come for. Well-placed Japanese-speaking Jesuits in the crowd kept the audience informed of the proceedings and the significance of what was going on.

Yasuke had seen many such plays on his travels with the Jesuits, but the lighting and number of listeners and their clearly rapt attention, made this one special. As he took to the stage to perform his part, the crowd gave a special cry of joy to see the giant dressed as a king in gold robes and holding the golden chalice which represented the myrrh. Balthazar performed his part and withdrew to more sounds of amazement.

For the next year, Yasuke's duties as Valignano's bodyguard and valet took him throughout the west of the southern island of Kyushu, meeting the chief Christian lords and headmen there and visiting dozens of missionary outposts. Aside from Kuchinotsu and Kazusa, Valignano engaged in countless short trips—to inspect churches and mission sites, and to conduct diplomacy among half a dozen minor lords and village headmen. The Jesuits were far from a secluded monastic order. Their business was rescuing souls for Christ, and they threw themselves into it with a frenetic, often fanatic, energy.

Accompanying Valignano, Yasuke learned a considerable amount about the country and its people, both from personal observation and from being privy to conversations between Valignano and assorted dignitaries and policy makers. The pace proved relentless. Although Yasuke had once thought they were done with sea travel for a while, it turned out almost all of these short sojourns were by boat. Roads were a largely impractical option in this mountainous, rainy and enemy-infested country; and almost everyone lived near the sea anyway.

Beyond basic security, Yasuke had not had much need to directly defend his Jesuit employer. There had been no recognized assassination or kidnapping attempts, and no direct skirmishes with enemy troops. Valignano, instead, had skillfully avoided the conflicts raging around them—though both Lords Arima *and* Ōmura suffered various small defeats and humiliations at Ryūzōji Takanobu's hands. The genuine power of prayer, perhaps? Or Yasuke's diligence was paying off. In any case, it had been a year in Japan and no Jesuits had been killed on Yasuke's watch; the mission seemed to be entering a period of stability. The African warrior took it as a well-earned measured victory as he'd remained thorough and on guard even in the quietest of weeks and sleepiest of villages.

Had they arrived a year earlier, however, the conditions would have been remarkably different. The Shimazu clan (another regional anti-Catholic force), with only thirty thousand troops, had routed the fifty thousand warriors of Christian ally, Lord Ōtomo Sōrin, when they clashed over the minor province of Hyūga. The Shimazu, who'd initially been curious about the first Jesuits, had long ago decided Catholicism was not for them, especially after their principle enemy, Ōtomo, had become Christian. And, as Ōtomo, Arima and Ōmura had burned temples and shrines during *their* Catholic-sponsored conquests, the Shimazu troops retaliated by torching the new Jesuit church. The Jesuits who'd been within had just managed to escape and fled through the night, beginning a chase that lasted several days. In the end, they only narrowly escaped with their lives, arriving starving and traumatized back in Ōtomo's territory.

The specter of these attacks and the loss of Ōtomo's army, whose shattered bodies stretched back twenty miles from the battlefield, had prompted Valignano's need for Yasuke's special services upon his arrival. Aside from the report of the torched church, however, the official reports that had reached Valignano in Europe, and as he traveled through India, Melaka and China to reach Japan, had all been highly, perhaps suspiciously, positive about what fertile soil Japan was for Catholicism. Those on the ground who'd written the reports, had somehow omitted details, or sometimes *mention*, of unfortunate setbacks and defeats. At such a distance from Roman authority, they were writing for posterity and promotion, and were unlikely to ever be found out. Now that he was here to see it for himself, the actual state of the mission in Japan, unsurprisingly, had not met Valignano's expectations.

In particular, there was significant discord between Japanese converts (especially the most senior ones not used to being gainsaid in their own domains), and the non-Japanese missionaries (who often behaved as if they knew everything). To meet the grievances of the Japanese community, Valignano quickly held consultations to identify ways in which the earlier mission had been mismanaged. These included overly strict discipline, racial discrimination in admittance to holy orders, an insistence on the superiority of European ways *and* a refusal to support the learning of the Japanese language by some senior Europeans—in particular the mission superior, Cabral.

Another problem Valignano faced was that most Europeans did not appear very civilized to the locals who saw them as, frankly, vulgar. By comparison, the Japanese were consistently well mannered. Valignano wrote, "even the children forbear to use inelegant expressions among themselves, nor do they fight or hit each other like European lads." Upper-class Japanese people, particularly, considered Europeans dirty, ill-mannered and ignorant of proper comportment. The Japanese were also used to daily bathing, and the ability to eat without touching food with their hands—both customs Europeans of the time customarily scorned.

Having identified these impediments, Valignano issued decrees on how Jesuits should conduct themselves and adapt more to local norms. (Though, even the relatively broad-minded Valignano still balked at bathing regularly and forbade his charges, including the Japanese *and* African ones, from doing so.) By the time he left Japan for the first time in 1582, he'd already opened three more seminaries with the aim of training locally recruited brothers and priests. The mission relied upon its native Japanese followers to help celebrate masses, marriages and funerals in Japanese, and for diplomacy in many cases. Until Valignano's arrival, Jesuit policy had forbidden Japanese men from becoming full members; they, instead, had to remain as semipermanent acolytes. One of the most important things Valignano would do during his tenure as Visitor was facilitate the first non-Europeans becoming full Jesuit members and ordained priests in Japan.

Then, to make Catholic priests' status more recognizable to the Japanese, Valignano reorganized the mission structure to more closely resemble that of the social organization of the Nanzen-ji Temple in Kyoto. Japanese religion at this time had become a fusion of imported Buddhist beliefs and native animist beliefs, hence, Buddhist "saints" were worshipped in the same places as ancient animist gods called *kami*. Sometimes *kami* and Buddhist saints eventually mixed in together and became one entity. Buddhism itself, was divided, sometimes violently, by sect, some of which, like Zen, had their origins abroad, and others, like Nichiren, which started in Japan.

Valignano copied their ranking system so locals would understand the social standing of the Jesuits and know which priests were more senior. Initially the priests had intentionally dressed poorly, marking their vows of poverty, but Valignano changed that, and they smartened up, or at least made sure their clothes were clean. This made the Japanese more open to the new religion, because it *looked* more like traditional ones, respectable, blurring the lines somewhat and gaining the Catholics more respect.

Valignano also directed the missionaries and other Jesuit workers to systematically learn as much Japanese as possible.

Only then, when they could speak directly to the locals in their own tongue, could they truly reach out across Japan for the Church. Perhaps influencing his plans, Valignano was particularly taken with the Japanese language, calling it "the best, the most elegant, and the most copious tongue in the known world," adding, "It is more abundant than Latin and expresses concepts better." What the European missionaries were particularly impressed with were the different ways ideas were expressed depending on *who* they were being expressed to. Language and culture were so intimately intertwined in Japan—and position and status such fundamental concepts—that their expression could be found in almost every sentence uttered. Thus, a person learned the art of rhetoric *and* good breeding along with the language as, in Japanese, one must know how to address the great and the lowly, the nobles and commoners, ensuring "decorum to be observed with them all." Ideal concepts for an aristocratic Jesuit hierarchy who wished to retain their nobility while encouraging the idea of the dignity of the common man.

Under Yohoken's systematic tutelage, a typical foreign missionary could attain conversational proficiency in two years, a considerable improvement on the previous haphazard approach, and they reported the language relatively easy to understand. Yet, to *speak* correctly proved far more challenging. The problem was the sheer number of ways to express the same concept. The language was, according to Valignano, "copious in the number of synonyms" with "infinite ways" to describe objects and actions.

Thanks largely to Yohoken's efforts, dozens of foreign Jesuits attained a language level that allowed them to hold their own in Japanese courtly circles. The western Jesuits were soon able to compose many volumes of accurate information such as Fróis's *History of Japan* and Valignano's *History of the Beginnings and Progress of the Society of Jesus in the East Indies,* which were circulated to a hungry scholarly class throughout the European world.

Information about contemporary Japan and knowledge of its

language were not yet even held in any measure of regard by its closest neighbors, Korea and China. The rare high-level communications between the three countries were conducted in written classical Chinese with which any educated Japanese or Korean was also familiar. Only a handful of unofficial interpreters or pirates in Korea or China could speak any Japanese at all. (In 1590, while Europe, thanks to the Jesuits, had plenty of detailed information about Japan, the Korean government was not even sure of the name of the Japanese hegemon, Toyotomi Hideyoshi—who'd soon invade them!) At its closest point, Korea is only thirty-five miles from Japan, while the sea journey to Rome around Africa is nearer to twenty thousand miles. Valignano, with Yohoken's help, made sure the Church wouldn't remain ignorant of Japan and its customs.

Valignano, however, by his own reckoning, could never quite get the language down. Although he wrote fluently in Italian, Latin, Spanish and Portuguese, the subtleties of the Japanese language remained elusive for the otherwise brilliant scholar, possibly because he was so busy. Despite his own linguistic shortcomings, in spreading the gospel—whether in India, China or Japan—Valignano continued to emphasize the need for cultural adaptation and local language study to his underlings.

Yasuke, meanwhile, was a natural. He became quite proficient at Japanese during his years in Kyushu, likely because he was waiting on Valignano during Yohoken's lessons. In twenty-some years, the warrior had lived in Africa, India and China, and spent countless months on Portuguese ships learning new tongues. This fluency had no doubt become a survival skill and a genuine gift. Japanese was simply the next language in a long list.

Whether Valignano specifically encouraged his bodyguard to learn the language or not, we cannot say. But learn it he did. And, with positive outcomes for all concerned—particularly Valignano, who would soon gain great favor and grace through introducing Yasuke to Lord Nobunaga and the rest of Japan.

CHAPTER FIVE
The Terms of Employment

Despite his willingness to work within foreign languages and cultures, Valignano had firmly held, and narrow, views on race.

Formed through hitherto-learned European stereotypes, the Visitor's perspectives were based primarily on how ready a people were to hear God's word and convert from "base heathens" or "Moors," to children of Jesus. His view was that Europe was "the most excellent of all the parts of the world, the part on which God with his most generous hand has conferred the most and the best good things." Although to the modern ear, many of his views are reprehensible, Valignano was not simply a white supremacist as we would know it today, and far less so than most of his contemporaries.

For Europeans of the time, the Chinese and Japanese people

were "white folk" and at the other end of the racial spectrum from the "dark folk" like Yasuke. So while Valignano admired China for its peace, tranquility, administration and wealth, he remained forever annoyed and puzzled by the low rate of conversion among these clearly reasonable, civilized, educated and rational fellow "whites." It challenged the very foundations of his, and Europe's, "rational" and carefully constructed arguments on both race and religion.

Although, over the years, some of his writings betrayed fluctuations in his long-held stereotypes, this racial ranking never really altered, and his first recorded views on Japanese people echoed those of the first Jesuit to reach Japan, Francis Xavier, who suggested the Japanese might be *more* civilized than Europeans; they "lacked in vice" and were ripe for the word of the Lord. Valignano wrote also that the Japanese were polite—"even the lower classes more so than Europeans"—and also capable of learning European academic subjects quicker even than Europeans. Quite capable, therefore, of accepting the one true God into their hearts. Thus, with these lighter-skinned peoples, Valignano was convinced the Jesuit mission was destined to thrive. Nothing could stop the inevitable victory of the Church; their skin color all but guaranteed it.

Valignano's world—one defined largely by skin color alone—proves even more troubling when the darker-skinned races are described. To understand his views, we need to know that modern skin color conceptualizations are the product of eighteenth and nineteenth century "scientific" classifications and Valignano's European Christendom had no such "ordered" racial world. To him, darker skin simply signified non-Christian. (This, despite the fact, of which much of Europe was largely unaware, that millions of Christians with "darker skin" had lived in North East Africa, the Arab world *and* India, for more than a thousand years.)

Beyond knowing that Islam (a spiritual foe and ongoing

threat) was largely dark-skinned, early modern Europeans also now associated blackness (perhaps in reaction to Islam) with the long-held misinterpretations of biblical texts, particularly the so-called "curse of Ham." Ham was Noah's son, and, after Cain and Judas, one of the most notorious sinners in Christendom. (The Bible says Ham's descendants were cursed by Noah because Ham gazed without shame upon Noah's drunken, naked state after the patriarch partook of too much wine.) While no document in which Ham is recorded mentions his black skin in any way, that Ham "was black-skinned" was a widely held view in Valignano's time. (In the way that most mistakenly think of Mary Magdalen as a "whore.") It is uncertain where this misconception came from, but it lasted long enough that even Martin Luther King Jr. addressed it in 1956: "Oh my friends, this is blasphemy. This is against everything that the Christian religion stands for." But for most Europeans in Valignano's era, dark skin was often associated with sin.

Thus, of Indians, Valignano initially wrote they were poor beyond measure and given to low and mean tasks. That they were of low intelligence and very ignorant. This was modified later—no doubt after seeing the reality of the learning and opulence found within Indian rulers' courts—to note Indians were *not* without intelligence and culture *but* still extremely "untruthful." He distrusted Indian Jesuits, even those of mixed-heritage Portuguese descent, and often seems to have openly hindered their promotion.

Valignano's views and writings on Africans—in conformity with most Europeans at the time—are even more damning. He wrote that Africans were people incapable of understanding Christianity. Strangers to all human refinement. Without talents. Of low intelligence. Lacking in culture. Given to savage ways. Born to serve. He'd reached these conclusions after only spending a very brief time on the island of Mozambique, mainly populated by Europeans and Arabs, during his journey from Portugal to India. These beliefs must have been part of the ideological baggage which he brought with him from Europe.

Despite this, Valignano still entrusted his life—and, thus, the security and future of the entire Japanese mission—to an African man.

The exact terms of Yasuke's employment are not known.

Around the age of nineteen or twenty he had entered the powerful Jesuit's service as a bodyguard-cum-valet. Likely trained in violence, as well as comportment and service, during his teenage years, Yasuke was already undoubtedly proficient and well versed in both professions. His was a high-status job among working folk. Serving a noble churchman such as Valignano would have brought considerable honor, and the opportunity for a comfortable living too. Selecting this role over other positions which might have been open to him was probably an obvious choice. If he'd *had* a choice. If a slave, he still was probably paid and had opportunities to get tips, etc.—a model of slavery closer to ancient Rome, where some slaves managed eventually to buy their freedom. Indeed, many Portuguese and Indian slaves of Yasuke's time did too.

The Jesuits were a new, radically modern order, and (despite the trade they occasionally played some part in for financial gain) *officially* condemned the institution of slavery. Over the following two centuries, in some parts of the world, such as Brazil and Paraguay, they even organized and armed native peoples to resist European slavers. Valignano himself explicitly disapproved of slavery; though as a man of his time, did nothing to stop it *and* was convinced God's work could not be done if the Jesuits did not have servants and slaves to do the more mundane chores. These everyday jobs included engaging in violence, something which priests could not be seen to do, and so this kind of unsavory but necessary activity had to be outsourced to someone like Yasuke.

Valignano, by all contemporary accounts, was somewhat lacking in the Jesuit spirits of poverty and humility and had little compassion for the poor or lower classes beyond the conversion of more souls for his balance sheet. His focus was on the upper reaches of society in general. Whatever Valignano's personal views—and, whether or not he inwardly lacked the spirit of poverty—he'd sworn the Jesuit vows and could hardly be seen walking around with an actual symbol of either wealth *or* slavery. Accordingly, Yasuke was most likely not a slave as we understand the word today.

The most likely scenario is that Yasuke had won his freedom in India, through service in war or through the death of a previous master. He could then have been recommended to Valignano by a Portuguese go-between or been individually recruited as a freelancer. Another possibility is that he was bought, then manumitted, as a specialist military slave from a band of mercenaries or a local broker. In any case, he would have been following in the footsteps of many other *Habshis*—as African soldiers were called in India—who'd entered Portuguese and Jesuit service to boost Portugal and her missionary order's ability to defend their interests around the world.

However, to be clear: Yasuke *had* been a slave once. There is little doubt he was violently abducted by other Africans as a child and brought to India to be a soldier by Arab, Persian or Indian slave merchants. Everywhere in the world in this era—Africa, the Americas, Asia and Europe—where people *could* be found to be enslaved, they were. Europeans, Arabs, Indians, Japanese, Chinese and Africans all sold, bought and/or abducted people for domestic bondage or export. These slaves underpinned ruling-class power and middle-class lives, providing every conceivable service from brute force to sex, music to manual labor; they carried goods, farmed land and sailed ships. The Portuguese esteemed Africans as being the most "docile and obedient" slaves and much valued them for their strength and size. In later centuries, the French professed to prefer slaves from the Indian sub-continent for similar reasons. Racial profiling at its most vile.

Japan, too, was also a slave-trading society at this time. Both domestic and foreign humans were available for sale. Foreigners were also able to buy Japanese slaves from Japanese or Jesuit brokers, most often girls destined for concubinage, in Macao, India or beyond, but also boys to work on ships or like Yasuke, as slave soldiers. A Jesuit source noted in 1598, "Even the [Indian sailors] and servants of the Portuguese are buying up [Japanese] slaves and selling them in Macao." Valignano himself wrote, "who can bear with equanimity that [Japanese] people have ended up scattered all over the heathen kingdoms of the world, [home to] abject peoples of false religions given over to vice. Not only must they suffer bitter servitude among black barbarians, but also be filled with false creeds," from which it can be inferred that African and Indian Muslims ("false creeds") on Portuguese ships, were among the buyers. Despite their indignant tone, the Jesuits, at least in the earlier years of the Japanese mission, were also involved in this trade, certifying that slaves had been taken in a "legal" way and possibly even selling war orphans in their care.

The existence of Japanese slaves was a sensitive one because of its illegality in Portuguese law (to which the Jesuits were subject due to their sponsorship by the King of Portugal), Valignano's immediate goals of charming rather than annoying his Japanese hosts, and because the Japanese were "white folk." He explained that Muslims were different. "Since they are barbarians, and enemies of Christianity, they remain in perpetual slavery if they are taken prisoner after battle." By Valignano's logic, it was acceptable to enslave non-Christians, especially Muslims, and enforce their conversion to save their souls. Better a chained soul saved for God's Kingdom than a free heathen, was Valignano's strongly held view.

Slaves were also regularly imported, brought back *to* Japan from pirate raids in China, Korea and even farther afield, as well as being purchased from Europeans who brought them from as far away as Africa. In 1613, a Spanish embassy to Japan ran out of funds and they decided to sell all their valuables to a Japanese buyer. This included *"a black man and the mattresses of his bed."*

The forms of slavery extant in Asia were perhaps different from the present-day image of what bondage entails, which is largely formed by the later Atlantic slave trade. While their bondage was equally degrading and tumultuous for the individuals concerned, slaves in Japan and much of Asia owned possessions, experienced a degree of freedom and were often even freed, adopted into the owner's family or married to a family member. Some eventually became rich and respected members of their local communities.

But what was it to be a slave or bonded to a master in an age where all mankind served someone else? Everybody had a master or a patron, and was beholden to others for promotion, work, shelter, food and even basic survival. The exact social structure differed widely among different cultures, but a person who was truly "free" in modern terms was probably someone who would starve quickly, die from exposure or be prey to violence. To "be free" was to lack a master or patron who extended his or her power—to feed, employ, physically protect or house—on your behalf, however self-interested the motives for said protection often were. (Thus the hopeless panic which results when characters like Romeo or Dante are banished from their cities and lords; they lose everything.) Although the Jesuits probably sold orphans who were in their care, they would have done this for what they saw as good reasons: once the child was grown, he or she needed a patron, and the institution of slavery provided a last resort in this respect.

The true evil in slavery though, was the abduction, the rape, cruel punishments, the enforced migration, the slaughter in Africa and other victimized regions of an estimated ten people for each slave who survived to be sold. And, in many parts of the world, the absolute lack of human status afforded those enslaved. To southern Europeans in particular, a slave was considered little higher than an animal and the slave's master had absolute right over whether he or she lived or died. Again, in the sixteenth century *everyone* had a master and it was highly undesirable to be "free" in the sense we understand the word today. But, to be a slave was something else. It was to be treated as subhuman.

While in Valignano's retinue, Yasuke was clearly not treated so. Most of his contemporaries counted themselves lucky to be fed, clothed and housed—often this sufficed as "wages." Yasuke himself did rather better than food and shelter. He was probably paid, and certainly well dressed, managed to receive some education, and was armed to the teeth with well-made weapons. And of course, all of these things made his ability to serve Valignano that much greater. A man of Valignano's stature could not be seen to be served and protected by a poorly armed beggar in rags. This was especially important in Japan, where, to protect their dignity and station, the servants of a lord were themselves supposed to look like great men so as not to dishonor their masters.

At the end of 1579, Valignano was making a brief visit to the Amakusa Islands, just south of Kuchinotsu, and one of the oldest and most successful missions, where most of the population had been peaceably converted. While there, a message packet arrived all the way from the mission in Kyoto, with news of a setback for the Jesuit mission.

Oda Nobunaga, the mighty warlord whose name tolled like a heavy bell over every conversation—and on whose interest and protection the Jesuits depended upon to keep the Kyoto mission going—had apparently suffered an embarrassing defeat. Rather, his second son had; defying his father to invade a tiny province, Iga, which the son had believed would be a pushover.

Done, the boy claimed, for the "greater glory" of his father and to add to the provinces that the family controlled. Instead, the wily peasants there had thrashed him, killing thousands of Nobunaga's troops and sending the powerful warlord's son scampering home again with his tail between his legs. Iga remained independent, they had seen off the latest in a long line of would-be invaders largely defeated by men and women called *shinobi*.

Or, sometimes known as *ninja*.

The Jesuits had heard rumors of these *ninja* from the locals and Japanese warriors. But this was the first time official correspondence recognized their existence. They were known to the Japanese as the deadliest of assassins. Some said they weren't really men at all or that they could change into beasts or ghouls during battle. Some called them ghosts, others claimed they could make themselves invisible and walk on water like the Lord Jesus. They specialized in stealth and mountain-forest fighting, although they were by no means limited to this terrain. They could, so it was claimed, swoop though the sky like bats, drop from treetops like spiders and burst from the undergrowth like deadly boar. They could make any item into a weapon, and the females specialized in using hairpins and chopsticks on their victims. They could see through walls and strike their opponents through doors. Most of the time no one knew they had attacked until after the dead were discovered.

Just the year before, it was told, a great lord in the north, Uesugi Kenshin, was killed by a ninja while sitting in his latrine. The assassin had supposedly waited in the stinking pit below all night for his chance and the killing stroke went straight up the sphincter and into the lord's stomach. Kenshin, they say, did not die immediately, but took four days to perish in extreme agony. The *ninja* had probably drowned in the cesspit after poisoning himself. That was what they did, rather than be captured. Unlike normal warriors, they allegedly killed only for money, and had no honor beyond what they were paid. They were a hardy folk who gave little value to their own lives or even those of their families. They served for the greater good of their clans, to earn money to survive in their inhospitable lands. Selling their services was the only way the community could survive; the individual meant nothing. Ironically, it was said to be Lord Nobunaga himself who paid the ninja's fee for that particular assassination.

The stories seemed absurd, but: *What if they were true?*

In the spring of 1580, Yasuke moved with Valignano to Nagasaki.

The port city, gifted months before to the Jesuits, was now entirely under Valignano's control. The teen lord, Arima, after months of waiting, had finally been baptized. Conversions continued apace. Now it was time to focus on the first Jesuit-controlled colony in Japan.

God may have made the earth and waters, but Valignano and his highly educated Jesuit team made Nagasaki. He worked with engineers, sailors, planners and local merchants to expand on the existing basic infrastructures and build a seaport which was to become the foothold and foundation of an entirely Catholic Japan.

Within a year of the Jesuits taking full control, the population had doubled—a hybrid seaport of Japanese, Chinese, Europeans, Indians and Africans. The town now comprised around four hundred households and was bursting with Christian-convert migrants from all over the country, a much-increased permanent foreign presence, and ever more refugees displaced by local warfare—warfare provoked, to a certain extent, *by* the Jesuits' presence in the region.

The port blossomed on imports of silk, guns, exotic and religious products, and exports of silver, sulfur and slaves. In turn, Nagasaki had become a thriving, multicultural port city, with all the problems and tensions that can ensue from such success. Community relations were often strained, and misbehaving mar-

Nagasaki Harbor in the early nineteenth century, by Phillip von Siebold. The central focus of the picture is the Dutch trading post of Dejima. Moored near it to the right are two European ships and numerous smaller Japanese vessels. To the left, near the Chinese warehouses (now Nagasaki China Town) are two Chinese junks.

iners fought indignant locals out of arrogant pride or because they were starving and had stolen food. The fact that, by law, only Catholics could live in Nagasaki did not, it would seem, guarantee Christian brotherhood. A still-graver concern was the threat of the fiercely anti-Catholic Ryūzōji (still besieging Lord Arima's castle) to the north and the Satsuma clan to the south; although a wary peace seemed to be holding.

Under Valignano's orders, Nagasaki had been fortified with wooden walls, with plans for stone replacements when more funds became available. The fortifications also now boasted bastions, moats and multiple gun ports. To man his walls and guns, Valignano needed men, and the process of arming the Christians of his new town proceeded apace.

In the less threatening atmosphere of walled Nagasaki, Yasuke's role changed slightly. One of his new jobs was to help train the militia, a ragged collection of refugees and masterless warriors, *ronin*, who'd recently embraced the Catholic cause. In time, these men would become a force to be reckoned with.

But Valignano, of course, conceived and planned far beyond just military matters. Several new churches also appeared. A new hospital, poorhouse and orphanage. And outside the city proper, there was even a burgeoning leper colony—the sign of a truly growing town. It was well-known that Jesuits looked after the poor and other unfortunates, so both groups flocked from afar. During certain festivals, the priests even symbolically washed the feet of representative paupers in public.

Outside the city walls, per Japanese convention, in Kawatama-chi across the Nakajima River, lived the outcast butchers and leather workers. (Outcast because their association with death meant their business was considered "impure.") As most leather came via Chinese ships, particularly from Siam (Thailand), Nagasaki quickly became a center of these industries. Here, Yasuke browsed and bought bows, armor, arrows and—most heartening of all, a staple from his childhood—meat. A rarity in Japan, though available widely in Nagasaki for Chinese people, other foreign residents and Christians.

Aside from the products in Kawatamachi, Yasuke would have found plenty to do in Nagasaki, already far bigger than other towns he had visited in Japan up to this point, to occupy his few spare hours. He'd have hunted for deer and boar in the mountains, fished in the sea and rivers, made and met friends of all sorts, and likely, again, enjoyed a romantic liaison or two. There were plenty of eating and drinking establishments and nonromantic sexual release was easy to find within the walls *and* across the city limits, in Ōmura's domain.

Surveying it all perched the new Jesuit headquarters, a well-fortified compound on top of a strategic promontory commanding the entrance to the entire harbor. It was essentially a fortress, separate from the main city fortifications, and guarded by Valignano's local Catholic militia. A final stronghold should the need ever arise. It was a beautiful building, constructed in the Japanese style, and something between a palace and a temple. At the foot of the promontory was a brand new pier jutting out from the beach beneath the stronghold, for easy access to the sea.

Not only was the mission compound a place of religious worship, but much of the town's business took place there as the Jesuits now had their fingers in almost every transaction in the region. One observer described it as being "like the Customs House of Sevilla" in Spain. Deals were concluded, contracts witnessed and signed, and merchandise—including humans—were bought and sold. It was the city hall, guild hall, corn exchange and main church all rolled into one and people were in and out all the time. A challenge for the bodyguard, but nothing Yasuke couldn't handle. He would have known the Japanese language well by now, better, in fact, than Valignano. More importantly, he would have felt he'd earned Valignano's trust. Knew his moods and habits, his disinclinations and even, now, some limitations.

The conflicts between Arima and Ōmura had calmed somewhat, as both lords—despite support from the Jesuits—had largely capitulated, kowtowing to Ryūzōji's overwhelming force. This meant a fragile peace, but as yet, Nagasaki remained unthreatened. Perhaps due to its shiny new defenses. Its population swelled further with refugees from the persecution that the anti-Catholic Ryūzōji Takanobu wrought, but despite his avid hatred of the Jesuits, Ryūzōji had agreed to a peace with Arima and seemed to be less of a threat to Nagasaki itself. He was biding his time.

Yasuke and Valignano remained in Nagasaki until September of 1580, watching the war orphans and dispossessed families tramp through the city gates to be looked after in Jesuit poorhouses or adopted into local families. It was a heart-wrenching sight, some of them were barely clothed and all looked skeletal. If they had been poor before, they had nothing except their faith now. A testament to the effectiveness and passion of Jesuit missionary activity; these second- and third-generation converts sincerely professed their belief in Christ and were willing to lose the little they possessed to preserve it.

Valignano had successfully solidified the Catholic hold on the various regions of western Kyushu which they'd visited. His inspection of Japan had been, thus far, an unmitigated success. Nagasaki—once a minor Japanese fishing village—was now firmly in the hands of the Church, a thriving base from which to claim an entire empire for Christ. Valignano and the Jesuits had successfully gathered another forty thousand converts under the wing of the Catholic Church, and planned to found four seminaries and numerous other institutions, and had placated and flattered several Japanese warlords to ensure better security for the Church moving forward. All for the greater glory of their God. Not bad work for only an estimated fifty-some Jesuits in Japan at this time.

This task completed, it was time to head to the other side of Kyushu to the home of Japan's richest Catholic lord, Ōtomo Sōrin of Bungo. From there, supported by Ōtomo's patronage, eventually they would head inland into the heart of the country, northeast toward Kyoto. The most important souls in the Japanese world waited there. The fiercest warlords, merchants with wealth beyond compare, and even the emperor himself. If *they* could be brought under the banners of Christ, then surely the rest of Japan would follow.

Each and every river punt, coastal barge, trail and step north or east was bringing Valignano, and Yasuke, ever closer to their greater destinies.

CHAPTER SIX
The Witch of Bungo

Lord Ōtomo Sōrin was the most powerful and wealthiest of the Japanese Christian lords by far. As ruler of Bungo (now Oita Prefecture), he controlled almost half of Japan's second largest island and commanded armies of up to fifty thousand men. His plush new residential palace in Utsuki had been, officially, constructed as a haven and fortress where he could pursue his new Christianity without being pressured to recant by his own people. However, Ōtomo's actual reason for his pecuniary support of the Jesuits and his seemingly pious gilt hermitage was far from pure. He'd embraced Christ and moved from his domain's capital, leaving his twenty-year-old son there to rule, primarily so he could divorce his wife of thirty-five years and marry his new, younger lover.

This ulterior motive had surprised no one as Ōtomo had

never been the most righteous of rulers, or men. When younger, he'd predominantly been known as a prolific womanizer, taking for himself the most beautiful women in his domain and routinely exercising *droit du seigneur*. Even when married, he'd spent vast sums bringing the most famous and refined courtesans, female *and* male, in from Kyoto for legendary court debauchery that lasted for days. Still, his most notorious deed was to have a soldier, whose wife he desired, put in the front line of battle. (Clearly, a time-honored scheme of smitten and cold-blooded rulers throughout history.) After the man was predictably killed, Ōtomo toyed with the wife for some time, destroyed her reputation and then moved on to other women. His senior advisors and frustrated wife—who came from another powerful local family, the Nata—protested against him, but he would not mend his ways. This, coupled with what many of his vassals saw as his sacrilegious behavior in embracing a foreign god, nearly brought his domain to civil war as his warriors took sides and squared off.

When the Jesuits arrived, with their new God and interesting gadgets, Ōtomo saw a way to rid himself of his tedious wife *and* the burden of power at the same time, while still retaining independence and comfort in his choice of residence and romantic partner. The Jesuits also saw advantage for themselves and were happy to oblige. The baptism of the most powerful Japanese lord yet would indeed be a coup and they advised Ōtomo that if he and his lover were to *both* be baptized, then all would be legitimate; as his first wife was not Catholic, his original marriage was invalid in the eyes of God. Thus, both Ōtomo and his younger lover, known in the Jesuit records as Julia, were soon baptized and the couple moved to Utsuki to live happily ever after.

Ōtomo's estranged wife, alas, had not taken the divorce well. Upon the news, she shaved her head, tried to kill herself, was kept on suicide watch and then routinely threatened to take her daughters out into the wilderness to die as punishment to Ōtomo. Far worse—as far as the Jesuits were concerned—she'd remained behind in the old capital, and wielded enormous influence over her son and his rule; influence she was using *specifically* to destroy the Catholics.

Ōtomo's first wife's Japanese name is lost to history, this despite her great power and important family ties (which reveals something of how history is written). The Jesuits simply labeled her "Jezebel the Witch," and for now, that is the only name we are left with. She had many followers through her family connections, but also—more damning in the eyes of the Europeans—enjoyed spiritual and political legitimacy as one of the high priestesses of the age-old Japanese god of war, Hachiman. As such, she was a well-known and celebrated devotee of ancient and esoteric magical rituals. She resented the Jesuits' challenge to both her spiritual and temporal power over the people of the domain and thwarted their efforts in the area incessantly. She'd already orchestrated several coup attempts against her husband and his new Christian friends, and routinely leaked rumors of plans to kill every Christian priest and brother in Bungo. When the Jesuits tried to take their mission to outlying villages, Jezebel sent messengers and hundreds of letters warning the locals to return to the old gods or face terrible consequences. Valignano and the other Jesuits were convinced she worked directly for Ba'al, the false and demonic pagan god, and Fróis wrote that the very Devil himself "had deeply taken hold of her mind."

With Valignano's relocation to Utsuki, a bigger and grander city than any they'd yet stayed at in Kyushu, Yasuke was entering another troublesome security situation. This Jezebel woman and her allies had good reason to want Valignano dead. Ōtomo's divorce was not recognized in Japanese law, and as far as most people were concerned, Jezebel was a deeply wronged woman with an unfaithful, work-shy cad for a husband. It was under *his* watch, a year ago, that the Bungo army of fifty thousand had been thrashed, the Catholic church burned by the Satsuma clan and the Jesuits had run for their lives.

Yasuke and the Jesuits needed to be on guard for both poison and rebellion. The threat of ninja assassins was, in this region, quite genuine. Such warriors were an ideal weapon for Jezebel to use: covert, lethal and deniable. If they remained undiscovered, she'd keep the moral high ground and do away with her troublesome husband and his new friends. Outright rebellion, political intrigue and family pressure had all failed. Perhaps the only option left *was* a swift and stealthy ninja strike. And her status as a high priestess and as a practitioner of magic meant she was ideally placed to ally herself with them. These assassins often disguised themselves as wandering esoteric priests, entertainers and magicians. For such people to associate with Jezebel would arouse no suspicion. Then all it would need would be an assassin's bullet, poison administered to the open mouth of a sleeping victim by thread in the dead of night, or a blow dart fired by a "begging monk" in a busy street.

Justifying the Jesuits' concerns, an unexplained fire broke out during their stay. Arson was suspected. The only house that burned down belonged to the married daughter of Ōtomo's new Christian wife. The couple barely escaped and everything else was lost. In solace, Valignano gave the young bride a rosary and she cried with joy (at least according to the later Jesuit report).

Jezebel the Witch was blamed immediately, or at least her devotees were. Security was increased, though Ōtomo waved off all concerns, claiming his former wife was "only a woman." (A comment which perhaps sums up Ōtomo's character nicely.) The Jesuits were not convinced. She was quite clearly a force to be reckoned with, and there was no room for any surprises. Despite the threat of arson and "witches," there were still three vital jobs which needed to be seen to in Bungo.

First on the list was the founding of the most ambitious seminary yet, one built in Utsuki, beneath the protective shadow of

Lord Ōtomo's castle and palatial residence, intended to attract students from throughout the Portuguese territories in Asia, as far away as India. It was to become the first truly "international school" in Japan. Of the first class of students, twelve were Japanese and eight were Portuguese. The curriculum was also more challenging than the seminary in Arima too, eventually including other European languages aside from Latin, philosophy and logic, as well as basic subjects such as liturgy, math and music. For several weeks, Valignano himself lectured at the school twice a day on Jesuit matters and morality—though to the Portuguese-speaking students only.

The second task was to organize a delegation of young Japanese noblemen for a pilgrimage to Europe. This was to be the next step in Valignano's grand plan for a Catholic Japan. With Ōtomo, Ōmura and Arima (all now baptized Christians) picking up the bill, these handpicked Japanese teens would tour and astonish Europe with their sophistication and dedication to the faith. And, upon their return to Japan, they'd then be able to impress the glory of Europe upon the rest of their country.

Yasuke likely assumed he would follow where Valignano led, and journey with the delegation west to Lisbon, Madrid, Milan, Florence and Rome. Places he'd only heard or dreamed of, but lands the Europeans hearkened back to again and again. The continent was their home and Yasuke was curious, if only for its contrast with the worlds he knew. The bustle of mercantile, multicultural Lisbon, the beauty of the palaces and gardens of Florence, the churches of the Vatican, the slowly growing glory of St. Peter's Basilica and the cosmopolitan finery of the papal court. As Valignano's man, Yasuke had access to the palaces and corridors of power and would meet the mightiest players of Christendom. He knew also that Valignano would have no less need of protection in war-ridden Europe. The regular reports from Rome, while a few years out of date, made it clear all was not well. The Dutch provinces were engulfed in the flames of revolution; unruly Protestants in France had been massacred, but the survivors still engaged in sedition; heretic English pirates were raiding any Catholic ships they could find after the Pope had declared their witch queen, Elizabeth, "the pretended Queen of England and the servant of crime" and then ordered her subjects to defy her under pain of excommunication. If that were not enough, despite the major victory at the Battle of Lepanto, the Muslim Ottoman Empire still threatened the whole of the Mediterranean, and North African pirates were constantly raiding and slaving along Christian coasts from Greece to Ireland. Ships and harbors approaching Europe were not safe. Yasuke's martial skills would be sorely needed.

The final task Valignano had to fulfill during his time in Utsuki was to fund and finalize a crucial, and decisive, trip within Japan. He needed to travel to Kyoto and appeal to the greatest power in the land, Oda Nobunaga. There, the Visitor would request formal permission to depart Japan and secure approval for his remaining colleagues to continue their missionary work. For the first time in nearly a century, there was a recognizable power in Kyoto who could almost be called a national leader, and obeisance to him was required. Nobunaga had, thus far, been more than tolerant of the Jesuits in Japan, but if Valignano could transform that tolerance into support or even conversion, all of Japan could be Catholic within a matter of years.

While Oda Nobunaga did not hold the title shogun—a position essentially meaning "military ruler" which he himself had abolished in 1573—the Japanese warlord now controlled the dominant center of the country and was making final moves toward pacifying, or as he called it "reunifying," the rest. He ruled primarily from his castle in Azuchi, not far from Kyoto, the ancient capital and still home of the in-name-only "ruling" emperor. Azuchi was close enough to keep an eye on goings-on there, but far enough away to not become embroiled in imperial court shenanigans. For that, Nobunaga had also secured quarters in a temple called Honnō-ji, on the outskirts of Kyoto. There, he and Valignano were to finally meet.

But reaching Nobunaga was no simple undertaking. Traveling to Kyoto, or Azuchi, from the Kyushu coast was at least a two-week journey, three hundred miles away and inland toward the heart of Japan.

The most established and safest route was to travel along the Seto Inland Sea, a sprawling waterway that split the largest Japanese islands down the center, peppered within by some three *thousand* smaller islands. Alas, the islands and surrounding waters were entirely controlled by ruthless bands of pirates.

Each day, ships traversing the passage were seized and burned, their passengers held for ransoms, sold into slavery, raped or killed. Unless, of course, they'd paid the "courtesy fee," in which case the pirates became the most hospitable of hosts and accompanied ships to ensure their safe arrival at the determined destination. Generous payment, with all its perils, afforded the best chance for safe passage. Alternative paths would take them by land and add *months* to their journey, crossing treacherous mountains *and* the territory of enemies of the Church. Not an option. It was pirates, or nothing.

The pirates, or Sea Lords, of the Seto Inland Sea were known and feared throughout Asia as far away as southern China and even modern-day Cambodia, well over two-thousand miles away. So much so, they'd harmfully influenced mainland Asian views of Japan. Many Chinese and Koreans assumed *all* Japanese behaved like pirates—violent, cruel, greedy and vulgar. (As if all medieval Europeans were judged solely by marauding Vikings.) Pirates had always been in these seas, but the breakdown in centralized political power during The Age of the Country at War meant there were no governmental forces powerful enough to control or suppress them. Coupled with weakness in the defense of the Chinese coastal provinces, whose riches they plundered, they became ever more powerful.

The Sea Lords—similar to the more traditional territorial lords found on land in that they were often hereditary and clan based—were ruthless in their practices and often desperate, a powerful combination. They had the reputation of being fearless in battle, sometimes fighting to the death against forces ten times their size rather than concede.

The disorganized Ming Chinese soldiers and officials, expected to defend their coast, often fled in dread fear of these eastern "demons," leaving the unfortunate fisher folk and farmers to face the onslaughts alone. No wonder, then, that many of the maritime Chinese communities felt it better to make friends with these marauders than fight them, hosting their clandestine vessels and providing them with crucial intelligence on neighboring provinces. As the pirates grew bolder, they ventured as far as Thailand and met, among others, English and Dutch sailors who also shared tales of their fearsome reputation. Their multicultural crews—including renegade Chinese locals, and Portuguese, African and other Asian mercenaries—often initially posed as legitimate merchants. And, indeed, some of them were. But if trade was refused, or terms were unfavorable, they would turn pirate again in a flash, looting the nearest coast to make good their losses.

In Japanese waters, however, the pirates were somewhat less badly behaved, as long as they were treated with the right degree of respect. While mainland Asia was typically offered no such protection, Japanese travelers and coastal communities could pay "tolls"—protection money—for security against "those other" nefarious, rival pirates. If the gratuity payments were not forthcoming, travelers, and coastal dwellers, faced an uncertain future.

When not transporting travelers and cargo for a "reasonable fee," these same men and women often acted as mercenary navies for one land-based lord or another. Only a few years earlier, a Seto pirate clan had constructed—for Nobunaga, no less—several of the world's first ironclad ships. These bore armor plating fastened across their hulls, and were loaded with small cannon and muskets. According to a Jesuit account, the ships were "the best and largest in Japan, being about the size of [Portuguese] royal carracks" which "shocked the eyes and ears of those who came to see them." The pirate-built ships had devastated their enemies with ease.

Even with the required levies paid, Yasuke and the Jesuits would have to sail through hostile and treacherous waters.

The Mori clan, on the northern Seto coast (modern-day Yamaguchi) had expressly driven the Jesuits away in 1557, two decades earlier. Then, the Mori had taken control of the region from their former overlords and found themselves disgusted with the, in their view, sacrilegious behavior of the foreign missionaries. This usurpation had not been without pirate assistance and some of the Sea Lords remained allies with the Mori. And, the Mori—in continuous warfare with Nobunaga who coveted control over their lands for his "reunification" project—were keenly aware of the armaments the Jesuits could eventually supply to the ascending warlord. The Mori clan bristled at the prospect of greater contact between these unwashed, meddling foreigners and their powerful sworn enemy in the capital city.

Therefore, to avoid the danger from the Mori clan, Valignano and Yasuke took a slightly longer journey than normal, one that avoided the safety of a northern coast-hugging voyage that passed through Mori-controlled territory. They, instead, island-hopped through the central, rougher, but less threatened, waters.

Fortunately, the Jesuits' gratuity payment had bought a powerful and well-armed escort: Lord Murakami of Noshima. The most powerful of the pirate lords.

Until recently, Murakami had been considered a virtual vassal of the Mori clan, but had lately been leaning more toward the rising power of Oda Nobunaga. This followed his fleet's defeat, and near destruction, in 1579 by Nobunaga's ironclad ships (vessels built by a competing pirate clan). Murakami, the consummate sailor, had read which way the winds were blowing. Escorting Valignano and Yasuke to his audience with Nobunaga was part of this shifting alliance.

Described by Fróis as "the greatest pirate in all of Japan," Murakami lived in "a grand fortress and possessed many retainers, holdings, and ships that continually fly across the waves." He was so powerful that "on these coasts as well as the coastal regions of other kingdoms, all pay him annual tribute out of fear he will destroy them."

It was *his* team that would escort the Jesuits and Yasuke throughout the journey. Who but the most feared pirate in Japan to get them through pirate-infested waters?

CHAPTER SEVEN
Pirates and Choir Boys

In early March, Valignano and Yasuke boarded a ship for Kyoto. They embarked from Ōtomo's Utsuki capital and were glad to finally be escaping from Jezebel's territory and the threat she posed. Every face in Bungo looked like a potential enemy and stepping onto a boat was, unexpectedly, proving a genuine relief.

Their transport was no deep-sea Portuguese galleon, but a shallow-hulled coastal beauty, constructed specifically by Ōtomo for this route and task, to transport Valignano and his party to Nobunaga. A gift to the Jesuits, the barge was decked out in red-and-gold cloth with parasols and gazebo-like structures to protect Ōtomo's guests from the elements. Autumn would bring the area typhoons severe enough to rip roofs from houses and sink boats, but it was only March and the sea was reasonably calm, though the sun could be severe and it often rained. Ōtomo had built the luxurious barge and also picked up the entire tab for the Jesuit's protection and safe passage with pirate lord Murakami. Ōtomo was closely allied with the Murakami pirates, and had used their muscle regularly as a mercenary force when he needed seaborne attacks. Here, he engaged them only to see that his charge, Valignano, reached the port of Sakai, from where they'd continue to Kyoto, and the great Lord Nobunaga, by land.

Ōtomo was also warily friendly with Nobunaga, largely because Nobunaga's power had not yet reached his borders and so he needed Nobunaga's friendship as a balance to the other major Kyushu powers—Ryūzōji and Ōtomo's lifelong foe, the Satsuma clan to the south. The sumptuous barge was to remind Nobunaga that Ōtomo's support for the Catholics was not only lip service, but something he would pay solid silver and gold for; an insinuation that the Jesuits were part of Japan's future and, with all due respect, Nobunaga should join the club.

Valignano and Yasuke traveled with three dozen fellow Jesuits and local Catholic representatives from the domains and missions of Kyushu. They boarded from a specially constructed dock, just below Ōtomo's castle in shallow waters at the mouth of the Usuki River. A short distance out to sea, as the forested hills and steaming sulfur pools of Kyushu shrank and the island of Shikoku and its smaller Seto brethren grew, several ships appeared against the horizon.

Murakami's Noshima pirates approaching to escort them.

Despite assurances of safe passage, and their weariness from worrying about Jezebel and her followers, the Jesuit party prepared again for the worst. Pirates were not to be trusted any more than witches, even if they were paid allies. Only five years earlier, these same men had been the allies of the Mori clan. *What if the Mori had offered a better price than Ōtomo?*

Lord Murakami's men proved a mixture of oarsmen and warriors. The sailors were dressed in rough-looking short kimonos, tied at the waist with hempen rope and were either barefoot or wearing rice-straw sandals; the warriors were slightly better dressed, some sporting items of armor and most wearing helmets of some sort, even if these only amounted to leather belts affixed with a steel strip and tied around the head. These men were obviously soldiers, but they were not rich enough to wear the ornate armor Yasuke had seen on the land-based warriors in Kyushu, and in a sea fight, armor would be quite the hindrance if you ended up in the water. Still, the sailors would be handy in a fight should the need arise.

The accompanying ships carried a floating arsenal of weapons; small cannon, spears, grappling hooks and chains, polearm sickles (to counter the enemy's grappling devices) and unstrung bows stowed in racks in the upper-deck cabin for easy access. They also had muskets and grenades; nasty little devils filled with shrapnel and explosives, launched, after the fuse had been lit, by spinning them around the head on a rope; these were kept somewhat safer in a locked storeroom. Their arsenal also included rocket launchers which shot clusters of fire arrows, a tactic recently learned from Chinese or Korean raids or allies, and not yet common in Japanese warfare.

Murakami's men had distinct flags to raise at the right moment when approaching associates and allies, so they could pass without hindrance. God willing, in just over a week, Yasuke's party would arrive in Sakai.

Valignano remained in state, working alone or in conversation with his colleagues, being briefed thoroughly on what he was seeing as they traveled. The opportunity for a double cross was high. Yet, the pirate sailors went about their business, largely ignoring their passengers, although happy to pass on the odd piece of maritime lore or even a joke. They were a hard lot, but the best at what they did, and they were used to foreign "guests," such as Jesuits who traveled regularly between Kyushu and Kyoto.

There were no sleeping facilities on the new Jesuit barge, as they stopped each night on a different island. There, they were put up in the best accommodations available—often small forts or pirate bases at strategic points along the way.

Beside the extortionate tolls, the Sea Lords' main industry was trade; providing and selling sea-derived products, foreign plunder and foods to landlubbers. In addition to fish and sea grasses such as kelp (essential for the ubiquitous Japanese soup stock), people who lived along this narrow and island-strewn sea also dived for pearls and coral. This job was done by the ama, or sea women, celebrated divers who harvested shellfish, pearls and coral for use in kitchens, medicine and jewelry. Yasuke had seen such work before in India, but the fact that the ama were all female and dove nude certainly added a new wrinkle. Ashore, when they stopped each night, he could see the ama more closely and noticed their teeth were rough from grasping shells as they rose from the sea depths, and blood-red eyes discolored from the high pressure of years of deep dives. Yasuke watched in wonder as, like mermaids, they submerged for five minutes at a time off rocks and moored boats, descending to great depths in the clear waters.

As usual, the large Jesuit party had brought their own supplies with them for meals, but also relied on the locals they passed to supplement their meals. Their new hosts each night entertained them. This entailed modest portions of food, but copious amounts of rice wine, sake, toward the end of each meal, to which the Jesuits contributed their prized and highly exotic European wine. As it was customary for any gift brought by the guest to be cracked open and enjoyed by all present straightaway, they had a jolly time and, afterward, a good night's sleep. The frolics likely did not extend to Yasuke, who'd have to remain alert for treachery; even paid pirates were still pirates.

After eating his own evening meal—enjoyed while Valignano presided over the prayers or while waiting for Valignano to retire for the night—Yasuke could finally doze just inside the door of the Visitor's room. Whereas a normal person could easily fit within the standardized six-foot length of the long tatami mats, Yasuke's great height meant his feet spilled beyond, onto a second mat. Each night, he slept to the gentle sound of the wind and waves against unseen shores, the hardworking villagers, fishermen and pirates having retired themselves after dinner, with the sunset.

The exception to this quiet life was in the "grand fortress" which was one of the effective capitals of the Sea Lords' domains, Noshima Island. In reality, a small, but easily defended pair of islands that Yasuke reached halfway through his voyage. The entire main island had been turned into a fortified stronghold, complete with walls which stretched to the beaches, numerous bastions and quays for the ships which docked there—like something out of a fairy tale—with high wooden walls that seemed to keep out the sea. It would take an audacious navy to even *attempt* an attack on the island fortress.

They came ashore at a wooden landing where Lord Murakami himself welcomed them warmly to his base. Murakami was dressed as finely as any of the other powerful Japanese lords Yasuke had met in the past two years. He wore a shimmering robe of deep blue Chinese silk and a black cap of stiff cotton on his head.

The climb up the steep island slopes to his small hall was a welcome stretch for Yasuke's long legs. And, it turned out, Murakami had all the trappings and servants of a land-based aristocrat. Far from the harsh conditions experienced by his sea people, he lived in relative luxury, and was only too happy to show off his fine porcelain and lacquered ware, no doubt stolen for the most part. This night, his important guests, having politely admired their host's treasures, were wined and dined late into the night as he regaled them with stories of various sea escapades and the provenance of his luxuriant possessions. (Mainly China, though he did not mention how, exactly, he'd come by them.)

Out of respect to the Jesuits' celibacy, the customary female entertainment had the night off.

The "Venice of Japan." That's what Europeans called Sakai, their final destination on this sea voyage. An attractive settlement of rivers, canals, mercantile spirit and international trade, and also a center of gun, sword and knife manufacture. It was a rich city with tentacles of commerce stretching far overseas, a city ruled, like Venice, by a merchant oligarchy. The Catholic mission there was the oldest in central Japan and perhaps the most successful. Several prominent Sakai families had already converted and made suitably large contributions to the Church. They were ecstatic about having the chance to host the Pope's direct representative and receive Valignano's blessings, however brief his stay was to be.

Having survived the pirates, the Jesuit party stepped off the barge onto a first-rate stone dock, an arrival far removed from their simple beach landing of two years before in Kuchinotsu. Compared to other ports he'd seen, Yasuke found this one clearly well-to-do and well cared for by its citizens. Warehouses lined the docks and wealthy merchants' compounds, and numerous vast temples, to spiritually safeguard the seafarers, formed the backdrop. Merchants and sailors did not have far to go to give thanks for the safe arrival of their ships and cargos.

This time there were no Latin hymns, but a welcome committee of leading citizens and a large corps of warriors sent by the region's primary Catholic lord, and one of Nobunaga's most senior generals, Takayama Ukon.

The dignitaries all knelt to kiss Valignano's hand and there was much rejoicing at his safe arrival. Valignano then gave public blessings again. Takayama himself was not present, but sent word by a senior vassal that he looked forward to hosting them in his castle in a few days' time.

Afterward, they were escorted straight to the home of Konishi

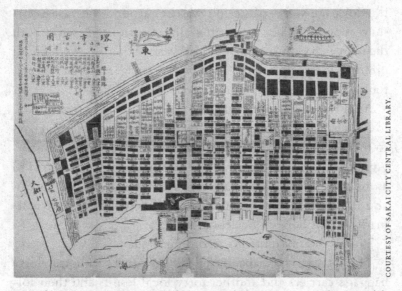

COURTESY OF SAKAI CITY CENTRAL LIBRARY.

Map of Sakai in 1704. The sea, and docks, are at the bottom. East is at the top, and the burial mounds (Kofun) of past emperors are clearly shown.

Ryusa, one of the Jesuits' first converts in Sakai, and a wealthy international dealer in tea and foreign medicinal products such as ginseng and rhubarb. The Konishi compound was large and comfortable, a most suitable place to recover from the trials of the sea journey. But Valignano could not be idle. The region's Catholics dignitaries had gathered to hear a private mass in the home of this generous magnate and were not to be denied. The crowd of wealthy Catholics listened, kneeling on tatami mats in the audience hall, in awed and ecstatic reverence as the Pope's personal Visitor blessed them and led them all in Latin prayer. Some of the congregation muttered the wording, pretending to follow along, but plenty among them could speak the true words. The Jesuit missionaries had been busy in these parts for years and the high-class congregants were well educated in Catholic ritual.

This area of the country was thought to be far less threatening than Kyushu had been, but the troops who'd met them on the dockside were still set on watches around the vast Koni-shi compound. They were, after all, in territory closer to both the mercenary ninja and Buddhist warrior monks, even if both groups had largely been quelled by Nobunaga. In either case, it took only a few, or even one, to end a life.

After recovering overnight, the growing Jesuit entourage was to head to Osaka, the next stop on their journey of several days to meet Nobunaga. The plan was to depart Sakai in a distin-guished procession to convey the power and glory of Christ to as much of this metropolitan region of Japan as possible. It was also Easter week, and processions and finest apparel were hall-marks of this holiest of celebrations.

Things did not quite go according to plan.

The party leaving Sakai included thirty-five packhorses, forty baggage carriers and another forty local Jesuits and their fol-lowers. The priests, brothers and Japanese notables from Sakai, Bungo, Arima, Ōmura and Nagasaki who'd accompanied Vali-gnano on his journey as a demonstration of their numbers and power, were all on horseback. They were dressed for the ride in drab clothing, but nevertheless, had still made an effort to stand out. Large crucifixes were displayed prominently on chains which hung around their necks and colorful devotional banners flew high. Valignano himself led the procession bearing a holy relic: an avowed splinter from Christ's cross encased in a jeweled casket wrapped up for safekeeping for the journey.

This procession was, as intended, one of the most peculiar and arresting things the locals had ever seen. In this war-ravaged age of disunity, Valignano's procession was an extreme specta-cle and novelty for those they passed. Many of the participants were of racial types never, or rarely, seen before. There was Ya-suke, of course, but also a number of pale Europeans and some Indo-Portuguese men too. Even some Japanese marchers wore strange-looking European robes and contributed to the out-landish effect. It's no wonder people of all classes converged from

The narrow streets of Sakai that the Jesuit procession struggled through. This photograph is from the early twentieth century, but besides the electricity pylons, would probably not have looked much different in Yasuke's day.

COURTESY OF SAKAI CITY CENTRAL LIBRARY.

miles around to get a peek at this once-in-a-lifetime experience. Hundreds crammed the narrow streets to see Valignano and his traveling Catholic show. The streets between Konishi's compound, near the docks, and the outskirts of town were crowded with onlookers, gawking and pointing, and the dignified procession soon became the center of a manic crowd.

Yasuke walked directly behind Valignano, and was meant to add to the awe as much as any relic in the Jesuit's possession. The decorated spear, the powerful, muscled arms. Here, in Sakai, he was probably the first African person they'd ever seen, and the locals were astonished.

A commercial city, Sakai's citizens had probably seen darker-skinned men—the occasional tanned Portuguese sailors, Indians perhaps—but, in living memory, no African man such as Yasuke. The combination of his size and shade fascinated the crowd; the women were gasping and grabbing at him as he passed, the men pointing at his muscles and comparing them with their own.

Had Valignano been planning to use him thus from the very beginning? Had Yasuke been selected by Valignano in India, knowing how the Japanese would respond? Or, was it merely an added bonus that the Visitor realized only after arriving in Japan? Perhaps the reaction to Yasuke took the Jesuit completely by surprise as they struggled to leave Sakai. In either case, Yasuke's appearance was a hit and the crowds clamored to get a sight of him.

Lord Takayama's men escorted the Jesuit procession. Along the way, more than a hundred of his soldiers, armed with swords and spears, lined the streets, attempting to control the growing, boisterous crowd. It was not enough.

In a world before photographs, television and the internet, people were still amazed at seeing something new for the first time and unapologetic about its thrill. Scraps and pushing broke out in the crowd straining to get their eyeful. People shouted out in glee and fell backward through paper and lattice windows into the houses and shops which lined the main thoroughfare out of the city. Guards were taunted and jostled. A shop collapsed when too many people crowded onto its roof for a better look.

"But nobody complained," claimed Fróis later; the sight of Yasuke, the African giant, made up for everything. The spectators could not get enough of him.

Mounted warriors from the head of the procession turned back to help the others through, half pulling and half pushing Valignano's horse through the press of people. But, again, Valignano was not the problem. The crowd wanted to see Yasuke. They *needed* to see Yasuke. And as soon as those on horseback were able to clear a space, the mob filled in behind them, surrounding and cutting off Yasuke and the other seventy porters and attendants who remained unmounted. Shouts and kicks and threats were to no avail. Valignano could either leave behind more than half his party—including all his gifts and supplies— *or* take three days to leave the city.

Neither option would work. A third remedy became clear enough.

Yasuke, who routinely walked in attendance beside Valignano, would need to mount up so the procession could continue and penetrate the throng. That was, at least, the easiest solution. Crowds tended to move out of the way of horses. Yasuke climbed into the saddle of a horse commandeered from a local guard with glee. It had been many years since he'd ridden. Then too it had been to escape, but in much different circumstances. On horseback now again, he was able to better forge his way through the crowd.

As they left Sakai and its masses behind, the line formed up again and the Catholic party and their attendants processed through the Japanese countryside.

Throughout the ensuing journey of several days, the experience in Sakai repeated itself. They encountered crowds on foot and horseback, all clearly eager to witness the amazing procession but primarily to gaze upon its most impressive figure, Yasuke. They openly gaped at the African colossus. His presence possibly spoiled the effect of religious awe Valignano had hoped to promote. (Or again, perhaps, Valignano knew *exactly* what he was doing.) Each village and town they traversed brought more and more spectators, after all. And, if they'd only come to gaze upon Yasuke's person in awe, Valignano would at least make sure they left having heard Christian hymns and something of the Good Word.

The holy week of Easter coincided with their last days on the road to Kyoto. They celebrated in the castle of the most truly dedicated of central Japan's Catholics, Lord Takayama Ukon, a powerful general and close senior retainer of Oda Nobunaga. Takayama was a man sincere and secure in his faith, unlike the lords of Kyushu who often seemed more interested in guns than God. He already had plenty of guns and easy access to even more

Sakai-manufactured firearms, which Nobunaga had long since, in theory at least, commandeered for his men alone.

Takayama's castle of Takatsuki was a good-sized but unassuming fortification, only one day's walk from Kyoto, on the flat plain that stretched south from the seas up to the mountains north of the capital city. Its central keep was surrounded by moats from which the walls rose steeply above the water. Its other enclosures, where the warriors lived, were defended by a series of exterior walls and waterways. Attackers would have to break through three or more fortifications to enter and take the central defensive bastion.

The seminary students from Azuchi, the capital of Nobunaga's domain and the latest of Valignano's seminary schools (he'd not yet been there himself and had trusted the Jesuits in Kyoto to establish it), as well as many Catholics from Kyoto and nearby missions, joined them for the holy rites that week.

The festivities included a flagellation procession of penance on Holy Thursday, scores of believers scourging their backs with thick woven and knotted hemp whips until they were a bloody pulp, reveling in the pain and religious euphoria it brought. On Good Friday, a young man who had missed the previous day's bloodletting while drinking and cavorting with courtesans in Kyoto, performed a public self-scourging alone in penance and gave generous alms to the poor. Such piety made the Jesuits feel like they'd somehow been transported back to Rome.

On Easter Sunday, an enormous crowd gathered around Takatsuki castle. Locals, eighteen thousand of whom had been converted by this time, mingled with Christian pilgrims from elsewhere and a large number of Takayama's warriors, all giving thanks for Christ's resurrection. Thousands more non-Catholics also came to see the great event, in awe at the exotic rituals and strange goings-on.

The Easter Sunday festivities themselves started predawn with a lantern-lit procession around the castle accompanied by the novel sounds of Latin hymns and violin music from the Japanese seminary students who'd been practicing for months in anticipation of this day. This was followed by a sermon from Valignano himself and then High Mass where communion was celebrated by thousands. Valignano was well satisfied. His plans were advancing nicely and the enormous turnout likely surprised even him. His eye for spectacle and awe was once again paying off.

After the enormous Mass, Lord Takayama treated the entire Jesuit party and senior retainers to a sumptuous feast of celebration with many speeches in the audience chamber of his grand keep.

Special news came with eventide. Just after the feast, a mounted messenger approached the castle riding hard along the Kyoto road. He dismounted at the stables outside the main fortifications and approached the gate at a run, his armor clanking with every step.

Yasuke, who was on duty outside the main reception chamber, watched him approach the main entrance to the hall and bow to the sentries before having a brief word with them, glancing sideways at Yasuke in surprise as he did. The sentries bowed back and slid the gilt paper doors apart silently on their smooth wooden rails; the messenger entered with dignity despite his obvious haste and strode down the hall, past the visiting foreigners and other guests of honor eating from low tables on the tatami floor to the head of the large room. The messenger waited for Lord Takayama to notice and beckon him, then approached, kneeling and performing a deep bow. He delivered his message in low tones, waited for the reply, bowed again and withdrew as quickly as he had appeared.

Takayama stood, a look of great satisfaction on his face. This was, after all, his show as he'd arranged Valignano's audience and acted as his patron in this part of Japan. He spoke loudly for all to hear and Father Fróis, seated next to Valignano, interpreted his speech in excited tones.

"Fathers and Brothers in Christ," Fróis translated. "We have been done great honor. Lord Nobunaga has requested us to proceed with all haste to his presence. You will be seen within mere days."

Oda Nobunaga had officially ordered them to speed up their visit. He could not wait to see them. The whole room held its excitement but the magnitude of the moment was not lost on anyone. While they'd all known they would be granted an audience with Nobunaga eventually, they had expected to lodge in Kyoto for weeks or months before being summoned. This swift recognition of Valignano's presence was an honor rarely extended.

Takayama, Oda's sworn man since his youth, announced in decent Portuguese that they'd all leave first thing in the morning to meet his lord. The audience would be in three days and if they moved fast they could be in Kyoto late the next afternoon.

Takayama's sincerest hope, just like Valignano's, was that Nobunaga would himself also become Catholic, and the signs were certainly positive if the man was demanding they move up their visit. Would he make some grand announcement? A request for baptism, a grant of further rights and privileges to the Azuchi seminary, or something even bigger—as Lord Ōmura had graciously done upon Valignano's arrival two years before? It wasn't beyond hope, although Valignano had no leverage over Nobunaga in the way he had had over Ōmura and Arima, and he knew it. Nobunaga kept people guessing, *divide et impera*, and kept his cards close. He was a master politician as well as warrior and also, for the most part, a man beloved by his people.

As evidence of that, Nobunaga had planned and funded a massive public spectacle involving thousands of troops and special horseback performances to be performed by him, his sons, an inner core of retainers and his horse guards for later in the week. This extraordinary event was the talk of the region and would bring in visitors from all over Japan, meaning many more listeners for the Jesuits to proselytize to.

Valignano addressed his hosts and colleagues and then, as Fróis interpreted, made a flowery speech of thanks for Takayama's hospitality, even adding some basic thanks and a blessing in Japanese at the end. (He had learned *some* Japanese in his two years.) The Jesuits made ready to leave before dawn the next day.

They left Takayama's castle for Kyoto, a mere twenty-five miles away, to arrive after midday. Humbly walking on foot, banners and icons raised high, candles and lanterns blazing in the cool of the early morning darkness, they processed like stars threading through the night sky.

Valignano led the finely dressed crowd in person bearing his holy relic of the Cross. Next came Takayama and the other Japanese notables accompanying them from Nagasaki, Bungo, Ōmura, Arima and Sakai, all displaying devotional icons. Behind the notables followed twenty-five choir boys from the seminary in Azuchi singing hymns of Christian praise in a joyful Latin harmony. The boys were exquisitely dressed, all carrying white candles, and the four at the front were garbed as actual angels in pure white with artificial wings made from bamboo and brilliant silk. Finally, came the fathers and brothers of the Society of Jesus dressed in clean surplices and stoles draped with long capes, bearing glass lanterns which shone in kaleidoscopic colors and designs. Perhaps for the first time in this region so distant from Christian Europe, the foreign words and notes of the choir did not seem so outlandish. They were being sung by Japanese boys and celebrated by local people as they approached a hegemon who seemed to welcome their message with open arms.

As they traveled together, Takayama proudly showed Valignano's party evidence of the changes in the human landscape; the clear carnage of burned-out and abandoned Buddhist temples and smashed Shinto shrines he'd ordered. In their place were new Christian churches by the side of the road and on the low slopes of the mountains. Man-sized crosses, and bigger, were planted everywhere. The procession had grown again. It was now accompanied by crowds of people, more than two hundred of Takayama's warriors and numerous faithful from his fief who'd joined the procession as it advanced to Kyoto. The scene was set for a most magnificent entry into Japan's greatest city.

It was March 27, 1581.

Only a few hours away, Kyoto and Lord Nobunaga waited. As did consequences that Yasuke, for all his vigilance, could never have foreseen.

CHAPTER EIGHT
A Riot on Monday

The march into Kyoto began as a well-planned Valignano production. The procession of priests, worshippers, seminary students, hundreds of Christian samurai and hired help had journeyed on foot and been followed all the way from Sakai. Sometimes by a dozen curious farmers, and sometimes—when passing through a decent-sized village—by as many as a hundred. But, expanded by a contingent of believers at Takatsuki, their pilgrimage had mushroomed quickly to include *several* hundred escorts and hangers-on.

Engineering large crowds was an old Jesuit tactic to generate publicity and new followers, and the exotic foreigners provided much entertainment as well as provoking sincere religious interest in the local populace. On at least two occasions in Kyoto before Yasuke, such crowds had been seen; one time, more than ten thousand had turned out for a Catholic funeral, something the Jesuits were very proud of. But never had the crowds turned violent and never had they truly threatened the physical fabric of the city. This time, however, was different.

Valignano had not at first realized his entrance into this city was atypical. But Kyoto was the most populous and boisterous city in Japan on any normal day of the week. And today, Kyoto was not having a normal day.

Tens of thousands had traveled to the city for a proper vacation tied to the impending *umazoroe*, Nobunaga's promised Cavalcade of Horses spectacle. The streets and alleyways were packed and heaving with life. Most of the crowd was drunk on a mix of alcohol and festival fever, and having a fabulous time. Pranks being played and good-natured wrestling between old friends and new. Old men were passed out in corners with their distinct conical *kasa* hats tumbled to the ground next to them and their undergarments parted to reveal all; dancers, their kimonos hoisted up, capering to the swift beat of hand drums and the whistle of bamboo flutes, were surrounded by teeming crowds of onlookers. Giggling giddy women, of all ages, arms akimbo, were flashing breasts, or even lifting their kimonos for a laugh, cheer, or even to dramatically pee in the street for still bigger laughter. Other ladies, the more dignified local aristocracy, looked on in horror, covering their dainty mouths and black-varnished teeth with pale hands while men, however dignified they pretended to be, openly ogled and pointed with their fans. In among the adults ran grubby shoeless children, playing tag, war games, nicking snacks from bamboo stalls and drawing gleefully with moistened charcoal on the faces of the passed-out elders while trying to suppress their giggles. The human soundscape was incessant, chatter and cheering interspersed with loud laughs, gasps, exclamations and the occasional gurgle of vomit.

By the time Valignano realized there were so many nonlocals in town, it was too late. Kyoto was unexpectedly, and uncharacteristically, filled with modest country bumpkins and rural warriors from all over Japan. Not from the port cities where Japan was meeting the world, but from the plains and mountains inland, where one could live for generations having not traveled more than a day's walk in any direction and having spent an entire lifetime looking only at the same dozen farming families. Few of these people had ever before seen Europeans— or even a Chinese man—let alone the remarkable wonder of a giant black-skinned warrior from distant Africa, a place most of them didn't yet know existed.

While a man with black skin may have seemed remarkable to them, the connotations were entirely positive. Japanese people seem to have had no negative images associated with dark skin at this period in history. On the contrary, many even revered it due to the fact that the Buddha was sometimes portrayed as black-skinned. Moreover, there was also Daikokuten, a Japanese manifestation of the Indian god Shiva, a deity of wealth and prosperity who is normally portrayed with ebony black skin.

A very good resource to judge the Japanese relationship with, and attitude to, Africans at this time is available to us today in sixteenth- and seventeenth-century Japanese artwork. Particularly in the form of folding screen pictures.

Such screens were typical of the upper class, and decorated lords' castles and the homes of the wealthy. Many have survived because they were so valuable. The black characters depicted in this artwork are representative of this point in history and the plethora of subjects available for the artists to depict. As with European portraits, and modern contemporary photography, the screens recorded what was going on at the time.

The artwork featuring Africans was produced initially by the Kanō School, drawn and printed by teams of artists led by master craftsmen, in the early 1590s, shortly after Yasuke's arrival in Japan. The screens show European, Indian and African visitors, and the

A folding screen representing the arrival of Portuguese merchants in Nagasaki. They are welcomed by Jesuits and local dignitaries. The entourage includes Africans as porters, parasol carriers and armed bodyguards.

This detail from the folding screen above shows both armed African bodyguards and porters carrying bread and a chair. All are sumptuously dressed.

Japanese people who associated with them. The earliest versions of such screens were created by eyewitnesses of these events, and include multiple depictions of dark-skinned men. Later, these *nanban byobu*, or "screens of southern barbarians," would become a genre copied in different forms for a century afterward, providing fascinating viewing for generations of Japanese art lovers.

The most common image adorning such screens is the African

African bodyguards with bows, spears and parasols serve a Portuguese merchant. Kano School, early seventeenth century.

man—always men, with no records of African *women* in Japan until the 1860s—as a sailor, slave or servant of Europeans or Japanese. These men are generally portrayed in lower social positions, carrying parasols, unloading ships, driving exotic animals or serving food. They are unarmed, bareheaded and without shoes or stockings, although often appear wearing smart doublets, shirts, collars, jackets and wide pantaloons. While the cloth of the merchants' apparel is notably shinier or with braid and trim, to show higher quality, the Africans are still well dressed and have several layers of clothing including white undershirts. This indicates they may be socially inferior, but not poverty stricken.

The next category of dark-skinned men is shown wearing skullcaps or sometimes a turban, with extravagantly styled facial hair. These would be Indian workers, or *lascars* as they were known. Both *lascars* and Africans seem to make up the majority of crews on the ships represented in these pictures and they can be seen scaling the rigging, manning the crow's nests and performing gravity-defying feats on the spars. The Japanese artists must have been truly astonished and impressed by this sight to continually paint it with such detail.

A third representation seen in art is the armed black man: Africans like Yasuke, working as bodyguards or mercenary muscle for the Portuguese. These men are typically better dressed, sometimes with shoes and socks in comparison to their barefoot brethren. The weapons they bear vary: sword, lance, trident, bow and arrow, among others. No lower-ranking Portuguese soldiers seem to appear in these pictures, and that is consistent with historical fact. Due to high mortality rates among the lower ranks and voyage fatigue most Portuguese rank and file could rarely be persuaded to board a ship again after their arrival in India. Hence, African soldiers in the pay of the Portuguese were common in the Far East. The inclusion of the trident as a weapon, rare in contemporary European, African or Japanese warfare, is significant and harkens back to Japanese people connecting the idea of Africans with the Indian god Shiva, who is typically portrayed carrying a trident weapon.

The final portrayal of black men, however, is Africans wearing the higher-quality dandified clothing and shoes characteristic of the wealthy Portuguese. A lacquered writing box created by the Rin School of art is one example of this representation, and shows clearly that not all Africans were slaves or indentured workers; the man portrayed here is quite clearly rich and prosperous. He is a very tall and powerful-looking bald man, dressed in luxurious garb, a short cape casually draped upon one arm, and wearing silk slippers. Hanging from his waist are two European-style swords, one on either side of his body. From his great height, he is talking down to a Portuguese merchant and there are two young black boys in evidence. One boy is dressed in expensive-looking Portuguese clothing, carrying a stringed

COURTESY OF PAULO DE CUNHA DONATION, FUNDAÇÃO ABEL E JOÃO DE LACERDA, MUSEU DO CARAMULO

Writing box with a very large and well-dressed African man, two African boys and a Portuguese merchant. This could possibly be Yasuke.

instrument, and the other in Japanese clothes, clearly a servant, carries the merchant's cloak.

This is all evidence of a particular fascination the Japanese of the era had for markedly dark skin as evidenced by the public reaction—and Nobunaga's forthcoming extreme favor toward Yasuke. The Portuguese merchant Jorge Alvarez once noted the Japanese would "travel fifteen leagues to see [black men] and would entertain them for three or four days." Africans were rare but became very respected, and indeed popular, in Japan.

The proof, if needed, was now in Kyoto.

The notion of perhaps exploiting Yasuke's exoticism to help

draw spectators to the holy spectacle of relics and divine singing had clearly turned on Valignano. Crowd management, in whatever form he'd planned or expected, was quickly revealed to be impossible. There were too many people. And they kept coming. The initial flock of several hundred had already doubled in only a few blocks.

The Japanese explicitly relished anything exotic and new and nothing as exotic as this outrageous procession had occurred in Kyoto for centuries. Here, on a morning where tens of thousands of tourists were drinking and looking for something different to see, were Jesuits from the semifabled lands of the south, choirs singing in strange tongues, children dressed as angels and a giant with black skin.

But getting *only* a look, for any crowd this size, is rarely enough. They tore at Yasuke's clothes and scratched his skin with gnarled peasant nails. One fearless woman reached out, shouting in joy, to yank away a piece of clothing as he passed. Not from hatred, but some odd form of affection mixed with a primeval trophy hunting; the urge for a souvenir of the moment when she'd crossed paths with the real live Daikokuten— the "black god of prosperity."

Their march through town had grown into a genuine riot. Thousands of people now. Mostly those seeking the unique experience of partaking *in* a riot. They had little idea, or care, for why they were actually rioting, or who they were pursuing. For most, the chase itself was enough. The shouting and shouldering each other as one corporeal "happening." The laughter and vulgarity and cheers. The spilled drinks and toppled food carts. The random punches and playful pokes. Some even breaking into genuine tears of excitement, while only those few at the very front truly knew who or what they were chasing.

Yasuke. Running mere steps ahead of the pressing mob of thousands. Running, by all later accounts, for his life. If he stopped, he knew they'd rip him into a thousand bloody trophies. His entourage and comrades now separated and spread over several blocks, no battle plan to follow, no orders to obey, this was surely a lone soldier's worst nightmare.

He hurried directly behind several of Takayama's warriors, who shouted warnings and cleared as much of a path as they could. Trying to get to their church in the center of Japan's most crowded city, they shouted at confused bystanders and vendors; those all too easily jostled and even swept into the roiling human tsunami which followed. Deeply regretting, no doubt, that they'd left the horses at Takayama's castle and decided to walk the last stretch for a more humble arrival.

The assigned guards, thankfully, soldiered on and kept at least the senior Jesuits from serious harm. Through the heaving streets and narrow alleys, past gilt shrines and walled temples. The shouts and murmurs of seven then eight hundred then thousands, the mob's sound engulfed the regular city clamor of temple bells, chanting priests, merchant chatter, vendors hawking their wares, workshops, musicians, dancers and street entertainers. Shoving past those on foot, the procession drove aside porters with frames on their backs, peddlers with wheeled carts or *tenbin* poles draped over their shoulders, and the alms bowls of begging monks. All went flying, their owners groaning or shouting out with anger and despair. Those wealthy few carried within their two-man curtained, lacquered, bamboo-curtained palanquins found their vehicles crushed against the sides of the streets, their carriers holding on for dear life.

Trailing along this violent line, now spread over several blocks, were the rest of the entourage. Valignano, the other priests and brothers, accompanying samurai, servants and choir boys. Mixed and lost within the ever-growing throng. The Europeans' outlandish garb and unwashed stale whiff, which normally allowed them a wide berth, was ignored by the ecstatic crush of those hemming them in—each of them secretly relieved that Yasuke, alone, had become the crowd's primary focus.

The next side street spilled into a shrine courtyard and the head of the maelstrom, Yasuke, as well as a Takayama warrior and a teenage acolyte, burst through, the brief moment of open air giving the African warrior's powerful legs and arms room to really work before a new flood of onlookers crushed close.

The rear of the shrine was walled with a flimsy-looking bamboo fence. It was either burst through the fence or be pulled apart by the crowd. Yasuke hoped it was as weak as it looked, kept his head down and smashed through the fence. The aged wood gave way in a shower of splinters and he was through. Yasuke charged onward, willing himself to be smaller, invisible, not so foreign-looking and obvious. In India, he'd encountered cities which seemed unending, alleys filled with people and animals. He'd fancied these Indian cities to have the most people he'd ever see in one place in his whole life.

With each frantic step he took in Kyoto, he was not so sure that was true anymore.

Kyoto had been the capital of Japan for almost a thousand years, and it had been among the largest, and most crowded, cities in the entire world. Before the hundred years of war, up to a half million people had thrived and labored in the city, a commercial metropolis larger than London, Paris, Moscow or medieval Rome. It was divided into two sections: the upper city to the north which was the province of the rich, powerful, divine and holy, and the lower city which was home to everyone else and where Yasuke now ran for his life.

When the African warrior arrived, the city was still recovering from the civil wars, during which it had been routinely fought over and often torched by the most powerful families and clans of the realm. With each round of destruction, more residents fled to other towns, or perished as "collateral damage" or as citizen members of the militias. By 1581, the population had reached around one hundred thousand again, its people coming from all over Japan and of all classes: laborers, artisans, samurai, beggars, aristocrats, hardened bandits, stoneworkers, blacksmiths, carpenters, dancers, weavers, pimps, tofu makers, cloth dyers, cooks, palanquin bearers, gunsmiths, prostitutes, diviners, sake brewers, soil carriers, doctors, moneylenders, water vendors, spies, apothecaries, entertainers, hawkers of locally manufactured goods and purveyors of fine wares such as ivory imported from as far away as Africa.

In a teeming mass, men and women peddled snacks, pickles, water, sake, sacred charms, dancing monkeys, brewed tea, dried fish, noodles, vegetables, tofu, sex, chirping cicadas in woven bamboo leaf cages, ear cleaning services, song birds and blessings through the streets. A more established merchant could invite patrons into his, or her, emporium where they might purchase polished rice, guns, swords, fine tea, lacquerware, fans, swords, silk products, clothes, wild animals, salt, oil, tools and implements for everything from carpentry to the tea ceremony, and singular delicacies, rarities and artwork from China, Luzon, Korea, Siberia, Siam and the far northern reaches of the realm and beyond. The upper class and better sort of merchant were gradually reestablishing their households here as Lord Nobunaga consolidated power and restored stability. Their money brought economic renewal and Kyoto had again become a magnet for high-class exotic consumables from abroad, particularly silk, art and medicine.

While out-of-work laborers and recent immigrants begged for alms and lay sprawled on street corners, business was booming, especially construction. Old guilds were thriving again. Artisans and laborers had rebuilt Kyoto block by block, its buildings walled with pine, plaster, bamboo and paper, and roofed with ceramic tiles for the rich and wood shingles for the less well off.

Most Japanese towns of any size were either merchant or warrior cities. The capital of each domain, where a lord based his army and court, provided a ready market to attract merchants and service providers. The towns grew fat on the stipends of their warrior consumers. Modern–day Tokyo and Nagoya are good examples of this type of city. Merchant-run cities, such as Sakai (now all but swallowed by Osaka), were dominated by business cliques which served whole regions with services and products—but Kyoto was the exception.

Kyoto was a city built not for war or commerce, but for political and religious power. Home to the emperor and his spiritual influence and, for much of Japanese history, home of secular governments too. In an age when there was a very fine line, if any, between religion and politics, the two went hand in hand and attracted everyone regardless of faction or sect. Pure Land Buddhism, Nichiren Buddhism, Shingon Buddhism, Zen, Shinto. A Catholic church. Kyoto's countless temples and shrines drew in hundreds of thousands of pilgrims, priests and studying acolytes. And its manors with their carefully manicured gardens once again housed Japan's aristocracy, senior samurai and diplomatic players.

Despite the city's recent renaissance, entire blocks once inhabited by sprawling estates remained barren and scarred. Vacant plots lay burned or blighted, with nothing more than the overgrown foundations of ancient mansions and once-venerable temples. Shoots of recovery were visible in land reclaimed by squatting smallholders or shanties. Lower Kyoto, aside from a few vast temple compounds, was ranged in blocks and consisted of mostly small buildings hedging in the streets and alleys, shops and houses wedged together with narrow frontages, enough for the doorway and perhaps a display of wares, no more. (Once inside, the buildings remained narrow but ran straight back quite a way to accommodate trade and living space.) Each block had gates which could be locked at night for security, and the well-organized residents' committees kept the streets and communal facilities such as toilets and wells in good order.

While not yet again the metropolis it had once been, Kyoto's proud inhabitants still believed their city was the center of the world, culturally and intellectually superior. It was a place of civilization and wonder, where people of quality from all across the empire appeared, national spectacles took place, and even the occasional Korean merchant, Chinese monk or southern islander garbed in exotic colorful attire could be found. It was a place well established and secure in its national prominence. And visitors like Valignano and Yasuke only confirmed its importance.

As they ran through the streets, Kyoto's only Catholic mission, a tall pagoda-shaped building, appeared above the many canted roofs. Yasuke could finally see their goal, maybe only two or three blocks to the church now. Over the surrounding roofs, reedy tendrils of black smoke lifted from workshops making everything from ironware to sake. Yasuke, heaving from his run, at last inhaled properly and noticed, then tasted, the bitter odor of charcoal smoke from stoves, bath houses and forges as he fought to grab a quick breath.

Another two blocks and the Catholic church fully appeared before them.

Dedicated to Our Lady of the Assumption, Mary, and built only four years before with donations from Takayama Ukon and other wealthy local Catholics, it stood above the neighboring buildings, a substantial three-floored structure, richly decorated in the Japanese manner, roofed in grey clay *kawara* tiles with gently curving gables. Nothing at all like a European church or any of those built in Portuguese enclaves around the world. It had been built in a Japanese style deliberately to make potential converts feel more comfortable and to best integrate into the rest of Kyoto.

To the atypical church's left was a large rice market where merchants traded from busy tables and carts. As Yasuke and the others entered the square, the crowd trailing closely behind, everyone in the market turned to see what the commotion was about. By the time Yasuke reached the compound which

The Jesuit Church in Kyoto by Kano Soshu, c. 1578–87.

surrounded the church, the roar of the throng swept over the whole block.

Several burly Japanese guards in service to the Jesuits stood waiting at the doorway. *"Kochira!"* (Over here!), a Takayama warrior in the lead shouted, waving the entourage ahead, his whole face and shaved pate shiny with sweat from exertion and fear. Yasuke and the others squeezed through the opening, then stumbled across the wide courtyard to the main building. Several attendants from inside the church had already rushed to the compound door and helped drop the bars into place. The pursued, slowing down, hurried from the side door, past outbuildings and deeper into the church compound to the central tabernacle.

The interior was in shadow, a stark contrast to the bright sun outside—some light entered through the wood lattice and paper windows and a few wax candles in silver candlesticks, looking oddly out of place in this otherwise Japanese-style building. The far end of the matted room was dominated by the altar, and a simple crucifix, and while the ceiling was not as high as a church's normally was, it was high enough even for Yasuke to stand up straight, a rarity in the last two years. Yasuke knew there were no real fortifications to help keep the clamoring crowd at bay, only a thin wooden wall topped with tiles. The gold rendering of Christ on the cross, forming the backdrop to the altar, stared down at them all, deep eyes full of compassion for their plight, starry in the fluttering candlelight. He would have to be their protection now. Yasuke perhaps dropped to his knees for a quick prayer of deliverance. By now, the rest of the party—those trapped farther back—had successfully pushed through the crowd and gotten into the compound also.

The superior of the Kyoto mission, Father Gnecchi Organtino, appeared. He was a thin, long-bearded European of maybe fifty, who calmly delivered orders as outside, the noise of the ever-growing mob hit the doors and walls around the courtyard, like something solid slamming against the bracings. The tatami on the church floor visibly quivered from the energy of the gathered horde, now numbering in the thousands.

Several Jesuit brothers and six more lay brothers appeared from various shadowed coves and raised walkways. Many of the Jesuits could not help first eyeing Yasuke's enormous size with the same curiosity as the crowd outside even as they rushed to help hold the compound gates.

As the first stones fell over the walls, a shout of *"Shutters!"* came in Japanese and the wooden doors of the ground-floor windows were slid firmly closed. Others ran to those compound doors yet to be attacked, racing to prop several long benches and a pair of low tables against them.

Yasuke was no doubt frightened. Trapped, having allowed himself to be so cornered. The church was not built for siege, and the continued rocking of the outer walls and the sound of more and more roofing tiles being hit by stones and other missiles hinted the structure would not last long under sustained assult. Also, forbidden to carry weapons, not one of the missionaries was armed.

"We heard the sounds of people wounded by stones," Fróis recorded later. "And, others dying."

The outer wall started to give and planks split in two, several gleeful faces shifted and pushed into the gap, searching for a glance of their quarry. More rocks thudded against the front walls. Something crashed upstairs, a rock hitting one of the windows, and then the lattice exploded into splinters. Outside, the crowd was unevenly chanting both to be let in and to let "him," Yasuke, out.

"Everyone agreed that if we displayed the man, we could earn eight to ten thousand *cruzados* (Portuguese currency, and a huge amount of money—about a third of the mission budget for a year) in a short time," joked Fróis later in his letters to Rome.

Outside, in front of the church compound, the mob stood shoulder to shoulder, spilling into adjoining alleys, and into the rice market—kids were on parents' shoulders for safety and old people, less able to keep themselves upright, were being trodden underfoot.

Then, suddenly, the crowd was moving *away* from the church. And quickly too. Staggering, stumbling to escape. A second stampede began as half a dozen cavalry and a dozen lightly armed foot soldiers marched slowly into the center of the mob. The mounted soldiers parted the crowd with wooden staves, beating men and women alike. Their clothes bore the *Oda mokkou* crest, a black flower blooming with five dark petals, each petal lined in gold; the crest of Lord Nobunaga's clan, the Oda.

The Oda soldiers shouted at those they passed, their words lost beneath the murmur of a retreating mob and the occasional scream of protest and agony as a staff hit home. Their horses pushed bodies aside, and human feet clad in straw sandals kicked out at any who got too close. Soon all that was left in the marketplace and around the walls were a few sorry trampled bodies. The silence was palpable, but shortly fists were again pounding at the church compound door. But they were different this time, somehow more assertive. Official.

This time, it was Nobunaga's men.

Father Valignano and the other priests and brothers, their

hearts still thumping, pulled themselves together, stood straight and assembled by the splintered gate, fingers laced together as if in prayer. Yasuke waited too. *What else could this horrendous day bring?*

The door opened to reveal Nobunaga's guards, the street behind them clear beyond a few bodies still writhing in agony, and the dust of thousands of retreating feet. The cavalryman—lightly armored for policing duties, not battle—bowed briefly to the priests from his saddle. There were two swords thrust into the sash at his waist.

"Yes." Father Organtino, the head of the mission, spoke in superb Japanese. "How may we serve, officer?"

And then the order: "Lord Nobunaga will see him."

"Yes," Organtino replied. "Father Valignano and I have been granted an audience—"

"No." The captain shook his head. *"Him."* He'd pointed behind the priests, and the missionaries all turned to Yasuke who stood at attention, rigid, eyes forward gazing only at the churning dust outside.

In shock at his deliverance, his heart finally slowing again, Yasuke was now calm enough to appreciate what had just happened. Thousands of people made to vanish in minutes by only a few sticks and the painted crest of one man.

"His Highness requests the pleasure of this man's presence," the soldier clarified again, more politely. "He desires to see what disturbs his peace."

Valignano kept his face serene, emotionless, unreadable. Their eyes met for only an instant. But was it triumph or jealousy Yasuke saw reflected there?

"Shouchi itashimashita." Yasuke shifted his gaze to meet the soldier's fierce stare and gave a deep judicious bow. *I hear and obey.*

CHAPTER NINE
Tenka Fuba

Lord Oda Nobunaga, as with many who've reached such heights of power, was not known for his abundant patience. Yasuke was expected immediately.

It was only a five-minute hike from the Jesuit church to Nobunaga's current headquarters at the Honnō-ji Temple—a large walled Buddhist compound Nobunaga had commandeered a year before after he'd piously, and shrewdly, donated his own lavish Kyoto residence, the Nijō Palace, for the use of the imperial family.

Valignano and the others had not been invited.

In short order, fresh clothes were found to replace Yasuke's ripped garments—not an easy task considering his size—and then just as quickly, Yasuke and Father Organtino, a longtime acquaintance of Nobunaga and the obvious Jesuit to accompany him, were hurrying out the gates and following Nobunaga's men.

In deference, as no guest dare appear before Nobunaga armed, Yasuke's weapons were left back at the church. It was the first time Yasuke had been unarmed in more than a decade. Replacing his customary spear was only his trust in Father Organtino, a man he'd met moments before.

Gnecchi Soldo Organtino, another Italian, had been in Japan for eleven years, living almost that entire time in Kyoto, and was now the lead Jesuit in the capital region. He'd seen the dawn of Nobunaga's power firsthand and had become the Oda lord's main Jesuit confidant. Their conversations were wide-ranging and deep; he taught Nobunaga the use of a globe and other innovative instruments and Organtino had always been impressed with Nobunaga's keen intelligence and ability to quickly grasp new concepts. Something approaching a real friendship had developed between the two men, and it was through the Jesuit that Nobunaga's respect for the western foreigners developed. When Father Organtino asked for permission to build a new seminary in Azuchi, Nobunaga's primary city, Nobunaga had agreed promptly and the Jesuit had transferred to Azuchi to supervise its building and act as its first head.

Along with his mentor, Fróis, Organtino was considered the best linguist among the foreign Jesuits. He was also the architect of the church in Kyoto and must have been very relieved to be escorting Yasuke, whose riot had nearly destroyed it, away from his life's work. The previous day, he and his seminary students had been in Takatsuki astounding their audience with Latin hymns for Easter. Now, Organtino walked the short distance to an audience with his friend, Nobunaga, escorting Yasuke. There could have been no one better to introduce the African warrior to the man whom the Jesuits called, somewhat erroneously, "The King of Japan."

Nobunaga's men led the way on horseback. The bright multicolored ribbons which entwined the horses' tails and then attached to the back of the riders' saddles seemed to lead Yasuke onward. A dozen more foot soldiers marched directly behind him and Organtino, their armor clinking and their straw sandals scuffing on the hard mud of the street. Ordinarily, a supplicant granted an audience was required to bring breathtaking gifts to be presented through an official go-between. But Yasuke had nothing, and there was no time to worry about formality. This was an official summons.

From the mounted soldiers, whom he'd dealt with many times before, Organtino elicited some more information in preparation for the audience. It was best, whenever possible, to know what Nobunaga wanted long *before* being directly asked. It seemed Nobunaga himself had noticed the commotion Yasuke's arrival had caused—difficult to miss the roar of a mob of thousands less than five minutes' walk away—and had, at first, demanded only to know who or what was disturbing his peace. Once the Japanese lord had gotten word of the cause from his informants, however, his focus had changed and he demanded to see this so-called "black man," not believing such a wonder existed.

Nobunaga had thus far tolerated the Catholic mission and its evangelists as interesting curiosities and purveyors of thought-provoking ideas. But the winds often changed fast in Old Japan. It was rumored—a rumor started by Father Organtino himself—that Nobunaga was thinking about becoming a Christian. This was entirely the missionary's flight of fancy, however. Nobunaga had little time for any god, and was more likely an atheist and certainly an iconoclast who tore down temples that offended him, mercilessly massacring their inhabitants without the slightest worry of eternal damnation. He had little concern for spiritual matters, except where they could render him temporal advantage or political legitimacy.

The inspiration for Nobunaga's ostensible support for Jesuit interests was twofold. First, to antagonize the Buddhist establishment with whom he'd been at war for years. He believed priests and monks should stick to spiritual matters, something that many Buddhist sects in Japan most definitely did not do. Many temples still owned vast estates, ran whole domains and manipulated temporal rulers with abandon, threatening them with eternal damnation if they were thwarted. They even retained armies of warrior monks, sohei, who fought with a legendary fanaticism for their abbots. (Ironically the Jesuits, whom Nobunaga was clearly fond of, were often accused of similar tendencies in other parts of the world, but Nobunaga did not know that.) These sects hated Nobunaga because he threatened everything they possessed and they fought him at every turn as he absorbed more of Japan. In turn, by this year of 1581, Nobunaga had subjugated virtually all monastic threats, and only a few remote holdouts remained to be dealt with.

The second reason he welcomed Jesuits was to harness European knowledge of the wider world and secure access to novelties such as cannon, globes, glasses and peculiar clothing, including European cloaks and tall wide-brimmed Portuguese hats. He regularly wore these foreign garments in public spectacles, triggering minor fashion crazes among the wealthy, and challenging the skill of Japan's tailors who'd never attempted such styles before.

Organtino briefly filled Yasuke in on a few stories to ensure he knew what was coming. Nobunaga could be the nicest of men, but he could also be an author of destruction and death.

The Jesuit priest recalled that as a fresh-off-the-boat missionary, he'd been present in Kyoto ten years before when the city was covered in smoke for days after the burning of the three hundred temple buildings on Mount Hiei, just to the north. The warrior monks who lived there had perished, along with their wives and children. All of them. Not even an animal had been left alive on that mountain. More than twenty thousand souls had perished in the fury and flames wrought by Nobunaga's troops. He was not a man to be crossed.

Yasuke and Organtino approached the Honnō-ji Temple, a former Buddhist sanctuary. It was one of the largest walled properties in the lower city, and so Nobunaga had commandeered it as his Kyoto residence. That the act would infuriate Buddhists across Japan was merely a bonus.

Although the daily services and devotions had long since ceased, the grounds still retained the trappings and outward appearance of a religious establishment and everyone, including Nobunaga, still called it "the Honnō-ji." The buildings within could still have been devotional, if not for the constant tramp of marching feet, and soft slow clop of cavalry mounts shod in straw shoes. (Horseshoes, in Japan, were made of straw, and on a long journey, horses would wear through dozens.) The inner sanctuary, devoted to Great Saint Nichiren—a Japanese Buddhist priest who'd lived three hundred years before and who some believed was a reincarnation of the Buddha himself—had been shorn of its customary decoration of the statue of Nichiren surrounded by arrangements of exquisitely carved lotus flowers. Nobunaga had removed their lustrous gold coating for "safekeeping," and the statue and carvings had been burned. The monks, of course, were also missing. Supplanted in full with well-armed samurai sworn to the Oda clan, grooms, cooks, maids and a plethora of other servants.

Yasuke and Organtino followed their escort through the main gate and bowed to scrutinizing guards. Word of Yasuke's summons and arrival had spread fast through the compound, and Nobunaga's people formed a small crowd around them. While Organtino attracted the odd, interested glance, most eyes stayed on Yasuke. This smaller assembly proved better behaved than the mob though, and Yasuke and Organtino's party slowly worked its way through the temple's inner courtyard, past numerous build-

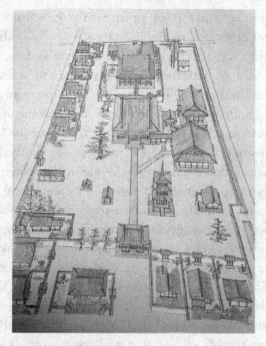

The Honnō-ji Temple at the time of Yasuke.

ings. They passed the vast incense-burning vats, but no smoke wafted within the compound that day. Nothing burned there in these godless times. They veered to the right directly before the main sanctuary to enter a large building, which served as Nobunaga's audience chamber.

Yasuke and Organtino were both nervous. Nobunaga was not past executing an entire village or burning a temple if he felt he had been somehow slighted. It was entirely possible he'd merely summoned the person who had breached the public peace for a swift execution. Yasuke did not want to end with the death of a criminal. And, of course, even if he was not to be executed, the fortune and official favor of the Jesuits now rested heavily on the African warrior's shoulders. Valignano had spent several years and a small fortune putting together this introduction, but now, with no warning or discussion, *he* was to become the chief emissary for the Jesuit mission in Japan. Yasuke was not unaware of his new obligations and he did not want to fail Valignano.

Organtino was also highly aware of these factors, and so it was with cautious steps they each approached their fate.

The antechambers within the temple building were crowded. An assembly of samurai and merchants mostly come on business of one sort or another. But as soon as Yasuke and the Jesuit had been announced, all other supplicants were summarily dismissed.

Nobunaga ordered Yasuke and Organtino to enter at once.

The doors of the main chamber slid open along wooden grooves to reveal a large tatami-matted chamber with a high vaulted roof and stout wooden columns along its length. Though the columns were plain waxed timber, the roof beams between them were decorated with assorted paintings of gold, silver and bright colors which covered the wood completely. These still retained their Buddhist themes: scenes of galloping animals and magnificent nature in vivid reds, blues and greens, illustrating stories from the life of Nichiren and other saints. The sliding walls also functioned as window shutters, and most of them were pulled open to let in the late-afternoon sunlight. Those that were visible were covered with simple inked scenes of nature, gold leaf coating the frames; perhaps it was the same gold that had once adorned the lotus flowers in the main temple sanctuary. More gold was painted onto the lattice screens that stood to the rear of the chamber. Behind the screens, in silhouette, petite shadows huddled together and drifted by like ghosts; the court ladies gathered eagerly awaiting this rare spectacle.

Organtino indicated that Yasuke, momentarily frozen to the spot, should enter, and he bowed his head to fit under the low door beam, reflecting as he did that from their vantage point, the people inside could have seen only his seemingly headless torso. Quite a sight. Still, no harm done; if he seemed more otherworldly, it would only add to his mystique.

Yasuke was announced, with slightly muddled geography,

by the head of the escort who'd brought them as "The Black Monk from *Christian*."

Yasuke crossed the threshold to enter the reception chamber, and immediately prostrated himself, bowing his head low to the floor. Organtino, slightly to his rear, did likewise. Two rows of Japanese men in formal robes knelt on the tatami, their backs straight and heads turned toward him, along the sides of the room leading up to one single man sitting on a slightly raised section at the far end facing the entrance through which Yasuke had entered. He spotted Takayama Ukon among the kneeling courtiers—the same powerful lord who'd paraded into town with them only hours before—and Yasuke tried to gauge Takayama's expression. Anything to give him a clue as to what was coming next.

Oda Nobunaga sat cross-legged directly in the middle of the dais, garbed in an opulent short-sleeved green cotton over-robe with an under-robe of shimmering white silk. He was well built. Tall, thin, sparsely bearded with a small moustache. Not yet fifty years old. Behind him was a sheathed sword, upright on a stand; no other piece of furniture was in evidence.

He was, and had been for a decade, the most powerful man in Japan.

Japanese systems of government had gone through a bewildering number of changes since the time of the semimythical first Emperor Jimmu, who is traditionally believed to have ascended to the throne in 660 BCE. For the next thirteen hundred years, emperors and empresses (several rulers were female) ruled as tribal leaders or semimagical shamans, until 645 CE when, borrowing systems from China, something similar to what we would now recognize as constitutional government came in to force. The emperor and imperial court became the center of power, forging a unified nation-state, and also establishing rites and a formalized spiritual role for the emperor as "Son of Heaven."

Over time, leading families among the imperial court nobles usurped power, and the person of the emperor became ever less relevant to actual governance. Legitimacy, however, still remained with the office until 1185, when, after years of civil war, the Minamoto clan seized the reins of government, moved the capital city from Kyoto to Kamakura (near modern-day Tokyo), and established military government under the rule of a shogun, what we would now call a hereditary military dictatorship. The emperor and imperial court were "released of the burden of government," and now "free to concentrate on their spiritual role" of ensuring that heaven looked beneficently on the Japanese realm, *Tenka*. Aside from a brief three-year period of restored imperial rule from 1333–1336, and a change of shogunal family to the usurper Ashikaga clan, the situation had not changed in the centuries that followed.

By 1581, the imperial family was impoverished and largely irrelevant to citizens' lives, and the "palace" in Kyoto—what was left of it—had deteriorated into moldering ruins. Bold sightseers could walk around its deserted grounds unchallenged, and unnoticed. One Jesuit, João Rodrigues, described it: "The walls surrounding the king's palace were made of wood covered by reeds and clay, and were old and dilapidated. Everything was left opened and abandoned without any guards, and anyone who desired could enter the courtyard right up to the royal palace without anyone stopping him, as we ourselves did."

Meanwhile, the institution of the shogun's government had undergone its own changes over the years. Until the fifteenth century and The Age of the Country at War, the shoguns had been the undisputed rulers of Japan. But, as the country descended into chaos and conflict, shoguns lacked the means to enforce their writ, and the country disintegrated into warring domains where regional powers became independent again. The office of shogun limped on, impotent and impoverished, until in 1573 the last holder of the title, Ashikaga Yoshiaki, was deposed by Nobunaga. Nobunaga took Yoshiaki's remaining

THIS PRINT KINDLY PROVIDED BY THE SANPO TEMPLE, YAMAGATA PREFECTURE, JAPAN, WHERE THE PICTURE IS KEPT.

Posthumous portrait of Oda Nobunaga, by Giovanni Nicolao, c. 1585.

power, but as yet he had not reunified the country fully, nor had he claimed the title of shogun for himself.

Not yet.

Nobunaga was a tangled knot of contradictions.

He lived in an age of war and thrived in this environment. Now in the prime of life at age forty-eight, he had both caused the deaths of tens of thousands and also effectively seen to the welfare and good governance of the territories that came under his control or submitted to his power. What he believed was that The Age of the Country at War would never end without one almighty, brutal leader to enforce peace through national conquest. Only afterward, could someone bring about the unity and national tranquility that would arise from the ashes and rivers of blood. Few others in the previous century of turmoil had his vision and none had succeeded as he had. (Why he is still today, perhaps, among the most popular personalities in Japanese history.)

Works of justice and mercy or huge displays of entertainment, culture, generosity and wonder came to him as easily as terror and fire. He was an expert military strategist but reticent about revealing his plans, sometimes shocking both his enemies and underlings by his audacity and covert tactics. He did not follow any rulebook, nor fear long odds. He was a natural with innovation, pulling off new ideas and stunts no one else would ever even dream of. He was highly sensitive about his own honor, but often acted as if he despised all others, speaking to even his most senior retainers as if they were lowly servants. His *actual* servants addressed their lord only with their hands and faces touching the floor, not one daring to raise his head, and when he dismissed them with an almost imperceptible nod of his head, they hurried away speedily, lest some error be found. Yet, conversely, he often spoke quite familiarly with them about details such as his new hawks or the weather. He was upright and prudent in his dealings. He disliked delays and long speeches. He was accompanied by escorts which often ran into hundreds or even *thousands* of troops, but most of all enjoyed riding out with a few pages and attending to his menagerie, especially hand rearing the fledgling chicks of his hunting birds.

Most importantly for Yasuke and the Jesuits, Nobunaga was a notable connoisseur of novelty. He carried a deep love for rarities from far-off places, promoting guns in his teens and, later, European-style body armor. He often wore a lion skin, sported Chinese or Portuguese clothing, flew hunting birds from as far away as Korea, and used exotic items such as globes, wineglasses and all manner of rare gadgets. He loved talking to foreigners whenever possible. Father Organtino was prime among his foreign conversants, but Nobunaga also employed and enjoyed the company of Chinese engineers and artisans, and—now for the first time—an African warrior.

He'd been born the son of a minor clan lord in Nagoya in 1534 and was, by all accounts, a spoiled, impetuous and eccentric youth. He preferred fooling around with his friends to studying for the future leadership of his people, and numerous stories are told of his scandalous actions, including disrespectful behavior at his father's funeral in 1551. Instead of reverence for the deceased patriarch, he threw incense at the altar and dressed as if for a celebration. His actions won him the name "Great Fool," and the clan despaired for its future.

Until one day in February 1553. As the snow still lay thick on the ground, one of Nobunaga's senior retainers, Hirate Masahide—an Oda samurai who'd been Nobunaga's tutor from birth, stood with him at his first battle and helped arrange his marriage—performed the rite of seppuku (ritualistic suicide via cutting open his own belly) to show his sincerity and dedication to his lord, and thereby draw attention to the perilous situation the clan seemed to face if Nobunaga continued in his erratic ways. And win his pupil's attention it did. Nobunaga was shocked, and grieved publicly, even building a temple to honor his dead mentor. Through this loyal samurai's death, Nobunaga seemed to grow up and focus. He had started upon the road to a greater destiny.

It is likely, also, that those earlier years of tomfoolery were simply a clever performance to convince his rivals for the leadership of the clan that he was little threat. Rivals, who soon became victims of Nobunaga's novel methods, and brilliant acting. One by one, Nobunaga dealt with his competing family members, eliminating them until he was the unquestioned clan leader. Then he turned his attention to the rest of Japan.

While he was still in his teens and only a few years after the first European firearms had been introduced to Japan, Nobunaga already had a permanent corps of five hundred disciplined gunners. (By comparison, Elizabeth I had no standing army at all,

The Battle of Nagashino at which Nobunaga is supposed to have refined the technique of volley fire.

and a corps of less than four hundred Portuguese musketeers in Ethiopia had recently changed the course of a decades-old war in 1542.) His fragile military position was more dramatically improved when, in his late twenties, Nobunaga inflicted a decisive defeat on the neighboring and hugely formidable Imagawa clan at the Battle of Okehazama in 1560. The plan that defeated a force of ten times his army's size was audacious and decisive. He ordered a lightning cavalry strike during a rainstorm straight into the enemy camp and Lord Imagawa was decapitated. With the enemy army's literal head gone, the Imagawa forces collapsed and retreated within days. This victory eliminated an enemy who'd threatened to overwhelm the Oda clan for years and left Nobunaga free to go on the offensive rather than merely look to the defense of his small domain.

Nobunaga continued to fight and expand the Oda territories in central Japan over the next two decades. His notable acts of violence and politics included the killing, or enforced seppuku, of various family members (he even turned his brother-in-law's skull into a sake cup) and thousands of militant Buddhist monks, as well as legions of enemy combatants. He acquired a reputation for ruthlessness and cunning in addition to martial prowess, generosity and an eagerness to promote retainers on merit rather than only birth.

The "Great Fool" epithet ceased to be used and his personal seal came to include the ambitious words: *Tenka Fubu*. Loosely translated to mean "The Realm United by Might of Arms."

In 1568, he made his mark on the national stage by intervening in the shogunal succession, enforcing his own candidate over a rival. This meant he could more easily manipulate the supposed military ruler in Kyoto. He also piously, or tactically, repaired a national shrine at Ise, socialised with the imperial family and bestowed funds upon that impoverished institution. These actions brought him honor in the eyes of all and grudging respect from his foes. The men and women under his rule adored him.

Throughout the 1570s, Nobunaga's military campaigns enabled him to consolidate his power and extend it to the east and north of Kyoto. He was able to do all this not only through deft use of traditional tactics and the skilfull management of his retainers, but also through his invention and decisive deployment of the massed musket volley—three ranks of gunners taking turns to fire and reload, thereby assaulting an enemy with a continuous hail of bullets—which was refined at the Battle of Nagashino in 1575 where he all but annihilated twelve thousand of fifteen thousand samurai of the formidable Takeda clan under Lord Takeda Katsuyori.

Nobunaga's father had been the lord of only one small domain; by 1581, when he met Yasuke, Nobunaga had taken full control of eighteen key provinces of the sixty-six, including the eight most important ones around Kyoto.

His generals were pushing ever onward in their efforts to reunify the realm, *Tenka*, under one ruler. Even now, he was putting exceeding pressure on the only other significant remaining clans and warrior monk bastions and, within a few years, his designs would have completed the reimposition of central Kyoto power over all of Japan. *Tenka Fubu*.

"Everything is under my control," Nobunaga once assured the Jesuits. "Just do what I tell you and you can go where you like." And, for this short time, almost everything in Japan truly was under his control.

CHAPTER TEN
Feats of Strength

It was not Yasuke's first time before an important man.

For most of his life, he'd worked and lived side by side with hugely powerful men: mercenary generals in India, captain majors on Portuguese ships, Valignano and now Nobunaga.

In Japan alone, he'd now spent two years with Valignano in or around the courts of minor lords and was used to the etiquette. Starting in China, Valignano had ensured that both he and his bodyguard were as prepared as possible for the protocols of Japanese courtly behavior. It wouldn't further Jesuit interests in the slightest if they offended their hosts. Yasuke knew the appropriate depth of bow for each status and occasion, the convention of giving presents through a third party, and how to venerate gifts received by touching them to your head. A Japanese observer, Ōta Gyūichi, specifically praised his good demeanor. Here, certainly, was an occasion when the deepest and most respectful of bows was needed.

Yasuke kept the position of obeisance he'd adopted the moment he'd entered the room, waiting under the courtiers' curious stares and hushed chatter until someone indicated in a loud voice full of laughter to Organtino they could both approach. Even with his face to the floor, Yasuke knew it had been Nobunaga. No one else in the room would have dared spoken.

The African rose slowly to his full height and walked upright and confident, despite his nerves, past the kneeling courtiers to the far end of the room. Father Organtino followed him and they approached together within a few feet of the great hegemon, Oda Nobunaga.

Directly before the mighty Japanese lord, Yasuke knelt again, bowing deeply. All chattering in the room had stopped, the courtiers seemingly stunned by his presence, and everything now took place in such complete silence that it seemed as if they were all moving in a dream.

Upon Nobunaga's gesture to approach further, Yasuke crawled forward awkwardly on his knees. Given his size, he found the act difficult to do in a dignified way. Keeping his head as low to the mats as possible, he moved to within a few feet of the raised dais and lowered his head again all the way flat to the floor. Nobunaga addressed Organtino again, welcoming him, and asking the priest to tell this new man that he was gratified by his presence too.

Yasuke understood the words well enough without translation as Father Organtino raised his head and returned to a bolt-upright kneeling position. Yasuke mirrored the position, but keeping his eyes to the floor as Organtino explained to Nobunaga that Yasuke was able to communicate in Japanese if Nobunaga desired to speak to him directly.

Nobunaga tested the claim at once. He welcomed Yasuke and asked his name and if he was comfortable.

Yasuke, after two years of hearing courtly Japanese, replied with customary deference, and in an acceptable manner.

The Japanese lord nodded, pleased, then openly chuckled. A big grin still on his face, he rose incredulously to inspect this unique, curious visitor. He asked Yasuke to stand again. He'd heard about the cause of the great riot, was interested, but had serious doubts as to whether the man was actually black-skinned. *Was this not in fact some public relations trick of the curious foreigners, concocted to grab attention from the populace and please him? Could he catch these Jesuits out in the joke?* He would get to the bottom of it, and it was, in any case, an ideal excuse for a fun break from the business of forging a nation. Now, the added notion that the strange man from alien lands half a world away could also converse in "civilized language" was an even greater surprise and delight.

Yasuke did as he was bid while focused on remaining respectful. Too much rested on the next five minutes.

Nobunaga viewed the African warrior close up.

Touched his skin. *Rubbed* it.

Black was the color—if one believed in such things, which Nobunaga did not—of gods and demons. Not men. Nobunaga had seen such a "god," Daikokuten, before in the Kiyomizu Temple, a short walk away. And, the protective guardian demons at the gates of most temples were often dark skinned: black or deep burgundy.

Nobunaga called for water and a brush. He commanded his guest, good-naturedly, with a quick smile to his vassals, to strip to the waist.

Yasuke bowed and promptly removed his shirt and doublet. The African warrior awaited the scrubbing and held a smile, little worried his skin would change color.

The servants quickly fetched a bucket and brush, and offered them up to Nobunaga in two hands from a kneeling position, their palms facing heaven and their heads remaining bowed to the floor. The Japanese lord snatched up the brush and scrubbed the giant man's skin himself. Scouring at Yasuke's exposed arm. Then his back. Along his neck. Throughout, Yasuke stood perfectly still and firm, flexed, muscles tensed and bulging. Knowing he was on display.

Of course, to Nobunaga's great interest, no matter where he turned, Yasuke's skin color remained stubbornly the same. He eyed his guest curiously, an incredulous grin returning to his face. With a big sigh of satisfaction, Nobunaga tossed the brush back into its bucket, still being held aloft by the prostrating servant. Nobunaga announced to the whole room he was at last convinced of the verity of this dark-skinned wonder. He nodded to the foreign warrior with a smile.

Yasuke bowed deeply.

Nobunaga called out for food and drink. There would be an immediate party to celebrate the coming of this miracle man to his court. Yasuke had become the new guest of honor.

The other men in the room burst out with cheers and genuine excitement. Only one of them—Takayama, who'd hosted Valignano and Yasuke the previous week—had ever seen Yasuke, or any black-skinned person before, and it was as much a surprise and novelty to them as it was to Nobunaga. Takayama was also visibly pleased. His friends, the Jesuits, had clearly already found favor with his lord, Nobunaga. It was not always easy for Lord Takayama to reconcile his loyalties to Nobunaga *and* to his faith, but Yasuke had just helped to bridge the two very nicely.

Nobunaga, oblivious to Takayama's reflections on faith and reconciliation, just wanted to have a genuine celebration and welcome Yasuke to his realm. He sent a man to fetch three of his sons who were staying in the nearby Myōkaku-ji temple. This, for them, would be an educational opportunity; a chance to see a new and fascinating kind of person never before encountered in the capital.

While they awaited the arrival of Nobunaga's sons and the

serving of the feast, Nobunaga made small talk. The other courtiers conversed interestedly between themselves as Takayama
quietly tried to address the various questions put to him by his
curious colleagues.

Presently, Nobunaga's three sons arrived. Nobutada, the eldest
at twenty-four; Nobukatsu, twenty-three, and Nobutaka also
twenty-three. They were invited together up to the dais. Yasuke
remembered the story he'd heard of Nobukatsu's humiliating
thrashing by the ninja, but he knew little about the other two.
As with almost everyone else in the room, they'd traveled into
Kyoto from their own fiefs elsewhere and were in town for the
umazoroe horse spectacle, in which they would play central roles.

As Yasuke was shown to them, servants entered with low tables and placed one in front of each man present, forming two
neat rows leading up to Nobunaga's dais. Some women entered
to pour drinks and help out with the serving, kneeling in the
customary Japanese fashion.

Nobunaga remained central on the platform, but Yasuke was
led to a spot, and table, directly at the Japanese lord's right and
his sons knelt at his left in order, the heir, Nobutada, closest
to his father. Organtino, his interpreting services no longer
needed, relaxed and conversed with the samurai lords of the
court. The servants quietly ran in and out of the room, bowing low as they did.

As everyone ate and drank, Nobunaga began a conversation
with Yasuke—or, perhaps, an interrogation—as the foreign warrior giant had unmistakably become the evening's main entertainment. Before the whole room, Yasuke was quizzed by the
warlord on a range of matters. From his size to his birth country. How long had it taken to travel to Nobunaga's court? How
was the hunting, were the animals bigger than in Japan? Did
his people eat rice? Who had he fought for and where? What
kind of weapons did they use? Did he yet know tea? What were
the people like in India (Nobunaga first assumed he was from

India)? Did they eat animal meat, as he heard some barbarians did? Did he know the teachings of the great sage Confucius? Did the pink Jesuit barbarians treat him well?

Yasuke answered all Nobunaga's questions with gusto. He replied he was from a land even farther than India. That many of the men and women of his people were as tall. Not all were as strong as he though, and of course, to the laughs of his audience, he claimed none as handsome either. And, no, his skin color was a permanent fixture, it would neither wash off nor turn pale if he stayed out of the sun. He had no idea why, but it was so. He told of his marshy homeland, the majestic glory of the Nile River, the hippos, the lions; of fish that were as big as a man, and of their succulent meat. He told of the cows his people revered, regaling Nobunaga and his court with tales of their beauty and grinning widely at the mention of their life-giving milk, knowing the idea of drinking another creature's milk would make his audience squirm with disgust. To get an easy laugh—the idea, he'd learned while traveling the world, would be so alien to his audience—he told how, as a boy, he'd even stimulated the cows' sexual organs with his own lips to increase milk flow. And of the warm showers in fresh running cow piss to dye his hair a deep golden color. He then told the room bitterly of the coming-of-age rituals and the facial scarification which signified a grown warrior that he'd missed because the slavers had taken him. He told of the mighty fighters his people became as men. Of his travels on ships, the terrors of storms and the boredom of the endless voyages. Of friends lost on the way. He enthralled his audience with tales of the palaces, cities and weapons of India, the spectacular beauty of her palaces, mosques and temples, the glory of Arabic art, and of the tall ships, stone churches and deadly cannon of the Portuguese.

Nobunaga called for his globe to the great satisfaction of all present. This was one of his newest and most treasured possessions and he brought it out whenever he found an opportunity. Yasuke had never seen such a contraption, but he'd heard of them; the earth was round and you could see it rotating on these "globes." Not only that, you could see all the lands and seas pictured too. Amazing. Most of the room had never seen one before either.

Nobunaga asked Yasuke to point to the place where he was born (a place the Japanese knew vaguely of as *Rimia*, derived from the name of the country we now know as Libya—it was sometimes mixed up with the name *Korobou*, a Japanese rendering of *Colombo* in modern-day Sri Lanka, and the blanket term for dark-skinned people), on the globe. This was difficult as Yasuke still had no real concept of where parts of the world were. Maps were the domain of navigators and nobles, and ordinary people rarely saw these precious and ever-changing records of geographical knowledge. The globe, a map stretched round a ball to represent the believed shape of God's creations was still a stupefying concept to most in 1581. These were the latest thing, cost a king's ransom and most people had never even heard of their existence, let alone seen one.

Luckily, Father Organtino was there, and he stepped forward to help. He'd only ever seen this one globe, but was now quite familiar with it. In fact, he'd personally presented it to Lord Nobunaga several years before as a gift on behalf of the Jesuits. At the time, and since, he'd attempted to explain its workings, where he came from, how he'd traveled to Kyoto and where Nobunaga's realms were indicated, but found himself lacking in much of the knowledge which the globe might impart. With the help of visitors to Kyoto who were more familiar with this modern invention, he'd come to grips with it over the years and was now able to find areas more easily. The priest pointed to the continent of Africa. More precise than that he couldn't guess and had no idea where Yasuke was born. Neither did Yasuke, and none of his various slavers had thought to ever tell him.

Nobunaga spun the globe until he was able to see Japan. The

Japanese warlord moved the globe backward and forward several times, wondering at the distance.

Yasuke was more than an entertainment for Nobunaga, he was a source of new intelligence and knowledge. Nobunaga had built a small empire on his ability to gather information and intelligence of friend and foe alike. He collected facts and stories like some rulers hoarded ornaments and jewels, and he archived them away in his head to mull over and modify until practical uses could be found. Intelligence had helped him win most of his battles, often against the odds, and astute use of every tool and weapon available to him had helped him win the rest. His men appreciated his inspiring leadership as much as the fact he'd invest in the best weapons and equipment to protect them in battle, never throwing away lives casually as some warlords were wont to do.

Well satisfied with Yasuke's answers, and openly pleased by how much Japanese the stranger spoke, Nobunaga was in good cheer. He'd learned much and been highly entertained; so had his sons. As the food came to an end and the drink continued to flow, the nature of the party altered from an intelligence gathering exercise to entertainment. Warriors did little dances around the room with Yasuke, they touched his skin reverently and joked with him. White paper lanterns emblazoned with the Oda crest were brought in to brighten the proceedings beyond the daylight hours.

Yasuke relaxed for the first time all day, and started to enjoy himself. His life appeared safe, he had a pleasant liquor glow about him, he'd done no dishonor to the Jesuits *and* he was now guest of honor at Nobunaga's, the "King's," feast. It was quite thrilling to be the center of attention in the company of such exalted men. As the drink continued to flow, the formality level dropped still further, and one of the plump serving ladies, giggling, hinted it would be interesting to see how much the giant could carry. Nobunaga laughed and told the woman, if she wanted to find out so badly, Yasuke could lift her. The shocked woman shied away, laughing uncertainly and reddening, but Nobunaga turned to Yasuke and asked him if he would be so good as to pick her up in one arm.

Yasuke had done this party trick before when he was away from Valignano; instead of picking up just the one girl, he picked up two. And easily. Then he balanced them one on each arm, and raised both arms in the universal pose of the weightlifter.

The room went silent again, but then almost as quickly burst out in rounds of excited roars of congratulations: "Hip, hip, hoa!" Later, one of the courtiers in attendance, Ōta Gyūichi, would record of the evening that Yasuke's "formidable strength surpassed that of ten men."

Throughout, Father Organtino, accustomed to close relations with Nobunaga, remained on his knees eating slowly and quietly at his table. He would surely have been praising God that things were going so well. He had much to report back to Valignano.

At the end of the party, Nobunaga brought the proceedings to a close by rewarding Yasuke for his role in this highly successful day. Through his favorite nephew, Nobunaga presented Yasuke with ten strings of copper coins weighing eighty pounds—a small fortune—not to mention a huge weight for the fellow with the "strength of ten men" to carry.

Yasuke accepted the gift graciously, and, bowing deeply once more, headed awkwardly backward on his knees with eighty pounds of boxed copper coins, out of Nobunaga's august presence before standing to head out of the audience room door slightly dizzily, due to the drink. The last quiet giggles from the courtiers provoked by his ungainly bulk moving backward like a crab echoed in his ears along with cheery drunken farewells and still-amazed whispers. Being slightly the worse for wear, and in a haze of happiness and confusion, Yasuke only narrowly avoided banging his head on the door frame.

Thus had a fateful friendship been born.

★ ★ ★

Several days later, Valignano had his own official audience with Oda Nobunaga. Yasuke was not present for this meeting, and remained back at the church. Valignano knew well what a hit Yasuke had been but didn't need the distraction for this first encounter.

The Italian first presented Nobunaga with a velvet-covered gilt chair, an ornate throne of state from Rome and then several sets of crystal glasses from Europe.

He also offered Yasuke.

The Jesuit Visitor was shortly to be returning to Goa and, he believed, thence to Europe, where he did not have need of a bodyguard, or could find a new one. He was unlikely to require Yasuke's intimidating presence in the meantime, especially if the Jesuits were now, more clearly than ever, under Nobunaga's protection. Placing his man, clearly a sensation, with Nobunaga would produce much long-term favor for the Jesuits, and perhaps even provide a valuable source of intelligence for them as well. He would have to talk to Yasuke about that.

Nobunaga agreed, delighted.

It was settled. Yasuke would join the powerful clan lord's immediate entourage as a weapon bearer *and* novelty. Any notion of him ever being anything more than that did not yet exist.

In a month, however, he would have servants of his own.

Property. Wealth.

The ear and trust of the most powerful man in Japan.

In a month, he would be samurai.

PART TWO
Samurai

PART TWO

Torment

CHAPTER ELEVEN
Guest of Honor

Yasuke's first days in the service of Nobunaga were filled with confusion and an unforeseen bit of pain.

For three days, he'd waited outside the mighty warlord's audience room, but had not been called back in. None of the Jesuits had yet returned either. His only company during the long wait were rotating pairs of silent and stone-faced Japanese guards. He remained equally silent between them, kneeling *seiza* in the time-honored Japanese manner to mirror their pose, legs folded underneath thighs, buttocks resting fully on heels. A painful position for beginners, but one he'd well mastered over the last two years. Or, so he'd thought.

Each day, they resumed that same, silent position. And, by God, did his legs ache. They'd be burning after forty minutes and then go numb after an hour. After two hours, he'd stand, sway uncertainly and even pace a little, all aspirations of honoring the "correct" Japanese custom discarded. *How could they kneel like that for so long?* The Oda guards, both lower-level samurai, eyed him, mute but clearly amused; when called upon to move, they'd spring up as if nothing at all was the matter.

An attendant had explained that Nobunaga was focused on the impending *umazoroe* horse spectacle, but planned to summon Yasuke again before long. Whether "long" was hours, days or weeks, no one seemed to know. During the wait, Yasuke had no real idea what his new role was to be. The notion that Valignano had gifted him as one might gift a globe or an ivory-inlaid gun was unavoidable. Perhaps his future role was merely to be trotted out as an amusement, performing various feats of strength at forthcoming dinners. Or were his skills as a bodyguard and warrior actually to be employed? He simply didn't know.

He noted the curious stares of all those who passed. This was decidedly not a port town with myriad international visitors. Beyond the Jesuits, there were no Europeans, Indians or other Africans in Kyoto. No Thais, or Koreans either, and only a few Chinese. He'd seen nothing but Japanese people for three solid days, making it his first time to be the only foreigner around, sticking out even more.

He'd been accommodated in a small private residence, one of many where guests stayed, with a view onto an enclosed garden backing up to the exterior wall of the temple. He'd never had so much privacy in his life, and it felt strange. The only sound to keep him company while eating or falling asleep was the lulling intermittent toking of a bamboo *sozu* just outside his window (meant to frighten away animals, but also an intentionally serene and soothing sound) and the regular tolling of the temple bell.

Each morning, he dressed, eyed his spear and knife with longing—as no one was permitted to bear them in audience with his new lord unless bidden—and slid open his door. Then he stood at attention, waiting for some form of direction. Passing servants and guards only gave wary glances as they charged by. Whatever orders they pursued, it was quite clear, had nothing to do with him. Eventually he'd be conducted back to the doors outside the audience chamber and the two guards to again wait in vain. After several more hours, Yasuke was ushered back to his guest room for a midmorning breakfast where his meal was served for him to eat alone.

After dining, it was back to his spot outside the audience chamber, only to remain, with aching legs and increasing hunger pangs, for the evening meal, when a pretty serving lady would bow to the floor, and utter the magic words *kochira e dozo*: "this way, please." Virtually the only words he seemed to hear anymore. *Oh, how things had changed.* He'd gone from head of security, someone who was at least seen and listened to, to an afterthought or an awkward guest. Worse, he even felt underfoot—a novel and uncomfortable sensation, and one he wanted to end. For more than four years he'd waited on Valignano seven days a week, overseeing various security matters, directly standing guard and still playing carrier as needed. The ten years before had been spent answering to former masters. Here, no one seemed to need him for anything. So he sat in his room or outside the audience chamber with the two young soldiers and waited. Trying to remain as unobtrusive as possible—not an easy thing for a man of his extraordinary size and appearance.

All through those three days and nights, he listened to, and learned, the systematic hustle of the compound and its hundred servants and guards rushing past. The brush of their straw sandals, the clop of clogs, the neighing of Oda horses, the grunt of men and the laughter of women; its energy, so very different from the all-male mission houses he'd been living in for years. Honnō-ji Temple was clearly a center of power, just as Nagasaki had become during their stay. A place that was alive with activity and purpose.

Morning started early, well before dawn, as servants cleaned out night soil, prepared fresh robes, opened shutters and removed bedding to be aired and stored until the evening. Others kindled the cooking fires, began serving meals in shifts and boiled water for tea. As this was a military installation, the changing of the guard was accompanied by weary but efficient banter. Their night duty done, these men were always the first to eat, then they retired to their barrack bedding. At the same time, the horses were exercised, groomed and fed, adding to the cacophony of noises and smells. The deep solemn peal of the temple bell marked the passing of time in the commandeered temple.

Throughout the day, Yasuke watched a hundred people enter through the audience chamber doors, then exit again in a hurry. Administrative monks with tonsured heads, barefoot servant girls, young warriors full of verve and energy with sparse beards on their chins, long-robed merchants with serious-looking faces, priestesses in ceremonial robes and townspeople with petitions. While waiting to be called in himself, he played a little game, a diversion he'd crafted while waiting for Valignano's various summits to end. *What judgment, directions, advice or reprimand had Nobunaga just given each man or woman?* The pastime proved easier than usual. Everyone's time, as he'd been warned, was taken entirely with the imminent military and horse spectacle. This *umazoroe* had become a national focus.

The preparations, Yasuke had gathered from discussions overheard outside the audience room, were under the immediate direction of a prominent and trusted Nobunaga general: Akechi Mitsuhide, whom Yasuke had met at his initial audience. He remembered Akechi, slightly plump of face and with a furrowed brow. Clearly a serious man who was, evidently, trying his best to carry out his lord's orders to the letter.

Akechi's arrangements had been speedy, especially considering that all the other lords, both major and minor from every province within easy reach of Kyoto, were to attend with their vassals. Some two hundred thousand viewers in all. The original order for the event had been given by Nobunaga only a month before! A nightmare timetable. But Akechi—it was agreed by the others waiting around with Yasuke—was a good organizer, a decisive man. A man who Nobunaga trusted to get the job done. Akechi had been the very first of Nobunaga's vassals to be awarded a castle by the warlord, and his career and service since had validated Nobunaga's early trust and good judgment. It had been Akechi who'd set fire to the Mount Hiei temples and put the twenty thousand inhabitants there to the sword, finally destroying the power of the warrior monks who threatened Kyoto. It was Akechi who'd been sent to overcome treacherous enemies in the Tanba and Tango regions to the north, and then set against Nobunaga's deadly rival Uesugi Kenshin—the same rival, Yasuke recalled grimly, who'd eventually succumbed to a ninja blade while on the toilet. And it was Akechi who'd produced many of Nobunaga's lavish entertainments and spectacles, including this imminent *umazoroe*. Despite the short notice, the elite of Japan dared not be absent. For they were, all of them, completely under Nobunaga's power now.

Rank was a complicated concept in Old Japan, and it was often difficult to determine who held power and how they could wield it.

Provisionally, since the beginning of his line the divine sovereign, the "Son of Heaven," was the *dairi*, male or female, and presided over a court of hereditary nobles called *kuge*. But, the *dairi* and *kuge* hadn't truly had any temporal power in almost four hundred years. Power had, instead, passed to the samurai, the warrior class who'd usurped power after an internecine war ended in decisive victory in 1185.

Since then, the most powerful among the samurai families had held the reins of national government in a form of military dictatorship called a *shogunate*, headed by a shogun. The prestige of the former imperial court, however, remained to the extent to which the title of shogun was still technically bestowed by the *dairi* (although they had no real choice about who it was bestowed upon). The shogun appointed his family and most trusted samurai retainers to positions of power within government, including the governorship of distant provinces. These governors were called *shugo*. Beneath the *shugo* were various tiers of samurai in a feudal-style system.

By Yasuke's day, however, The Age of the Country at War had finished off the shogun/*shugo* partnership. With increasingly weak shogunate government in the fifteenth century, the *shugo* had gradually become independent rulers of their own provinces or parts of them, domains. The shogunate's influence shrank ever smaller until, like the *dairi* before, it ceased to exist in all but the historic prestige of its name. The last shogun, Ashikaga Yoshiaki, had been officially exiled by Nobunaga from Kyoto in 1573 and his title was abolished. Yoshiaki lived on in the protection of regional lords and temples opposed to Nobunaga, but remained unable to do anything about it.

The newly independent rulers of provinces, now essentially small states, were called *daimyō*, "Great Names," or sometimes *ryoshu*. Men such as Nobunaga, Arima, Ryūzōji and Ōtomo, were all *daimyō*. So powerful that the Jesuits referred to them as kings. These *daimyō* and their samurai—they *themselves* were samurai also, just at the top of the tier—largely kept the former hierarchy in place. A *daimyō* would portion out fiefs to his vassal samurai, the most important vassals getting the largest fiefs. The samurai, in turn, would distribute fiefs to *their* warrior vassals all the way down to the lowest-ranking samurai. Samurai income depended on the production of the farmers who lived in the fief to which they'd been assigned.

The biggest *daimyō* in Yasuke's time was indisputably Nobunaga. He controlled Kyoto, where he "took care" of the imperial court and any last vestiges of the former shogunate regime. His alliances with other lords did not always hold (though it was rarely he that broke them), but had spread throughout all of Japan. Nobunaga could command up to two hundred thousand troops on various fronts and in effect had military control of the country. He was *shogun* in everything but name.

On the third morning, Yasuke finally received his orders.

He was to take part in the military parade, the *umazoroe*, the next day. A last-minute addition. He too, now, was completely under Nobunaga's power.

They'd built the arena for the event from scratch only weeks before. Thousands of men were brought in from the provinces by Akechi as labor. Supplies of wood were not an issue because Nobunaga had, for years, commandeered virtually all timber in central Japan for his numerous building projects. The arena was a half mile running north to south, marked out by tall posts enfolded with colored felt, which ran the length of the field. The outer limits of the riding area were indicated with a short stockade of crisscrossed bamboo. Behind this low fence most spectators craned their necks, pushing and shoving to get a glimpse through the crowds in front of them. Small children squatted and squinted through the holes in the woven bamboo while their elder brothers and sisters filled nearby trees. The roofs of any buildings with a view were crowded with the inhabitants and their friends, exquisite *bento* picnic boxes prepared, and parasols blossoming to protect the ladies from the bright spring sun. This huge crowd was much like the one which greeted Yasuke only days before—the same people, treated to yet another wonder, all in the name of Nobunaga's glory.

On one side of the arena, Nobunaga had erected a viewing pavilion directly outside the east gate of the imperial residence. Although a temporary structure of simple cypress wood, it was painted in gold and silver leaf. Here on a gilt tatami mat sat the sixty-three-year-old, white-haired *dairi*, Ōgimachi, the 106th Emperor of Japan and his *kuge* nobles. Befitting their exalted positions, and following etiquette and court protocol, none appeared pleased nor displeased, excited nor bored. As they solemnly processed to their gilded pavilion, they merely stared straight ahead, as if they'd seen events like this every day of their lives. With them sat principals of the old shogunate regime, who, like the court nobles, still retained a veneer of respectability and influence, but were the powerless puppets of Nobunaga.

The more common nobles, the *daimyō*, and their senior vassals, rulers of taxable provinces and wielders of military might, were accommodated in wooden stands to the right and left of the nobility, symbolically fulfilling their role of "protecting" the emperor.

The guests of honor, the nobles and courtiers in the pavilions and stands, were all in formal dress. The sweet smell of perfumed *sokutai* robes, preserved from ancient times especially for such occasions—held once a century, if ever—pervaded the air. The court costumes were multilayered flowing waterfalls of silk, extremely limiting to movement and very heavy. The outer garments, *osode*, were red, black, gold and deep blue, and embroidered with mythical creatures from a dozen different ancient myths. Kirin, Shishi and long wingless Chinese dragons. Beneath these robes lay visible undergarments of shimmering pure white silk, white due to the diet of white mulberry leaves purposely fed to the silkworms. The whole was topped off with the *kammuri* cap of black-lacquered silk with a scorpion tail–like peak. Each courtier seemed more gorgeously dressed than the next, all intent on outdoing one another, perhaps a sign they were more excited about the day than their passive expressions revealed.

Several hours after sunrise, the first horses and their riders entered the showground in tight procession.

Leading their carefully chosen cavalry of senior retainers, Nobunaga's chief vassals paraded in, foremost among them Niwa Nagahide, an old and trusted retainer who had the honor of holding a fief and castle near Azuchi. Even their horses were dressed in Oda clan finery, leather and lacquer saddles and ex-quisitely embroidered scarlet-and-purple bridles and reins. Next came several lines of warrior monks of the Shingi Shingon sect, their presence a symbol of Nobunaga's dominance over the pre-viously formidable militant Buddhist schools. After these two groups entered the man who'd stage-managed the whole day: Akechi rode tall and proud, his plump face impassive, astride his chestnut-colored mount. Beside him rode his mounted con-tingent and their allies.

Following Akechi, came Nobunaga's three principal sons and other close relatives—including his nephew, Tsuda, the man who'd presented Yasuke a small fortune only days earlier. No-butada, the eldest and acknowledged Oda clan heir, led eighty horsemen. Nobukatsu, his younger brother, and the others each led ten.

Behind them, among fifty other Nobunaga retainers, rode Yasuke.

A last-day addition, the brief order dispatched through an or-derly the previous evening had informed him of his place and tasks. There'd been no time for Yasuke to be part of the main equestrian events. His role was to simply ride as part of the pro-cession, as one of Nobunaga's retainers, and then fall out with other nonperformers to watch the later proceedings. They'd mounted him onto a beautiful grey steed. Like Nobunaga's other direct attendants, he'd been dressed in starched sleeveless white robes layered over undergarments of red silk, and trousers of black leather. The night before, a pair of anxious seamstresses had feverishly stitched together *three* outfits into something that might fit the giant man.

Surely, thousands recognized Yasuke from the day of the riot or from the rumors since. This wonder from the south! And now all of Japan saw who this man served: Lord Nobunaga.

A mild hush seemed to descend on the crowd as he entered the arena, but the buzz of chatter and cheers soon picked up again, plenty more to talk about this day, the extreme novelty just five days earlier was now only a footnote to this glorious parade.

Behind Yasuke trotted Nobunaga's other retainers, another hundred mounted archers, and the hegemon's highest-ranking clerics, men who operated as Nobunaga's administrative officials. Armed contingents followed—another eighty horsemen, and one hundred mounted archers, their bows on the string, ready for action.

As the horsemen completed their first ceremonial circuit of the riding ground, they fell into formation in the corner opposite the imperial party's gilt pavilion. The precision throughout was impeccable, and Yasuke had never before seen military discipline like it. Even the horses stood at attention. Next came four men bearing an empty chair garnished in gold and made of velvet. It was the same throne-like seat of state gifted by Valignano to Nobunaga at his audience three days before. Another visible sign of the homage being borne to Nobunaga from distant realms. Even the European barbarians put him on a throne.

Then Nobunaga himself finally entered, mounted on his war horse, Daikoku. Named for the god of great blackness. Daikoku personified how Nobunaga desired others to perceive him: a destroyer of evil, a transformer, a nation-builder and a person in whom renewal was personified. Yes, today's event was held "in the emperor's honor," but everyone, including the emperor, understood who the event was truly honoring. Oda Nobunaga rode completely alone, in glorious isolation. The other thousand riders and soldiers surrendered the main stage to him. The attendants bearing his gear rode a good distance behind him, symbolically there to serve, but not to take anything away from the day's main focus. Their lord.

Nobunaga's garments displayed symbols of his reign, signifying his successful establishment, for most Japanese, of peace and prosperity over a war-torn country. His under-robe was of plum-colored silk. Above it, a second layer of vintage silk vermillion rested, cuffs ornamented with twisted gold thread. His sleeveless robe was made of crimson damask, sporting the stylized crest of a paulownia flower and trimmings of silk and corded gold thread. Even the saddlecloth of his horse was made of the finest Chinese fabric, as were the mudguards with cloud motifs elaborated in crimson. His trousers were also crimson damask, his shoes of Chinese brocade and his gloves of untanned chamois leather. On his head, Nobunaga wore a cap symbolizing the power of a demon. Thrust into his belt were a pair of swords sheathed in gold-encrusted scabbards. And, over it all, he wore court robes made from cloth of gold. The fabric was said to have originally been made for a Chinese emperor, the Lord of heaven himself. Before war had consumed the islands and cut the trade routes, it had somehow made its way to Japan and had miraculously been kept safe from flame and damp until the realm was at peace again and it could be used for a suitably glorious occasion. Today was just such an occasion, the culmination of all history, and all present knew they were within the aura of a divine, heavenly being. Seven hundred mounted warriors had now trotted into the riding ground in perfect lines, all clad in the finest robes in Japan. But, none could match Nobunaga. And none dared.

Valignano, in the audience, and well familiar with both European and Indian pomp and pageantry, later remarked that never in all his days had he seen "such a resplendent and magnificent affair, on account of the great quantity of gold and silks with which [the riders] were adorned." With that comment, Nobunaga's wish for international renown was officially realized. Accounts of his grandeur—thanks to events like this and the Jesuits' presence at them—soon reached India, Africa, Europe and even the New World.

And the actual horsemanship exhibit had not even started yet. The parade was only the opening ceremony. After the last of the procession passed through the impromptu arena, things really got interesting. Yasuke had no role in the actual riding performances. Instead, one of Akechi's stewards guided his borrowed horse to a groom, Yasuke dismounted and was led to a ringside position where he knelt, his new norm, on the tatami in the stands.

For the next six hours, cavalry galloped or trotted back into the showground, performed acrobatics on horseback, swapped rides midgallop and performed horseback archery, shooting from all angles at increasingly distant targets at ever mounting breakneck speeds.

Nobunaga himself, and his sons, dominated the show.

The top Oda men excelled at every turn; they performed synchronized movements and mounted tricks—changing their rides in midgallop, standing on their saddles, even jumping as the horse raced beneath them—for the crowds of Kyoto citizens and visitors from the countryside. Nobunaga also swapped horses midgallop with his pages, in a continuous whirlwind carousel, which brought them eventually back to their original mounts.

The crowd mostly watched in awed silence, breaking into earth-shattering cheers only when each performance finished, before once more descending into rapt attention as a new act began. Throughout, Akechi's stewards buzzed about, dealing with any problems, but there were few. Akechi himself had seen to it that no possible issue had been left to the vicissitudes of fate; everything had been anticipated, every detail planned.

Palpable joy filled the arena. The enormous multitude felt unconquerable, chosen, truly alive. *And, why shouldn't they?* Their part of the realm was at peace for the first time in more than one hundred years. The common people now prospered and kept their hearths lit and bellies full. They were grateful to be alive in such a blessed age.

Finally, Nobunaga led the whole body of mounted soldiers in a full stomach-churning gallop past the imperial pavilion for one last royal inspection.

A week later, at imperial request, the entire show was repeated.

CHAPTER TWELVE
Treasures Old and New

Days after the second performance, Nobunaga departed Kyoto to return to his own clan capital of Azuchi. Yasuke went with him.

It was early May, a time of misty mornings and new colors. Vivid pink and purple flowers filled the world underfoot and from on high, the soft violet wisteria tumbled over branches like flowing water. They left the Honnō-ji Temple, heading east, trotting through the city streets before crossing the shallow depths of the Kamo River. Early morning fishermen and herons alike, standing in the sparkling waters waiting for the silver flashes of small sweet fish, watched them pass. A scene of tranquility and peace on the edge of the busy metropolis.

At Nobunaga's request, Yasuke rode at his side. It felt strange to be riding alongside his new lord, as he'd almost always walked in attendance behind Valignano. Nobunaga was clothed for the ride, simple dark robes and loose *hakama* trousers. Yasuke had no new outfit for this jaunt, so donned his pre-Nobunaga garb. The miles slipped by easily as the warlord explained various landmarks, newly gifted lands, and then recounted the key moments of his recent horseback spectacle. Yasuke—having learned from Valignano—praised the event with the perfect balance of exuberance and reverence. That he did so in more than-passable Japanese only delighted Nobunaga the more.

Behind them rode Nobunaga's pages. A group of thirty or so boisterous well-born samurai, still teenagers, who rode fast and flashy—joking bawdily with the warlord and each other in this informal setting, pulling simple pranks and competing in impromptu races. Yasuke could see these young men clearly kept Nobunaga feeling youthful, transmitting their vitality and levity to him. That they'd also, without a moment's thought, fight to the death to defend their lord and protector, was also a given. These youths had been handpicked from the sons of Nobunaga's chief retainers and allies. They were the best in his growing kingdom. Alongside his own sons, they would be the future of his dynasty, a future shaped under Nobunaga's tutelage and fond eye. With each mile crossed, Yasuke could only wonder what his role might be in that same future.

By midday, they'd traversed the heavily forested passes through the mountains which flanked Kyoto. On the other side waited Lake Biwa—an enormous body of water more than two hundred fifty square miles, almost an inland sea, that nearly split the main Japanese island of Honshu in two. As they traced the lake's rightward shore, along a well-trodden dirt road, the huge Hira Mountains loomed far away on the other side of the wide expanse of water. To their immediate right ran a wide plain stippled with freshly planted green rice shoots just peeping from the muddy brown water. On this side too, the Ibuki Mountains rose again in the distance where a white ribbon of snow gleamed like a moving stream within the highest woods. Nobunaga explained to Yasuke that his original capital, Gifu—which he'd recently gifted to his first-born son Nobutada now that he had his own new capital in Azuchi—lay just beyond those grand mountains.

The rivers they crossed at fordable points were shallow in this dry season, but would prove swollen and nearly impassable in the snow-melt months ahead, when the water poured down from the mountains. The horses splashed through the cropped scrub before crunching over gravel at the dry edges of the flow, and then waded into the gloriously cool water. The cold flow cleansed the horses' sweat and splashed refreshingly over the riders' legs. Then it was back to the dust of the dirt road and the clomp of rice-straw shod hooves. They rode hard, and easily covered the thirty miles in a day.

The pace was exhilarating, but likely served as a bracing reminder for Yasuke. The last time he'd been on such a fast gallop, some eight years before, the feeling had been quite different.

At the time, Yasuke had been in India with a unit of seven hundred other African mercenaries. They were fighting for the Persian lord Ibrahim Husain Mirza, who controlled a southern district of Gujarat. How much longer Mirza controlled the territory was now the question at hand.

The Mughal emperor, Akbar, had been campaigning throughout much of northern India for nearly two decades, adding to conquered lands he inherited mostly from his grandfather Babur. Gujarat in northwest India was next on Akbar's wish list.

Akbar had just caught Mirza's troops by surprise at a village called Sarnal on the Mahi River. There, Akbar's elite force of only two hundred horsemen, handpicked from his main force of seventy-five thousand, led by the emperor personally, had charged across the river at sunset and deployed into the narrow village streets where Mirza's greater numbers meant little. Taken by surprise, the Persian defenders and their African mercenaries had attempted a counterattack, but when their Persian commander turned and fled, they'd also wisely scattered to the four winds.

Yasuke and his comrades had joined the headlong retreat into the gathering night, bolting on horseback more than one hundred miles to the city of Surat, a great seaport on the coast. The ride had lasted all the next day, and night too. Gunfire spurred them on their way, but they'd managed to evade their hunters. When, at last, they arrived at the gates of the huge walled city, men and horses alike had collapsed from thirst and fatigue.

India—as Africa, Japan and Europe—was a region plagued by ongoing wars and internecine fighting. Foremost being the ongoing Mughal conquest and consolidation of northern India. How India's wars differed from Japan's was largely due to its geographical position at the center of the world's maritime, steppe and desert trade routes. This centrality attracted combatants from everywhere: Afghans, Turks, Persians, Africans, Arabs, Mongols and Portuguese all flocked to the Indian subcontinent to make their fortunes in war.

Even so, the need for soldiers far surpassed the influx of voluntary global mercenaries. As a solution, African boys like Yasuke were forcibly brought to India and trained to become slave soldiers.

Many *free* Africans also made the journey, seeking the same opportunities as the Turks, Arabs or Portuguese. But the vast majority were children captured in Africa, as Yasuke had been, and sold to foreign slavers in coastal ports, most often Zeila (now in northern Somalia), or Suakin (in modern-day Sudan). Here, their young lives were traded for salt, Indian cloth or iron bars along with other commodities such as guns. If not immediately put to work on dhows or galleys, they were taken on Arab, Ottoman or Indian ships, north toward Egypt, Arabia, Turkey and Europe, or east toward Persia and India.

During the voyage, slave traders often chose to invest in their slaves, educating or even mutilating them to gain more profit at the next stage of sale. For instance, while some were taught their letters, many more young boys were castrated. Handsome eunuch slaves fetched astronomical prices partly because only 10 percent of the victims survived the cut. By the time the captives reached northern India, almost a fourth of those who'd boarded ships in Africa had perished. On arrival in India, the Africans found themselves in slave markets, where they were again sold and taken farther afield to wherever trade routes and eager customers waited—places like Gujarat, the Gulf of Cambay, the Deccan, Cochin (modern-day Kochi), and to Portuguese Goa.

First arriving in Gujarat in northern India, Yasuke and the others had been herded into underground cells, with only street-level barred windows for light and air. The conditions were dark, airless, cramped and horrific. (On the ships, they'd been kept above deck and out of chains, doing simple maritime chores.) He was thirteen now; the voyage from Africa had taken almost a year—as the ships he traveled on stopped to trade or take shelter from adverse weather on the way. He'd been stripped, subjected to a full body examination and checked that he'd not been overly damaged by punishments or abuse on the way from Africa. The slavers who inspected Yasuke were themselves of African origin, perhaps having passed through exactly the same slave cells years before. Their appraising eyes summed up the young Yasuke, observed his size and growth potential and purchased him on the spot.

He was now a member of a military caste called *Habshi*—African warriors, often horsemen, who fought for local rulers or were loaned out by a mercenary band leader to whomever was willing to pay. Some of these bands numbered in the thousands, but most were only a few hundred strong. The Indians called the Africans *Habshi*—a word derived from "Abyssinia," the ancient name for Ethiopia—because a large majority of the Africans destined for India had started their sea journeys there. During the span of recorded history, it is estimated that as many as eleven million Africans were trafficked to India as slaves, primarily to be used as soldiers. During Yasuke's time, when soldiers were in peak demand, estimates reach into the tens of thousands.

Yasuke spent his first years in India training to use weapons, to ride a horse, to kill and fight. Too valuable to be used as mere fodder (the weakest slaves, who were judged to have little military worth, were often used as human shields, driven before the main force to absorb bombardments), he took the field only after training. Throughout, he would have been both brutalized and baptized into the cult of the killer, through actual battle, but also by carrying out commissions such as executions for his new masters. In his teens, he'd likely supped with assassins, marched and fought beside fifty thousand men, helped slaughter entire villages, joked and bet as comrades fought to the death in camp over some village girl, missing token or misheard comment. He also grew taller and his muscles hardened. He learned to kill with his hands. To ignore the gore and screams of new friends and foe alike. By eighteen, he was a valuable warrior. Now training young boys, as he'd once been a lifetime ago. His body a chronicle of ever-fading scars, a book written in blood.

Gujarat, where he was then stationed, was a region in chaos, fought over by feuding noble families and factions (some among them of African origin) with significant fiefs and military might. Into this turmoil entered the all-conquering Akbar, the third of the Mughal emperors who'd been fighting over northern India for the last half-century, seeking to add to their previous conquests. Since ascending to the throne at age fourteen in 1556, he'd defeated foe after foe, often by surprising them with his speed and daring. He'd consolidated Mughal rule in northern India, and been pushing south into the center of the subcontinent, as state after state had fallen to him. Sarnal, where

Yasuke had escaped from, was merely one of many places that had fallen to Akbar's lightning-attack tactics.

Surat, walled and on the coast, would not fall so easily. (Or, so Yasuke and the other *Habshi* hoped.) The city was nominally ruled by the last shah of the Gujarati Muzaffarid dynasty, the twelve-year-old Muzaffar Shah III. As his forces fled before the Mughal onslaught, some took refuge in Surat, which was under the command of an officer named Hamzaban. Hope and resources were dwindling fast as ever more news of widespread Akbar victories elsewhere throughout Gujarat arrived in the walled city. In the chaos of retreat, the invaders had also managed to take possession of the Muzaffarid elephants and much of their treasury. The desperate besieged army had appealed to the Portuguese colonial government in Goa, offering handsome rewards to come to their aid. However, upon seeing the size of Akbar's army, the Portuguese soldiers got cold feet and, behind the backs of the defenders trapped in Surat, tried to simply buy the assaulting Mughal emperor off with gifts of valuable novelty articles from Europe. They failed—Akbar detested double-dealing—and the siege continued.

Yasuke stood resolutely on the walls and watched Akbar's huge army deploy below. It was January 11, 1573. Beyond its numbers, the sheer diversity of the force was astonishing. First arrived the cavalry bands, armored in mail, their shiny silver-colored helmets, some with turbans wrapped around them, peaked in long spikes which could run a man through. They carried long spears, curved swords and round shields. Some had short composite Mongol-style bows, a nod to the Mughals' Mongol origins, in quivers attached to their saddles. Next arrived the infantry, sporting muskets, spears, swords, and led by their officers on highly decorated minicastles perched on the backs of elephants. Yasuke had seen these before in war, both on his side and in enemy attacks. Fearsome animals that could crush a man with a stamp of their huge feet or throw the same man into the air, impaled on a sharpened tusk.

The Emperor Akbar trains an elephant.

Akbar's foot soldiers acted quickly, erecting a palisade to defend against missiles from the city, and then started to dig tunnels toward and beneath the city walls. As the weeks went by, Yasuke and the other defenders could hear the scraping as the miners' shovels grew nearer, the soil piling in huge telltale mounds outside the wall. The work had now advanced as far as the gates and the huge "impregnable" walls felt less and less secure to Yasuke and his comrades who manned them. They were surely going to collapse imminently, either from the undermining of their foundations, or from the massive explosion Akbar's men would eventually set beneath them in the hollowed-out tunnels.

By February, Hamzaban and his Portuguese colleague Antonio Cabral (no known relation to the Jesuit superior who'd welcomed Yasuke to Japan, but the Cabral *were* a famous Portuguese noble family who got themselves all over the world) trapped with nowhere to go, despaired of assistance from other quarters. The walls would soon be breached. They surrendered and Akbar, in an unusually merciful gesture, agreed to spare their lives and send the besieged troops—Portuguese *and* African—back to Goa, having now established diplomatic relations with Europe for the first time.

Akbar was less merciful when it came to the commander Hamzaban. After the behind-the-scenes negotiations were concluded, Surat's leaders rode out under a flag of truce to offer their surrender before both armies. And Emperor Akbar had received them with the respect, dignity and pomp due to their station. He'd even erected a gilt awning a short distance from the wall where he welcomed the enemy commanders and their official surrender in the heat of the day. Yasuke and his worried comrades had no choice but to watch impotently as the city, and perhaps their lives, were bargained away. As far as he knew, the besieged troops would be killed or handed over as slaves to a new master.

Akbar was flanked by his lords and two huge African eunuch bodyguards clad in white with golden turbans. All seemed to go well. The Portuguese commander remained with his head bowed to the ground, prostrated in defeat. Hamzaban, however, seemed to be carrying on at length, giving an impassioned speech. Suddenly, Akbar motioned and the two *Habshi* bodyguards seized the prostrated Hamzaban and held him between them. With another Akbar command, a man walked from behind the awning with a large pair of tongs and yanked Hamzaban's tongue from his mouth by the roots. Hamzaban collapsed forward, blood spraying the ground, as Akbar gestured that proceedings were at an end. Cabral raised himself quickly, walked backward out of the emperor's presence and, when he judged it safe to do so, turned and marched quickly back to the city gates. Inside, he gestured to his own men, and they raced to pack up and return to their waiting galleys; they were leaving posthaste.

Word spread quickly. Yasuke and his African comrades were *also* free to leave the city, but had to depart immediately. They all had one day. Midday tomorrow, Akbar's men would enter the city and kill any soldier remaining; Akbar's mercy was clearly far from unconditional. One day was plenty but *where should they go?* They were masterless, their commanders killed or fled. Rumor spread that the Portuguese were looking for rowers to replace some men who'd perished during the voyage from Goa, and many of the *Habshi* grasped desperately at this escape method.

Yasuke was one of them. He joined the Portuguese galley crew as an oarsman and rowed the four hundred nautical miles to Goa. The winds were weak in the spring, but the galleys propelled by their oarsmen sped through the waves. Behind them, thousands of soldiers had been stranded in Surat. Their end, he never learned.

In four days, Yasuke was in Goa on the west coast of India, starting a new life.

The Portuguese had conquered Goa in 1510 and the city quickly became a center of trade, missionary and military activity: the jewel of the Portuguese Indies. Along with a majority Indian population of more than sixty thousand, Goa had a thousand civilian and military Portuguese inhabitants and some *ten* thousand slaves, most originally from Africa. Goa's lifeblood was trade—in Indian slaves, gold, silk, other fabrics and spices. The eventual exports to Portugal and the wider European market were paid for with the profits from the China-Japan silk trade, which was dominated by the Portuguese middlemen ever since the Chinese government had banned direct Chinese trade with Japan. While other forts and missionary stations also appeared along the coasts of the Indian subcontinent, Goa remained the key to Portuguese India.

Many of the Portuguese emigrants settled and married in India outside of Portuguese-ruled areas. They were called *casados*, the married ones, long-term settlers who'd escaped the poverty of their European homes to make their fortune. Portugal was a hard land with few routes to riches outside maritime service, and conditions under Portuguese rule were harsh and pay was poor. Europeans had come east to sell their services as mercenaries, or deserted the Portuguese Crown permanently to the highest bidder. Local Muslim or Hindu rulers would often pay a premium for renegade *casados*, who brought intelligence about the infidel Catholic enemy and crucial military skills, such as gunnery. This pattern of desertion by the Europeans meant that Goa, and other Portuguese outposts, cried out for manpower and depended heavily on men like Yasuke, who were commonly seen as more loyal to the individual Portuguese officers and dignitaries who employed them than a European would be.

Yasuke and several of his comrades were happily employed by a Portuguese merchant who gave them food, shelter and additional training in combat and domestic service. They worked in Goa in various merchants' homes as security and valets making a stable but meager living.

Roughly a year later, Yasuke met a tall Catholic man named Valignano.

As they neared their destination, Nobunaga's capital, a foothill rose out of the trees. Yasuke's thoughts returned to his immediate surroundings. Closer, the setting sun caught the top of the mount and reflected off something shiny, giving it a golden aura like an angel's halo in the weakening sun of the dying day.

Azuchi Castle.

The first of the style that now gives us the image of a "typical Japanese castle," Azuchi was an ornate palatial fortification of multiple floors with intricate gables and turrets. As much mansion and political power center as war-orientated stronghold, the full complex covered the low mountain and was topped by a massive gilt keep, Nobunaga's *yakata*, towering above the local landscape—reminiscent of the buildings seen in classic Chinese paintings, which had been imported in great quantity to Japan for more than a millennium and had set the new standard for beauty and splendor. The outside of its top floor was clad in gold and truly seemed to beam its light across the whole of Japan.

The party of horsemen entered through a huge gate under burning reed torches at the foot of the hill on which the castle was built. The sun set to the west over the lake. The speed they'd traveled had not allowed for much rest or food, and now that they'd arrived, Nobunaga called for refreshment before the long climb up the hill to the living quarters.

Fifty servants emerged from every direction to tend the horses and supply the demanded food. The group of weary riders sat to a snack of cups of bitter, vividly green, almost luminous, tea and sweet rice-flour dumplings dyed green, pink and yellow. It seemed to Yasuke that everything about Nobunaga was colorful—a far cry from the somber, austere world of the Jesuits. After eating, they rode up the hundreds of steep steps to the living quarters, passing through the hustle and bustle of evening castle life. As the horses climbed, Yasuke realized the whole hill *was* the castle. The keep atop was only the highest of multiple levels of buildings, palisades and defenses.

Some three thousand people would have lived there at that time. Yasuke and the rest of the entourage passed hundreds of soldiers on sentry—those on palace garrison duty while Nobunaga and his personal clique were away—and the busy houses of senior retainers and, again and again, through ever more ranks of bowing samurai guards. Today was not a day for huge ceremony and decorum. No one, besides the castle's thousands of residents, Nobunaga's own people, was watching. No outsiders to impress here. They pressed on up the hill, joking and in a fine mood to be home.

The inner precincts comprised multiple palaces of senior vassals and family members, temples, large administrative buildings, grand chambers, guest quarters and accommodation for those on duty. Retainers such as Yasuke were accommodated on lower levels of the mountain or in the warrior quarters of the town of Azuchi below.

Nobunaga had designed it to inspire awe and be compared favorably with the faded grandeur of the imperial family's quarters in Kyoto. On a clear day, Kyoto, only twenty-five miles away, was almost visible from the top floors of Azuchi Castle, which commanded unrivaled views over Lake Biwa and the vast plains which reached eastward toward the distantly looming mountains from whence Nobunaga had burst onto the national scene so many years before.

Following the lead of the others, Yasuke trailed Nobunaga to his private chambers. Then he and the others were dismissed, and retired to their own quarters. Nobunaga's favorites were accommodated in their own houses nearest to the keep. For example, Ranmaru, the most handsome of the amusing young samurai who'd ridden with them and clearly the group's star, had only a very short walk to the private residence that he had been assigned, immediately below the donjon (the innermost keep).

Yasuke, meanwhile, was initially only a short distance away, but by no means assigned a private house; he was to lodge with the guards for now. This did not lessen his elation, but he was still puzzled as to what to do. Nobunaga had talked with him at length, but still given few clues as to his new ongoing role. He bedded down on a futon beside the nine other men in the room. The others kept quiet and to themselves, acting as if he weren't even in the room. He soon discovered his roommates would change throughout the night as they took turns guarding the castle. These were temporary quarters for those on duty. Yasuke waited for his turn on the watch, but it never came. Worse, his dutiful offers to join the watch were ignored with curious stares. Had Nobunaga already forgotten him again?

A day behind Nobunaga and Yasuke rode a party of Jesuits led by Valignano. The Visitor was eager to see Azuchi Castle and the new Jesuit mission in town. The town of Azuchi, below the main castle, proved bustling and wealthy—living off the needs and excesses of the stronghold's inhabitants and Nobunaga's court. Daily, new buildings were being constructed and old structures were replaced with improved ones. The people and fruits of the countryside—food, timber, silver, gold, charcoal and weapons—flowed in and merchants from afar brought exotic wares they'd purchased in Japanese ports from international go-betweens.

The mission in Azuchi now comprised a Catholic church, residential quarters and a seminary, all housed in the same building. Father Organtino, who oversaw the mission in Kyoto and who'd first accompanied Yasuke to Nobunaga, had been assigned leadership of the seminary in 1580 and now rode alongside Valignano. They'd arrived the day after, having taken the journey at a much more serene pace.

The seminary was originally intended for Kyoto, the materials first bought and delivered to a building site directly next to the Kyoto church. But an inspiration of Organtino's changed all this. If they could build a church in Nobunaga's own city, Azuchi, it would demonstrate to all they were directly under the warlord's protection. The Jesuits would be untouchable by their enemies all over Japan. No one would dare. Nobunaga had proved happy enough to comply and saw this request as foreign validation of his rule and global reach. He offered the Jesuits several places, including a vacated temple precinct and newly reclaimed land, freshly drained with a brook flowing alongside it.

Father Organtino had chosen the reclaimed site, near the

foot of the castle mountain and around the corner from No-bunaga's horse ground, Matsubara. A place where Nobunaga went to ride daily, he would therefore see the Jesuit buildings every day and be reminded of Christ's presence in his city. Nobunaga had declared the selected plot too small—they needed more land to build an edifice grand enough for his Azuchi—and the warlord ordered the two neighboring properties va-cated and destroyed, adding the newly empty land to the Jesuit plot. The missionaries saw it as yet more proof of God's divine vision for a Catholic Japan. More than one thousand workers supplied by Lord Takayama (the most powerful Christian lord sworn to Nobunaga) lugged, carried and sailed the building materials from Kyoto.

The seminary-cum-church and Jesuit residence was a large three-storied building of European style with a prominent and highly un-Japanese bell tower at the front. It stood out. Around twenty-five boys were already students there and they'd come from all over Japan, from the noblest Christian families. These young men were Rome's greatest hope in Japan, nothing less than an elite who'd pave the way for Christ's entry into his new-est earthly domain.

That Valignano had, as he arrived in Azuchi, glanced up to the magnificent castle looming high above and wondered what Yasuke might be doing there, is unlikely. Unless, of course, he hoped Yasuke would prove a valuable source of information in the heart of the warlord's home.

For the next few weeks, Yasuke stayed in the guard quar-ters. He was, however, given the honor of his own room so as not to be disturbed by the changing of the watch. Still, life felt temporary and transient. To pass the time, he spent hours ex-ploring the castle.

Azuchi Castle was a true monument to Nobunaga, built *by* Nobunaga. He'd tasked his loyal retainer lord, Niwa Nagahide,

Azuchi Castle painted in the nineteenth century and based upon descriptions from Yasuke's time.

with overseeing the building, but attended to many of the intricate details himself. It was the first of its kind, and Nobunaga's inspirations drove the resulting masterwork.

On the peak of the mountain, upon a colossal stone foundation, nearly eighty feet in height, there were seven stories to the donjon. The five lower floors were lacquered black, but the sixth rose up a brilliant vermillion red and then the seventh and last was pure gold. They shone in the light of day and would have looked almost as if the top of the castle itself were the sun, emanating glory over the surrounding world. It took the breath away.

The finest artisans available from Japan and China had labored on the intricate details, the carvings and the tiling using only the best materials known, challenging and pushing their skills to the extreme. Other castles Yasuke had seen had been affairs of plain undecorated wood and mud, simple weapons of war, strongholds to control tiny domains. *This* ostentatious magnificence was a palace as well as a weapon, built to make the world take a step backward and then genuflect to its lord. A building to control men's minds as much as the landscape. The mountain base was encircled with moats and walled in smooth stone ramparts topped with tiled walkways which were repeated at different levels ascending the steep slopes, providing multiple levels of defense but also space for dwellings, storehouses, stables and temples. It had been designed with guns in mind. The location of the castle above the lake and upon the open plains afforded better visibility and lines of fire, while the numerous inner citadels gave defenders ample protection to shoot down on exposed attackers even if they had scaled the first wall and were forcing their way up the mountain. Yasuke was told that teams of *thousands* had toiled to pull the largest of the stones into place.

All visitors ascended the mountain up the wide steep, twisting stone stairs and cobbled roads, surrounded by high stone walls. Periodically each side opened on to wide plots of land, also surrounded with high stone ramparts, where senior lords kept their households. Each did their best to display the ultimate in elegance and beauty, thus forming a gilded winding avenue up the mountain to the peak where Nobunaga himself had his residence. Nearing the top, the plots of land became smaller; these were the abodes of Nobunaga's wife, Nōhime, mother, Dota Gozen, sister Oichi, various concubines and high-ranking pages. Close at hand should the lord desire to call upon them. Here also were the administrative buildings and temples that served the center of power, the main stable and the guardhouses to protect them.

And above it all was Nobunaga's *yakata*, his seven-story palace.

Upon entering and climbing through the first five floors, visitors were faced with myriad rooms decorated with gorgeous artwork. Each chamber had a different theme. Some were scenes of nature: wild geese on the wing, doves, pheasants feeding their young, a roundup of wild horses, trees, cliffs, bamboo and pine.

Others were adorned with images of ancient Chinese scholars and stories from legend. Still more were painted in gold, while the tea room was simply done in flakes of gold dust on white. These were the audience and dining chambers where Nobunaga received guests and supplicants. Different themes for different moods, seasons and occasions. The most strikingly unusual and novel feature of these five stories, however, was that the center was hollow, a great chamber that stretched from bottom to top with a stage at the bottom from which audiences on all five floors could be entertained.

The sixth floor was in the shape of an octagon, about thirty feet in diameter. It was painted vermillion red with scenes from Buddha's life and gorgeously carved balustrades adorned its curves. The top floor was a square box of gold, both inside and outside. An audience chamber for the emperor on his planned future visit. The pillars were painted with dragons and the walls decorated with pictures of ancient rulers and sages from Chinese history. It told the world Nobunaga was a modern-day incarnation of those revered sages, the wise, virtuous and rightful ruler of this Japanese realm, and perhaps the world.

After the first week, Yasuke was finally called to work.

Standing guard, his old weapons in hand for the first time in his lord's presence, he watched as Nobunaga got on with his public business and diplomatic machinations in whichever of the audience rooms suited his mood and his guests' rank that day. Yasuke's responsibilities were not so different from much he'd done for Valignano. He was a bodyguard, yes, but even more so than with Valignano, his role was to impress, entertain and intimidate as much as fight. Nobunaga wanted someone who would, by looking good, make him look good too. Yasuke was the ideal choice.

He did his job well. He could see it in the eyes of the men who came to kneel before Nobunaga, whether they wished to curry favor or whisper of war. He could feel their startled gazes, hear their hushed questions. He might never be called to defend his master in his own home, but that didn't matter. He was a strange and mysterious blade Nobunaga could wield at will, and that was enough.

When Nobunaga took the fancy, Yasuke and the warlord's young samurai pages accompanied their lord on trips, galloping along dusty roads; visiting castles, temples and shrines; bathing in rivers, lakes and hot springs; and feasting on succulent trout, ice-cold noodles, and luscious fruits of the forest. They also hunted with hawks and bows, and engaged in contests of strength and speed. And, where strength was concerned, wrestling or weightlifting contests, Yasuke swept all opposition before him, no matter what he was challenged to do.

But Nobunaga clearly not only took a delight in Yasuke's strength and body, but also his brain, speaking with him of the Jesuits, far-off lands, foreign tongues, hunting, and—Nobunaga's favorite topic, and Yasuke's foremost trade—warfare. Yasuke was an adept warrior with myriad stories and experiences to share with Nobunaga. His talks with Nobunaga grew over informal meals, or on walks and rides, when Yasuke was taken ostensibly as a weapon bearer, but fulfilled the role of confidant and tale teller as much as carrier. Nobunaga clearly reveled in talking with him. As the weeks went by, Yasuke became less of an exotic accessory and, increasingly, a useful member of Nobunaga's inner circle. He provided a kind of companionship and familiarity that a Japanese person, tied up with cultural obligations and etiquettes, could not readily perform.

He was able to give Nobunaga a new view of warfare in terrains wildly different from Japan's, with exotic tactics and weapons such as elephants and great cannon. Nobunaga, who'd successfully pioneered handheld guns and massed volley warfare in Japan, was just starting experiments in large gun use, having obtained a few pieces of ordnance through the Jesuits, and ached for a chance to try these out. However, the need for Nobunaga to *govern* his newly won territory as much as fight, meant most actual battles these days were done by senior underlings, themselves commanding many tens of thousands of their own retainers.

As Yasuke understood it, Nobunaga's only serious competition for mastership of the main Japanese island of Honshu was now down to two enemies. First, there was Takeda Katsuyori, the lord of Kai in the extensive mountain ranges to the northeast, leader of a clan who'd fought the Oda for generations. Weakened by a decade of Oda victories, but never fully defeated, Takeda was one lord Nobunaga genuinely hated. And, second, was the powerful Mori clan in the far west, who were slowly being pushed back by Nobunaga into their distant heartlands. Yasuke could tell Nobunaga was itching to follow them all the way home. The Mori—the same clan who'd made sailing with pirates the best option a month earlier—were wealthy and had a formidable military. To continue his rise, Oda Nobunaga next needed these two foes out of the way.

One morning, two months after he began serving Nobunaga, Yasuke was called again to attend his new lord. They'd met in the second-highest level of Azuchi Castle, and Yasuke was not surprised when Nobunaga suggested a walk. Often, the warlord enjoyed a stroll to clear his mind and hear another of Yasuke's childhood stories or of the things he'd seen in India. By now there was nothing Nobunaga didn't know about him—his strength had been proven, his mind and body had been tested, his loyalty was unquestioned. But Yasuke stood straight and still, ready for anything his lord commanded.

Yasuke and Nobunaga ambled out together, descending several flights of stairs until they were halfway down the castle mount. A fresh lake breeze cooled the early summer air and flowering plum blossoms and magnolia lined their path in snowy white. Wild hawks drifted overhead. And the warlord seemed in a peculiar mood today.

Nobunaga suddenly turned right, down a short path, through an opened gate and into a courtyard. Inside was a modest house, brand new, judging by the sweet resin smell of the fresh timber and the pale look of the unaged wood. Into the house they went; it was dim inside, but they passed from the entrance hall into a large room. Tree-dappled light entered through the open shutters and shone on the tatami nearest to him, so that it looked like dim gold.

The first thing Yasuke noticed was that the ceiling felt higher than normal. He hardly needed to stoop at all.

At the far end of the room was a simple black stand with a short sword cradled upon it. The artwork on the scabbard was exquisite, lacquered black with a gold inlaid Oda *mokkou* crest.

Yasuke wondered what it was all about. He imagined for one terrible moment that he was about to be ordered to take his life, but quickly disregarded this. He'd done nothing wrong. He and Nobunaga were still conversing pleasantly. How would he—

Nobunaga lifted his hands and indicated the dwelling. "It's yours," he said simply, and then laughed at Yasuke's stunned expression. "I had it specially made for a giant!"

Yasuke eyed the building again, as if seeing it for the first time. He could hardly breathe.

"You are my black warrior," Nobunaga said, reaching up to grip Yasuke's shoulder. "The demon who will ride beside me into battle, the dark angel who protects me and my family up in my home." He pointed. "The sword is a symbol of this. You are my samurai now. A member of the Oda clan." And with that, he turned and walked out, indicating Yasuke should stay where he was.

Yasuke was dumbfounded. He did not yet fully understand what was going on or how to respond. But Nobunaga was already gone, so it was too late to say anything anyway. The African warrior just stood there.

In what may have been a long time later, or it may have been only a minute, the near silence of the room was broken by the rustle of socked feet on the tatami. Yasuke awoke from his dream to discover two people kneeling, their heads bowed and faces to the floor.

Servants. Awaiting *his* command.

A member of the Oda clan?

And something else Nobunaga had said. The words turning in Yasuke's mind like someone trying to pick a lock somewhere deep inside him, fumbling for the precise combination.

You are my samurai now.

Seven thousand miles from the village he'd been born, it seemed he'd finally been welcomed home.

CHAPTER THIRTEEN
The Way of Warriors

After Nobunaga's decree, and the short moment with the sword, there were no other formalities required. Yasuke was a samurai.

The samurai today are one of the most renowned groups of warriors in human history. Famous throughout the world for their fighting might and artistic prowess. At Yasuke's time the samurai formed the ruling class and almost anybody of note in Japanese society *was* a samurai. The rest, for the most part, aspired to be. Much more than warriors, samurai also oversaw vast estates, commissioned huge building projects, engaged in learning, philosophy, social innovation, made laws, wrote poetry, enforced justice and patronized the arts.

Their status afforded them great respect from the people, and when a samurai rode or swaggered by, his two swords thrust into the sash at his waist, peasants and rich merchants alike would kneel and bow in the mud of the street.

The samurai, as an identifiable class, had not started out this way.

They began around the tenth century as guards and armed bailiffs tasked with guarding imperial property, and enforcing rents and servility among the tax-paying peasants. The tax code in Japan, based on a Chinese-inspired legal system depended, for the most part, upon rice collection. Other crops were generally not taxed, and wily farmers tried countless ways to reduce their rice crop and increase production of other useful plants, such as cotton or vegetables. That the peasants were not even allowed to eat their own rice—as it was too valuable for mere farmers to enjoy—only increased the need for enforcement of constant production and levies.

It took a nationwide conflict for them to advance from simple instruments of war and tax collection muscle to the ruling class. From 1180 to 1185, Japan fought the Taira-Minamoto War (or Genpei War), named for the two imperial court factions who formed the primary belligerents. The Minamoto samurai clan emerged victorious and managed to turn their victory into a total usurpation of national power. The imperial family became rulers in name only, and entrusted governance to a shogun, a samurai military ruler. The samurai were now on top.

From 1185, the samurai formed the elite in society, but the word itself means "to serve," and so the samurai warrior was taught from birth to, above all else, serve his master (also a samurai, though of a higher rank). The relationship between samurai great and small was characterized by rigid codes of honor and obligations from both sides in a codified system known as Bushidō (the way of warriors). The samurai expected himself, his family and his own retainers and household to be provided for by their lord, through income from fiefs worked by farmers, as well as other income sources such as trade, mining and toll collection. In return, the samurai would obey their lords' summons to arms and, the higher the samurai's income, the more men they were expected to bring along with them when their lord's call for warriors went out. They also served as their lords' civil servants, overseers and governance advisors.

During the fifteenth century and The Age of the Country at War, the endless battles took their toll on the limited ranks of the traditional samurai families, and many *daimyō* lords decided they needed to expand their armies. Gone were the days when a few hundred highly trained, magnificently attired samurai squared off against each other with swords in battle. By Yasuke's era, the armies were tens of thousands strong and the need for cheap soldiers had provisionally overridden the need to keep peasants exclusively growing rice. Many men now regularly dropped their tools and lofted spears when they were called upon, leaving the women, elderly and children to work the fields until they returned, if they ever did. Eventually, as the wars expanded in scope, the distances covered made returning home regularly an impossibility. Many of the peasants now found themselves receiving regular wages and better arms from their lords and they held an ambiguous dual status as farmers and lower-ranking samurai, known as *ashigaru*. (The key difference from traditional samurai being that *ashigaru* were not normally permanently retained, nor did they hold fiefs.) This development led in many areas to a more assertive lower class with a sense of their own power and military utility. These farmers had now *also* been to war, and held a spear or fired a gun. No longer would they be so easily bullied around by the samurai. They wanted a bigger portion of the proverbial rice bowl, perhaps even with some real rice in it.

Thus, following The Age of the Country at War, there was no shortage of "samurai" in Japan. Hundreds of thousands, perhaps up to half a million, could have claimed the epithet, though few would have any real family pedigree beyond the last couple of generations in the elite warrior world.

A *daimyō* could call upon both direct personal retainers such as Yasuke, and part-time *ashigaru* warriors to swell his ranks. The direct personal retainers could be classified into four groups. Family members, hereditary vassals, officers of the levies and *hatamoto*, who were the lord's personal attendants. Family members and vassals who held their own fiefs were expected to bring their own samurai and *ashigaru* with them when called upon to fight.

It is not known exactly which rank Yasuke held, but it would probably have been equivalent to *hatamoto*. The *hatamoto* saw to the lord's needs, handling everything from finance to transport, communications to trade. They were also the bodyguards and pages to the warlord, traveling with him and spending their days in his company.

As the realm became more and more divided during the great age of turmoil and the collapse of centralized power in Kyoto, other players saw an opportunity to enter the world of politics and war the samurai had dominated for hundreds of years. During the first half of The Age of the Country at War, these interlopers included monks from militant temples, self-governing peasant warriors such as ninja, pirate Sea Lords, and even merchant guilds and oligarchies in cities like Sakai. Sometimes these regional powers controlled whole provinces and seas. Vast areas, ruled not from Kyoto or from a samurai *daimyō*'s castle, but from an island fortress, a temple complex or a mountain village.

By Yasuke's era, these competing players included the Jesuits, carving out their own domain in Nagasaki. Accordingly, when men like Nobunaga had amassed great power, one of the samurai's primary responsibilities became not only to fight each other, but to reassert their authority upon these upstart lower-class and religious entities.

Prime among these threats to their power were the warrior

monks and their followers. Temples had always been powerful, controlling vast estates of tax-exempt land donated by long-dead nobles or samurai wishing to buy a better deal in the afterlife. Similar to the Catholic Church in Europe, temples engaged in politics and intrigue, and eventually became embroiled in wars. These warrior monks, *sohei*, might be compared to religious orders like the Knights Templar or Teutonic Knights in Europe, strong warriors who believed they were fighting for their faith. At first, different temples fought among themselves, but as their strength became obvious, samurai would sometimes ally themselves to a temple and make use of the institution's armed forces. The temples were well rewarded for their military assistance and became richer and more powerful.

During The Age of the Country at War, the ranks of the *sohei* swelled with co-religionists, often peasants, who saw salvation from the chaos of the troubled times in religion. The temple organizations who'd never before seriously challenged samurai power, now became major players in their own right, seizing whole provinces such as Kaga in the north, and vast swathes of land elsewhere. Any samurai who declined to join them were eliminated.

The samurai in general, and Nobunaga in particular, could not allow this rival power, and the campaigns to reunify the nation also became a de facto war to reestablish samurai rule.

Yasuke had become part of that war, fighting for a new future, one that promised a return to the past.

One can imagine the popular image of a traditional senior male samurai easily enough. In full battle regalia, on horseback with his sword or bow at the ready, but also the warrior off the battlefield. The long *hitatare* kimono, with jacket sleeves so long that the hands are ensconced within, and the wide flowing *hakama* trousers held up with a thin *obi* sash knotted at the front. Two swords in lacquered scabbards, one long, one short, thrust into

Samurai in armor photographed in the 1860s shortly before the caste was abolished.

the left side of the sash. The samurai holds a fan in his right hand, both to keep himself cool and to use as a pointer when indicating something or giving an order. The distinctive topknot hairstyle *(chonmage)* and dour expression, the status-conscious swagger of an unassailable warrior in the presence of his inferiors and his own servile kowtowing in the presence of superiors. He walks behind his lord, passing through an exquisitely groomed garden, judiciously discussing plans of warfare, artistic spectacles or public policy. Or he surveys his lord's fields from horseback, the stallion trotting slowly past lines of peasants bowed to the ground.

There are, and were, many romantic stories surrounding such warriors. Some based in truth and others inspired by later nostalgia for simpler and "purer" times.

One of the most well-known tales concerns ritual suicide. Samurai on the losing side in battle *did* sometimes cut their bellies—seppuku (sometimes called hara-kiri)—to atone for failing their lord. Not all did, especially if there was hope of living to fight another day. While, periodically, laws to forbid seppuku were promulgated to stop useless waste of life, most samurai would have found it a hard deed to fulfill; they were human, after all. And, after the major battles of the sixteenth and early seventeenth centuries were over, hundreds, probably thousands, fled abroad rather than kill themselves, fighting as mercenaries for foreign powers. Most who stayed in Japan escaped to remote areas to become *ronin,* masterless warriors who sold their services on a piecemeal basis, settled down to other jobs or became bandits. Others of the vanquished armies were simply drafted by the winning side.

Another romantic legend involves the samurai's unswervable loyalty to his master unto death. But again, while there is some truth in this legend, there are also cases where samurai changed sides, even in the midst of battle, or staged coups for money or personal advantage. In Valignano's view, treachery was one of the main flaws and limitations of the Japanese, an opinion that bears the hallmark of outside prejudice, but also perhaps one that reflects the perilous times Valignano observed rather than a dominant theme in Japanese history.

During one 1587 siege in Kyushu a senior defending commander, Hebaru Chikayuki, was persuaded by the enemy to engineer the death of the castellan and therefore end the siege. He asked a disgruntled samurai, Usono Kurando, who'd been passed over for promotion, to kill the commander while he set his own castle alight. It was a success, but the stain of treachery ensured the killer ended his life a beggar.

Another samurai story, less well-known but grounded in historical fact, is that of the female samurai, fighting to the death to protect her family or home, or even leading troops on the bat-

Female Samurai, by Utagawa Kuniyoshi, 1848.

tlefield. Women traditionally trained extensively in martial arts, especially with the *naginata*, a wooden staff with a sword blade attached to the end. Many also trained at archery, and all women were taught to use a knife to cut their own throat, and those of their children, rather than surrender to the enemy. When her husband was away or dead, the samurai wife had charge of the castle or manor, and ultimately, became the last line of defense.

One famous story is that of Ōhōri Tsuruhime, the chief priestess of the Ōyamazumi Shrine on the island of Ōmishima in the Seto Inland Sea (on the route Yasuke took to Kyoto in 1581). Ōhōri's island was attacked in 1541, but she and her warriors drove the invaders back into the sea. Four months later, they attacked again, and Ōhōri used a grappling hook to climb aboard the invaders' flagship and confront the enemy general face-to-face. He mocked her but got his comeuppance when she cut him down. Her men then pounded the enemy ships with grenades and the invaders retreated. Two years later, battling the same foes, Ōhōri's betrothed was killed in action. Brokenhearted, she flung herself into the sea and drowned. She was eighteen years old.

Yasuke had joined the exalted ranks of the direct retainers to the most powerful man in the land. But he hadn't fully grasped the magnitude of the gifting of the short sword marked with the Oda crest. He'd been given various weapons before in battle, even presented with beautiful tools of war by other lords and commanders, but they'd never come with such life-changing status, titles or esteem. Only after a period of reflection and noticing the different way he was treated by the *other* samurai of Nobunaga, did Yasuke fully understand. The bestowing of a sword from a Japanese lord to his vassal was a major occurrence. It was the symbol of his new status and virtually the highest honor Nobunaga could bestow on another without actually giving a fief. Nobunaga's own sons, Yasuke was told, had wept when they'd received similar ceremonial blades from their father.

With the sword came the modest house tucked within the trees off a path leading up to the castle, a beautiful garden with a spring-fed rock pool, and the two attendants—an old married couple given by Nobunaga to provide for Yasuke's every need and keep his residence immaculate.

Yasuke also received a moderate stipend. The salary was to feed, clothe and pay his new staff, and to provide Yasuke with the funds to entertain, and be entertained, in the manner appropriate to his new station.

Nobunaga, something of a sartorial dandy himself, then provided several sets of clothes. Based on the wardrobe alone, Yasuke would clearly be required to dress in different ways depending on whom Nobunaga wanted to impress. Not only a formal *kataginu* jacket with wide loose-fitting *hakama* trousers and *kosode* under shirt, but outfits the warlord had ordered from tailors in Azuchi and Kyoto—those with recent experience of copying European dress. Due to Nobunaga's own notorious taste for European garb, particularly cloaks, body armor and hats, and the inevitable fad which followed his example throughout Japanese court circles, this fashion was fast becoming a specialty of some clothiers. Wide trousers, doublet, collared shirt and short manteaus, as you'd find in any court in Spain or Portugal, all custom made to Yasuke's great size and neatly folded in intricately decorated wooden drawers in the back room of the house.

Nobunaga's inner circle were used to hosting Chinese engineers, artisans, experts and consultants, visiting European priests and exotic visitors from the realm's farthest borders, but they'd never been asked to accept an outsider as one of their own. It was a truly unique event in their history, in world history, and many would have been puzzled, possibly even mutely offended. However, Nobunaga's word was final.

Besides, this foreigner was affable, mannerly and worthy to train with. Sparring with Yasuke, Nobunaga's other samurai had already picked up new and unusual moves, and so had he. The process strengthened all of them and the Oda clan was all the more formidable for it. Nobunaga—curious about the bigger world beyond eastern Asia, and hungry for much that it offered—had never been a man overly concerned with tradition or protocol. If anything, he brandished his disdain for convention as other lords clung to their pedigrees and heirloom swords. It was he, Nobunaga, who'd create his own traditions and decide what was acceptable in a new Japan truly of his own making. Advancing Yasuke, a foreigner from thousands of miles away, to a samurai of his close household would be one of many "firsts" with which Nobunaga would challenge "Old" Japan.

Bestowing a foreign-born warrior with the title of samurai was, to Oda Nobunaga, one more way of establishing himself as the creator of a New Japan.

CHAPTER FOURTEEN
His Lord's Whim

As a samurai retainer, Yasuke's primary role had changed. His main job was no longer security detail; there were plenty of others to handle that. Instead he became something of a consultant. "Nobunaga never tired of talking to him," wrote the Jesuit Mexia, a key colleague of Valignano's who was with him in Azuchi. "As he was also strong and entertaining, he pleased Nobunaga." Which gives us a glimpse of not just his skill in warfare but also Yasuke's affable side. It is easy to see him training with Nobunaga and his comrades, and providing the odd laugh with a comic turn or feat of strength.

His life had changed considerably. He was now a man of independent means, a householder and employer with all the new responsibilities that came with those roles. He was not just a samurai, he was a citizen, and a highly ranked one at that. People in the streets did not only gape at him, they bowed, heads to the earth, as they addressed him. His servants took over all the household chores he'd been accustomed to performing for Valignano, and Yasuke was free to fulfill his various roles for Nobunaga, train for the battles which would surely soon come, and get to know his new sparring companions.

Like everyone else in Nobunaga's service, Yasuke was subject to the warlord's every whim or order. If Nobunaga wanted to spend the day at court dispensing judgments, then Yasuke would be in attendance as backup security, *and* novelty. If Nobunaga wanted to take his horse out, then Yasuke and the other pages would ride with him. If he wanted to engage his pages in competitions of strength for fun, then Yasuke would partake and be the winner. If Nobunaga wanted to put an idea past a new ear, then they would talk. That his residence was only a short distance away from the palace made it all the easier for his Japanese lord to summon him.

As companion, Yasuke had officially joined the two dozen handsome and spirited young men he'd first ridden to Azuchi with. These were Nobunaga's pages, the cream of the clan youth from ancient Oda family vassals' households.

Chief among them was clearly Mori Ranmaru, Nobunaga's favorite, renowned for his beauty and bravery. Then there were Ranmaru's youngest brothers Rikimaru and Bōmaru, too young to have domains of their own as yet, but already blooded warriors. And Ogura Matsuju and Jingorou, sons by another father of Nobunaga's concubine Onabe no Kata. Takahashi Toramatsu, a rising star. Otsuka Mataichiro and the others had been in this band of samurai for years and would soon be given fiefs of their own, replaced by their younger brothers and cousins in Nobunaga's immediate entourage.

All of good birth, with promising future careers. Following Japanese warrior custom, Nobunaga allowed these young men to enter his close service and promoted them through the ranks.

Mori Ranmaru, by Utagawa Yoshiiku, 1867.

They also, Yasuke soon learned, engaged in sexual relations with Nobunaga and other older samurai.

The samurai had adopted, supposedly from practices within the Buddhist monasteries, same-sex pederasty—*nanshoku* or *shudo*, the "way of adolescent boys"—as a way to promote to-the-death loyalty among warrior bands. Samurai boys in training were commonly apprenticed with an older warrior to learn martial skills, the samurai code of honor and formal etiquette. And, very often, the instructing male would take the boy as a lover until the apprentice became an adult. The older lover was expected to reflect on his role as a mentor through this benevolent love and become a better adult in the process.

In this arrangement, both parties, with their families' blessing, generally agreed to be exclusive as far as male-to-male sexual relations were concerned. Either male was *also* permitted to take female lovers. This codified system became role-defined; the adult male was the active, desiring penetrative partner, and the younger, sexually receptive boy was considered to submit out of love, loyalty and affection, rather than mere sexual desire. One Jesuit explained that, in Japan, committing sodomy with a boy did not cause the boy any discredit or dishonor, as sodomy was not considered a sin and boys "had no virginity to lose" anyway. Sex between the couple ended when the boy came of age—and then normally went in search of a younger lover himself. The original relationship would, ideally, develop into a lifelong bond of friendship and loyalty which would transcend to the battlefield.

As might be assumed, the Jesuits took a rather dim view of these homosexual relations. Valignano wrote that the Japanese "are much addicted to sensual vices and sins, a thing which has always been true of pagans." He went on to say, "Worse is their great dissipation in the sin that does not bear mentioning. This is regarded so lightly that both the boys and the men who consort with them brag and talk about it openly."

Nobunaga was known to be involved in relationships with many of his pages at one time or other. During Yasuke's time, Ranmaru was clearly his favorite. And after much sake, the brazier embers glowing, Nobunaga and his chosen one would quietly make their way to the warlord's sleeping room and a white silk futon glimmering in candlelight.

Homosexual relationships among warriors were not unique to Japan. There is a rich history of fierce warrior bands who would fight to the death rather than let down their lovers beside them. The most famous is perhaps the Sacred Band of Thebes, an elite corps composed of one hundred fifty pairs of male lovers who fought for Thebes, a city-state of ancient Greece. Friend and foe alike considered them invincible. Their end came at the Battle of Chaeronea, where Philip of Macedon effectively completed his conquest of the Greek states. While the rest of their army fled the field, the Sacred Band refused to surrender and were annihilated to the last man.

Closer in proximity to where Yasuke spent the key years of his life were the *Hwarang* (Flower Knights), warrior elite of the ancient Korean Kingdom of Silla. Although the evidence is not conclusive due to the loss of many historical sources, they seem to have been cavalry bands of young men taken from noble houses who were trained in the martial arts and enjoyed great success in battle. They also formed romantic attachments as Silla became the most powerful kingdom on the Korean peninsula for several centuries during the first millennium.

It's been suggested in recent academic studies that Yasuke was engaged in this kind of *shudo* relationship with Nobunaga. Yasuke, in his mid–late twenties, was far older than a normal youth role in *shudo* allowed; in fact, he was old enough to be in the senior role. However, the handsome giant certainly appealed to Nobunaga on multiple levels, so one of these may well have been sexual. Perhaps one of the clues to his swift rise is that he submitted to Nobunaga's advances.

Had Nobunaga attempted any kind of sexual relations, it is unlikely Yasuke could have resisted. He now owed everything to his lord and was entirely in his power. If it occurred, however, one wonders what Yasuke would have thought of it. He was from a very different cultural background, and although it is hard to find information on sexual practices in ancient Africa prior to Church missionary activity, there is a long history of acceptance of transgender individuals and homosexuality. Many African languages did not even have a word for homosexuality until loan words came from outside, indicating its probable unremarkable part of their human experience.

Traditional Dinka society was family orientated with a strong emphasis on all males having children, but their society also revolved economically, and socially, around cattle and cattle rearing, which would have left a lot of man-to-man time while herding. Young adolescent men, whose job it was to mind the herds, had time on their hands and few girls nearby, so close male relationships would have thrived. Perhaps, as with the samurai, any sexual relationships were considered as a "life stage." The ultimate aim of male adulthood was marriage and children to keep the family line going.

Near the Dinka lands were the Azande warriors, in what is now the border region between South Sudan, the Central African Republic and Democratic Republic of Congo. Similar to the samurai, elder warriors formally approached the parents of teenaged boys to ask for their hand in marriage. The boy then lived with the warrior as his wife until he came of age and was released to find a male teenaged wife of his own. The boys accompanied their husbands into battle and looked after their weapons. Azande means "the people who possess much land," giving an idea of their prowess and success as warriors; they battled regularly against the Dinka people.

Slightly to the east of the Dinka lands, among the Oromo people of modern Ethiopia, male-to-male relationships, *midiisa-i*, were common, as they had a rigid system of grades or ranks which were determined by age, and which did not allow men to marry until they reached the age of thirty-two. This meant an outlet other than sexual relations with women had to be found. The Oromo were also a warrior-orientated people, excellent horsemen, who spent a lot of time on campaign, and therefore even after reaching marriageable age would have spent a lot of time away from female company.

However, Yasuke had been in Jesuit employ for many years, and prior to that, had inhabited a Muslim world. Neither Catholicism nor Islam permit homosexuality, so there would have been tremendous societal pressure against the realities of human nature.

Nobunaga was said to be merciful and loved by the people whom he ruled; Yasuke understood he was also feared. Servants in Azuchi often literally shook when he approached. Visiting lords sometimes stammered when they spoke. One even fainted. In the town below the castle, shopkeepers casually retreated back into their stores as Oda warriors breezed by on horseback. In the middle of June, Yasuke finally discovered why, firsthand.

They'd traveled on horseback and then by boat to a shrine in the middle of nearby Lake Biwa. An excursion to revel in nature, something to get them away from the tiring business of government and diplomacy and a good excuse, after, to pay a visit to Nobunaga's senior vassal Hideyoshi's nearby castle at Nagahama. Nobunaga had been preoccupied with events in the north where restive peasants, warrior monks and local lords had taken advantage of the distraction caused by the cavalcade of horses to rise up. Nobunaga's men on the ground, and the reinforcements which he swiftly dispatched, had dealt with the rebellion, but it had been a worry. The ride out after to visit Hideyoshi's castle was another chance to forget such concerns.

Yasuke and five other pages had accompanied their lord and expected to stay the night in Nagahama before returning to Azuchi in the morning. Nobunaga, however, had changed his plans and, in an exhilarating lightning ride that used multiple horses per rider, the group returned to Azuchi Castle within the day, a distance of thirty leagues, much farther than anyone would ever have imagined was possible.

Word of his change in plans, alas, had not reached his servants back home. When Nobunaga returned, he found nothing was ready for him; worse, many of his female domestics had gone to worship in a nearby temple, and were still absent from the castle. Word of his fury spread across the entire town. Those servants caught unprepared at Azuchi Castle were bound and taken to the courtyard for immediate execution. The women at the temple were rounded up—together with the abbot who begged for their lives to be spared—and brought back to the castle.

Yasuke watched the whole episode from a platform in the yard. Nobunaga sat on the ornate chair he'd received from Valignano and Yasuke stood beside him. The warlord gazed on impassively as each woman knelt without resistance and the warrior assigned to the duty lifted the victim's hair. Another soldier poised his sword and then swiped. Each woman's head was parted neatly from her neck, the corpse toppled forward, and the first man held up the trophy for inspection while the swordsman wiped his blade clean on the dead woman's robes. Blood pooled on the ground. Each time Nobunaga nodded curtly in acknowledgment of the sentence carried out. After all the women had been killed, the Buddhist abbot was executed also.

Yasuke thought back to other mutilations, decapitations and executions he'd witnessed. One he'd seen in India had taken five swipes of a blunt sword; and another he'd seen on his journey to the African slave port as a child had been prolonged purposely by the perpetrators, who'd bet on how many strokes it would take. These, in comparison, had each been delivered in a single stroke.

His samurai brethren often used the bodies of recently executed criminals to test and hone their blades during training. The headless body would be fixed to a standing position with stakes so that warriors could swipe at it in different rounds of testing; the calves, then the thighs, the arms and so on to the torso until there was nothing left but mincemeat for the village dogs and waiting birds. At each round, they'd inspect their blades for notches and nicks, testing and proving the quality of their swords to themselves and each other. It was a chance to practice killing blows *and* to assess their swords on real human flesh and bone. Yasuke—who'd received similar training in India to inure him to killing—joined them.

They did not use the executed women in this way. Instead, their bodies were cleared away, the lesson at an end.

* * *

A week after executing a dozen members of his staff, Nobunaga stayed true to his mercurial style and held several extravaganzas for his people and numerous visiting lords.

Foremost among these was a sumo wrestling tournament.

Nobunaga was a well-known connoisseur of sumo, and many of today's codified rules and conventions can be traced back to him. Sumo had originated in the mists of unrecorded history as a form of worship, where humans performed ritual combat with spirits in shrines. It had continued as one of the many ancient martial arts warriors used to keep fit and battle ready.

This occasion was a minor spectacle compared to some of Nobunaga's earlier sumo tournaments. Those had attracted men from throughout all of Japan and lasted through days of feasting, drinking and—thanks to the local women *and* courtesans brought in especially for the event—a good deal of fornication. Prizes then had included ceremonial weapons, silken gowns, minor fortunes and even residences. The prize this time was, as always, handsome. The winner was to be awarded one hundred *koku*, enough rice to feed one hundred mouths for a year and the income of a medium-ranked samurai, something akin to what Yasuke himself might have enjoyed. It could be sold for a small fortune by the winning combatant. Even sumo wrestlers, Nobunaga joked, could not eat *that* much food.

Nobunaga, who routinely exhibited and challenged Yasuke's strength, eventually had his newest samurai enter the *dohyo* for a go. While the crowd cheered and his fellow samurai shouted encouragement and laughed, Yasuke played along. He first took off his shirt, then climbed the steps to the *dohyo*. The ring was edged in rice-straw bales on top of a platform made of clay mixed with sand. The *yobidashi* (bout announcer) called Yasuke forward and indicated the two white lines on the floor with his fan. Yasuke, grinning, took his spot and squared off against a foe who stood more than a foot shorter. The crowd—drunk, and understanding this match was just for fun—yowled with expectant laughter.

Yasuke understood his role well enough and when the *gyoji* (referee) motioned with his fan for them to start, the two began their slow dance around the ring. The tournament's many spectators would have seen the religious symbolism in the dark-skinned giant, who many equated with a semidivine spirit, engaging in the ring with a human foe.

Yasuke placed his hand on his opponent's head and held him back as he would a child, several feet away, to raucous laughter and Nobunaga's delighted grin. Next, Yasuke stood still and allowed his opponent to try and move him out of the circle for the win; but he could not be moved an inch. Finally, the African samurai grabbed his opponent and pushed him out of the ring in a single dismissive move; his expression, however, apologized to his defeated opponent. It would not do to humiliate his new comrades.

The crowd roared in approval. Yasuke was soon facing off against two, then three, at the same time. It took four to eventually drive him from the *dohyo*. The four triumphant Oda warriors drank and joked together with him for the rest of the day. Nobunaga joined them for a round or two.

A week after the sumo tournament, Nobunaga held another horseback spectacle—this time for the people of Azuchi. *What was the point of soon ruling the whole of Japan, if he couldn't bring such delights to his own people?* They needed to get used to this. The Romans called it "bread and circuses," perhaps "rice and spectacle" might have been more appropriate for Nobunaga.

The time for his final victories, after all, was near. Campaigns were going well against the Mori in the west and in the final planning stages for taking on the Chōsōkabe clan on Shikoku, Japan's third largest island. Nobunaga's troops were on standby, ready to attack the other minor factions, prime among them Nobunaga's old foe, the loathed Takeda clan ensconced in the mountains around Mount Fuji to the east. When the opportunity revealed itself, he would pounce.

Yasuke held his left arm firm as the bronze-speckled hawk approached for landing, harking the gentle ringing of the *suzu* bell attached to the bird. Its underbelly and wings were mostly white, but it was the narrow streaks of bronze at the tips of each feather which flashed in the midday sun. It landed heavily on the glove made from soft deerskin which extended up his left forearm, and after Yasuke had secured the jesses, he turned to grin at his amused audience.

Nobunaga smiled back, nodded in approval. Two of his sons— Nobutada and Nobukatsu, on a rare visit from their own fiefs to see their father—and Ranmaru stood behind him, their own hawks perched and waiting on their arms. Insects gossiped in the tall grass surrounding them.

A party of twenty had ridden along the Lake Biwa coast and then inland for an afternoon of hawking. Nobunaga had recently received six exceptionally rare Korean hawks as a gift from a lord seeking favor, and after weeks of lavishing care and training on the birds, he'd wanted to try them out.

Throughout the summer, supplicants from all over Japan had brought Nobunaga gifts to acknowledge his suzerainty. Jewels, expensive fabrics, horses, domestic falcon chicks and most prized of all, these hawks from Korea, a rarity that had not been seen in Japan for a century. Hawking was a prestigious pastime, introduced from Korea over a thousand years before, the sport of lords and princes. The best and rarest birds could change hands for a king's ransom and to be able to partake in the activity showed extreme status and wealth. Still, every large city had a *Takajo* street, a "hawkers" street, supplying trainers and necessary gear for the exclusive few who could afford to shop there.

Yasuke gave his bird a wedge of duck meat from the black-lacquered food box proffered by Nobunaga's hawk trainer, who'd stood close at hand, offering quiet encouragement throughout. The *suzu* bell tinkled as the bird chewed away at the meat. It had been a short flight for the older hawk, a hundred yards at most, but had adequately provided Yasuke his first hands-on experience with the sport. It was a great honor to be permitted to take part; normally even the hawk trainer was not allowed to touch the bird directly, using a special kind of blanket wrap called a *muchi* when needing to handle his charges.

The party stalked forward again, Nobunaga preparing his own bird as a pair of young retainers ran out ahead to scare up prey. Today's catch had been a dozen quail and a pair of cranes. Soon, another crane took to the sky and Nobunaga's hawk was set loose. It raced across the sky and struck the larger bird directly at the neck, clinging tight as the crane worked to shake the hawk away. Often it required a second hawk loosed to take down such large prey, but no one—not even Nobunaga's sons—dared loose his own bird unless asked directly by Nobunaga. They waited and watched while the hawk's vicious beak struck and the crane dropped from the sky, talons still clinging to its throat.

As they walked to the catch, Nobunaga recounted the story of a famous samurai poet who'd given up hawking forever after meeting a "samurai-like" crane who'd held off two hawks for some time before finally falling and then shown no signs of fear when the hunters had come to slit its throat. "The crane was resigned to the inevitable," Nobunaga explained. "And met its death with a calm and reconciled defiance."

Yasuke and the others reflected on his meaning while Nobunaga's youngest son finished off the dropped bird and handed it to one of the servants. The hawk, already landed and tethered with the jesses as the warlord fed it morsels of moist duck meat from the ornate food box hanging from his belt, shuddered at the scent of the fresh kill.

Over lunch, Nobunaga asked Yasuke about hawking in other parts of the world. Yasuke had heard of it, even seen it in India, but never partaken. He felt a bit disappointed in himself; normally he had an answer for Nobunaga's questions. Still, Valignano's former bodyguard had clearly become a trusted voice in Nobunaga's inner circle due to his wide knowledge and experience of foreign life and warfare.

In early August, Valignano—who'd been visiting Christian Japanese allies in the north—again passed through Azuchi. He was on his way back to Nagasaki, where he would prepare for his long-planned return home to Rome, a voyage which now included four young Japanese samurai emissaries he planned to introduce to Europe.

Before leaving, he visited Nobunaga one last time to bid his formal farewell. A necessary courtesy. They met in Nobunaga's sixth-floor vermillion chamber, Yasuke kneeling to the warlord's right. Yasuke and Valignano exchanged nods and cordial smiles as their gazes met. Valignano had been kept up-to-date with Yasuke's trajectory, but the sight of him, at Nobunaga's specific request, in his Japanese robes and samurai swords, was surely something to behold.

Nobunaga, naturally, granted Valignano's wish to depart his realm, but also made clear his desire the Jesuits continue their work in Japan. Their presence pleased him a great deal as he found their conversation fascinating, their music beautiful, and they always brought interesting new contraptions with them as tribute. Also, of course, they annoyed the established ecclesiastical powers.

The Jesuits had already had a significant influence on Japanese warfare. The order's patrons, the Portuguese, had been the first to bring handheld European firearms, more efficient than the older Chinese-style weapons. But now the Japanese had mastered their manufacture and even improved upon it by making waterproof guards for the ignition taper, and standardized bores for bullets. But who knew what newfangled gadgets the Jesuits would bring next?

Paying Valignano a particularly notable honor, Nobunaga gifted the Jesuit a huge pair of priceless gold-encrusted standing screens featuring paintings of Azuchi Castle. These were as tall and long as Yasuke and created by the most renowned artist of the day, Kanō Eitoku. Through Valignano, now word of Nobunaga's power and glory would be transmitted around the world. Valignano promised to deliver Nobunaga's gift directly to Pope Gregory XIII himself.

The Jesuit Visitor took the opportunity to bid a final farewell to his former manservant and bodyguard. The emotions for both men were complex, without words. They wished each other "Godspeed."

Yasuke had been promoted further than he ever could have imagined and stood rich, proud and tall beside his new lord. And, Valignano's visit to Japan had also been a wild success, his expectations exceeded, and it was with triumph in his heart that he paid obeisance to Nobunaga and, bowing deeply, took his final leave.

Yasuke was now alone in Japan.

Days later, all the streets and houses were hung with lanterns. For three long hot summer nights, the townspeople and garrison walked the streets, both out of devotion to the dead and in joy at taking their holiday. It was the annual Festival of the Dead, *Obon*, a Buddhist-inspired holiday in which ancestors' spirits pay a visit to the family home and after three days return to the spiritual realm, seen off with fire, song and dance. It is an occasion for family gatherings and also the time to clean the ancestral graves and think upon sacrifices made by dearly departed loved ones. Offerings of water, incense, sake, rice and fruit are made to the dead, new babies are introduced, family updates shared with the unseen ancestor ghosts as they rest for the long journey back whence they came.

Nobunaga was not going to miss the opportunity for a spectacular show. On the third and final day of the festival, he had the whole upper castle, surrounding woods and Lake Biwa below lit with thousands of flaming torches and lanterns to show the ghosts and spirits of the ancestors the way home. For those spirits slow to take the hint and leave, children tossed rocks onto the roofs to drive them away.

Nobunaga's horse guards even stood holding burning torches in boats far out from the shore. The light reflected and shimmered softly like the spirits themselves in the calm lake waters. Below the castle, a huge blaze formed the center of the festivities. The great war drums, *taiko*, were brought out and manned by teams of bulky samurai, setting the tempo for the *Obon* dance, as the people of Azuchi formed circles around the huge blaze and lost themselves in the hot sweaty evening and booming, mesmerizing beat of the *taiko*.

It was a sight to behold and as Yasuke took a break from the dancing, he drank it all in. It was a time to reflect on ancestors and the past. All that was lost to him so many miles and years away, somehow so close again. But he also reflected on the present, and even wondered about the future, a future now so much more filled with possibility than at any previous time in his life.

He was a samurai, within the closest entourage of the most powerful man in Japan. He had servants, his own home, weapons and fine clothing. And he not only invoked fear in those around him—something he had likely always received, grudging or blatant—but respect. His trajectory was moving ever upward, rising like the many *Obon* flames reaching toward the sky.

CHAPTER FIFTEEN
Oda at War

Yasuke's new clan, the Oda, were always at war.

Under Nobunaga's leadership, the Oda had expanded from their home province of Owari during the 1560s—when they conquered the neighboring domain of Mino—and hadn't ever looked back. Their strategy and pattern remained consistent for the next twenty years: giant strides forward in territory, followed with occasional small steps back in consolidation. In short order, they'd grabbed the whole of the central plain area of Japan, and the most important cities of Kyoto and Sakai.

As was common across the world in Yasuke's time, before resorting to war, Nobunaga often tried using his wider family as pawns in his political machinations. He himself started playing this game early by marrying Nōhime, the daughter of a rival in 1549. She proved unable to have children, so all of his offspring were born by concubines, although Nōhime officially adopted his direct heir, Nobutada. Twenty years later, Nobunaga married off his younger sister, Oichi, to cement an alliance with warlord Azai Nagamasa, and then married another sister, Oinu, to a rival lord, Saji Nobukata. Both marriages ended after only a few years when the warlords betrayed Nobunaga by foolishly allying with enemies against him, and soon died at the hands of Oda troops. Still, the marriages had bought peace for a few years and allowed Nobunaga to continue his spread across Japan.

Nobunaga's children also proved useful tools in his plan for domination. In 1570, he forced another clan, the Kitabatake, to adopt his second son, Nobukatsu. Within five years, Nobukatsu had taken control of the clan and disposed of all its senior members, and hence the territory came under Oda control. Nobunaga's third son, Nobutaka was foisted on the Kanbe clan. Other sons were adopted by senior retainers to foster alliances, loyalty and the growing Oda empire.

And then there were Nobunaga's daughters. His eldest, Tokuhime, was married to the eldest son of the lord of Mikawa, Tokugawa Ieyasu, to cement their alliance. However, Ieyasu's wife, Lady Tsukiyama, did not take to the Oda girl, and made her life a misery. The frustrated daughter-in-law told her father that Ieyasu's wife had been scheming against him and although nothing was ever proven, Ieyasu valued his alliance with Nobunaga more than his wife (putting his people's peace and prosperity above his own self-interest—Nobunaga would have wrought havoc had he been defied). She was executed, and, just to be safe, their son, Tokuhime's new husband, was forced to perform seppuku.

When such alliances weren't easily available, Nobunaga had his army. Fielding fast-moving divisions of up to fifty thousand men, many toting Japanese-made muskets, they were now completing the seizure of Japan by expanding their influence along both coasts and into the mountainous areas that remained resistant or semiautonomous. All of his sworn troops, pulled together now, surpassed two hundred thousand men.

Beside the reviled Takeda Katsuyori in the mountains of Shinano Province to the northeast, Nobunaga's only other real competition for national domination was now limited to the Mori clan in the far west, also well armed with locally made firearms. All other serious players had been defeated, weakened beyond recovery, or effectively become his vassals. The Mori remained wealthy, well armed and defiant.

Nobunaga's most formidable general, Hideyoshi, was tasked with ending that particular matter once and for all. Hideyoshi was a nationally feared and celebrated military genius who'd risen quickly from the lowest levels of society to become one of Nobunaga's most trusted advisors. Nobunaga knew how to use the men fate brought his way; Yasuke was the latest in a long line.

Hideyoshi had started life in a family so poor and humble he didn't even have a surname and he'd escaped a drunken stepfather to wander the streets as a child doing odd jobs. He'd entered Nobunaga's service as a sandal bearer in 1557 and, as the story was now told, had distinguished himself from the other Oda attendants by keeping Nobunaga's sandals warm under his own shirt. This simple innovation prompted a promotion, and as he proved himself further, by Yasuke's day, he was one of the most powerful men in the country with a castle near Azuchi, Nagahama, and command of tens of thousands of his own retainers.

While successful, Hideyoshi had not been graced with good looks, and Nobunaga even called him "monkey" in public. One of the Oda lord's little jokes.

Regardless of his appearance, Hideyoshi's advance across central and western Japan under Nobunaga's orders had been swift and bloody. Each year of the late 1570s brought a new campaign and new domains under Nobunaga's power. As the lower-lying regions on the Seto Inland Sea coast fell, Hideyoshi pushed north through the mountains and descended toward the San'in region on the other side where the Mori-controlled domains stretched along the coast awaiting his attention. In 1581, only one thing now stood in Hideyoshi's way: Tottori Castle.

The castle was perched upon an almost unassailable, thousand-foot-high mountain. Even the widest paths to the top were essentially vertical, and men could only scramble up in pairs. The majority of the town's population had speedily taken shelter from Hideyoshi's hordes in the mountain castle, and the only way to take the fortress was to break the defenders' resolve through siege. Hideyoshi duly engineered one of his famous solutions, one of the engineering feats that secured his reputation as an unbeatable foe. He flattened the outlying forts and ordered a moat and palisades, of both mud and bamboo, thrown up around the entire mountain. The fortifications were interspersed with watch towers so that none could escape. On the other hand, no relief force would be able to come to the Mori's rescue either; ramparts were built facing outward and the whole Oda army was accommodated *inside* the two walls. Now, the mountain castle and all its people within were truly surrounded and the garrote tightened; food soon became scarce as Hideyoshi's attack had been so fast that few of the besieged had time to haul in supplies. The Oda troops, on the other hand, could transport all the supplies they wanted by sea, and also secured all the local granaries, ports and rivers so Hideyoshi's men could feast while the defenders starved. The smells of steaming rice and the sweet aroma of the plentiful and delicious Tottori fish grilling, must have wafted up the near sheer sides of the mountain, torturing those starving within.

It was only a matter of time.

Yasuke was in attendance on his lord in late autumn, when a messenger came speeding on horseback up the Azuchi Castle mountain, drawing to a halt before the doors of the tall and imposing keep. Lord Nobunaga bid him enter and climb the stairs to the audience chamber at once as the message was from General Hideyoshi. Mori reinforcements were rumored to be pouring through the mountains to the south of Tottori in numbers enough to break the siege and send the Oda troops packing back home again.

Nobunaga's reaction to the threat was swift and decisive. He dispatched two other top generals, Akechi Mitsuhide (who'd orchestrated the horse spectacle in Kyoto) and the Christian Takayama Ukon, at the head of their armies to immediately reinforce Hideyoshi. If the rumors of a rescue attempt were true, there would be overwhelming power ready to crush any such rescue attempt. Takayama, at whose castle Yasuke had spent Easter with Valignano, was to assess the situation at Tottori and report back in person as soon as circumstances allowed.

Yasuke, the other samurai and the horse guards were also all put on standby. If need be, depending on the reports, Nobunaga himself would take the field. The African mercenary hadn't been invited as a curiosity or companion this time. He'd been ordered as a warrior. If they went, it would be his first battle for Nobunaga; his first fight as an Oda samurai. A chance to prove he was more than strength and stories and differently toned skin.

But the weeks went by and, while there were indeed rumblings of riding off to war, Tottori Castle was not the cause. Apparently, the siege was not the only matter at hand that fall. There was also an opportunity to forever eliminate another old enemy.

For centuries, the peasant warriors of Iga (a wild mountainous land where unwelcome outsiders often got waylaid and murdered) had proven rebellious mercenaries who, when they weren't working their inhospitable land, sold their specialized military services to the highest bidder. These were the same people who'd, two years previously, disgraced the Oda name by thrashing Nobunaga's second son, Nobukatsu. To make matters worse, the warriors of Iga were famed ninja, the most famous in the land, to such an extent that special operations operatives like the ninja were often called "Iga men" whether they were from Iga or not. If Nobunaga couldn't buy or control them—which seemed the case despite a decade of trying—these revered and unpredictable fighters would forever remain dangerous enemies, a thorn in his side. He needed to eliminate or subdue them entirely and redeem the wounded Oda pride.

Nobunaga had already endured ninja assassination attempts on at least three occasions. The first had involved a musket sniper from Iga's close neighbor, the Koga domain, in 1570. The ninja lay in wait for days for Nobunaga to pass, and his two shots from two-dozen yards had been true. Nobunaga had been blown from his horse, but the shots had been narrowly deflected by the edge of his collar armor. A very lucky escape, and the furious warlord ordered the would-be assassin hunted down. The hunt took three years, and, once caught, the ninja's end had not been swift. Nobunaga personally supervised his agonizing execution: the man buried up to his neck, his head removed in small horizontal slivers, top to bottom, by very blunt bamboo blades over the course of an entire day.

The second attempt had been by an Iga ninja named Manabe Rokuro, who'd somehow sneaked into the place where Nobunaga was staying. He'd been intercepted by the guards, but ran. Upon being cornered and, knowing the fate that awaited him if captured, he'd turned the assassin's knife on himself. Nobunaga ordered the fresh corpse displayed in the middle of the marketplace as an example to all. The third attack was by an infamous ninja-cum-thief, somewhat reminiscent of Robin Hood in contemporary folklore, named Ishikawa Goemon, also said to be from Iga, and took place in 1580 directly before Yasuke met his new lord. He'd hidden in the ceiling above Nobunaga's bed and used a thread to target drops of poison into the sleeping warlord's mouth. That he did not succeed was self-evident, but how he escaped and why he wasn't successful in his attempt is lost to history.

Fortunately for Nobunaga, two disgruntled Iga men had recently appeared at his castle. For a substantial payment in gold, these traitors offered to guide the Oda troops into the mountains to destroy the rest of the Iga warriors once and for all. It was the break Nobunaga had been waiting for. He would eliminate their threat for good and reclaim the clan's lost honor. The mighty Oda would not be defeated by mere farmers in arms—even if they *were* professional killers, they were still "only peasants" who took payment for their services rather than fighting for the higher principles of honor and the glory of one's lord.

Nobunaga jumped at the opportunity of razing Iga, but not personally. He remained in Azuchi that fall, Yasuke and the others at his side, still awaiting news of the Tottori siege. But, following his orders, by late September, fifty thousand troops under six commanders had massed on Iga's borders. And, to be certain of absolute victory, they were to all attack at once from different directions and take and hold specific districts. (His son's previous attempt had been with only ten thousand men, of which more than three thousand had perished in the humiliating disaster.) This time, the mountain ninja—which included the elderly and children—numbered, at most, one fifth the size of the Oda force. And, with the traitors' guidance, the Oda troops would not get so easily lost in the dense mountainous forests, nor would they be victims of ambush, trickery or, dare one even think it, magic. (An attribute often connected to the ninja in folklore.) To make Iga's doom perfectly clear to all sides, Nobunaga declared that for every day the local people resisted, the heads of three hundred to five hundred Iga residents should be collected, combatant or no.

At the end of September, the Oda attacked from all sides as planned, splitting the already meager defending forces and wiping out resistance ruthlessly. The stipulated number of heads were collected; Nobunaga's troops even exceeded their targets. Farms, villages, castles, shrines and temples went up in smoke and fathers put their families to the sword rather than have them slaughtered or sold as slaves after the inevitable defeat. Of one temple it was written, "When the smoke died down, inside and outside were dyed red with blood. The corpses of priests and laymen were piled high in the courtyard or lay scattered like strange autumn leaves lying deep of a morning." Wives and children were "fleeing hither and thither from this place to that place, but because of the attack they were cruelly slaughtered, mown down like blades of grass." Despite desperate, and strong, resistance, a concentration of ninja at one of the main strongholds in the province, Hijiyama Castle, was dealt with quickly; fire was the weapon of choice, as it so often was in Japan. Few, if any, of the inhabitants survived and it is recorded the embers remained hot for months afterward.

The Oda armies were now fully engaged on two fronts.

Yet Yasuke remained cloistered in Azuchi Castle. Waiting.

He took some comfort in knowing Nobunaga was *also* eager to personally join either campaign as soon as possible. Every day, the warlord grew more impatient, but he needed more information before deciding which operation to join. Little matter that winter approached and the late fall wind and rains blew hard against Azuchi Castle. Orders were given, provisions gathered, campaign tools readied and weapons polished and sharpened again and again as the days passed in inaction. If only Takayama would report back from the Tottori siege, Nobunaga could go south to join his son and generals in battle against Iga knowing his northeastern flank was secure.

Then finally, Takayama's report arrived.

Nobunaga and Yasuke climbed the stairs to the vermillion audience chamber at the top of the castle keep and awaited the anticipated report. Whatever happened next, Yasuke was going to war with his new lord. The only question that mattered now was *where?* General Takayama climbed the steep stairs with the answer.

His account started well. The several thousand Mori troops sent to break the siege had been beaten back easily before they could get anywhere even near the castle. General Hideyoshi's forces were, as they spoke, chasing the enemy west along the wild northern coast, taking heads and, also, Mori-held castles with abandon. The San'in front was secure. Takayama had pulled out a huge map to demonstrate the joyful news. However, the tale he told next chilled the warm, brazier-heated room to its core.

It was the siege. The thousands of men, women and children inside Tottori castle had endured the cordon better than could have been expected. Six months. Having been taken by surprise and not prepared for any blockade, they'd had no supplies, and soon consumed all the edible vegetation on the mountain and then eaten their oxen and horses. In a matter of weeks, there was nothing left. The faint and starving people within were dying in droves, their bodies lying stacked and frozen like waiting logs in the deep winter snow. And any help, the hope of which had first sustained them through their terrible hunger, had been driven back, routed with ease by their tormentors.

Yasuke knew what it was to be hungry. In Africa, on the long journey after being captured as a boy, the slavers had hardly fed their captives at all on the way to the coast. By the time Yasuke arrived at the boat that was to take him eastward, most of his comrades had collapsed and been speared by the slavers rather than have food and care wasted on them. Only the fittest survived, and even Yasuke had seen his own ribs and felt the gnaw of almost unassailable hunger.

General Takayama had watched from the wall top as an emaciated crowd of entrapped townsmen—their ribs prominent, eyes bulging from their skulls, legs ready to snap—begged from the wall to surrender, to escape the palisades Hideyoshi had built, to get food. They all looked like ghosts in the dirty trodden snow, the cold giving their bodies an otherworldly tinge. But Hideyoshi was immune to their pleas. The only way any of them were leaving was if the castle commander, Kikkawa Tsuneie, surrendered unconditionally; but he had no intention of that, he knew his duty, so those trapped in Tottori starved. Takayama next shared an incident he'd observed. A crowd of people had been begging to escape, and whether because an Oda guard took pity upon them or he was just fed up with the wailing, a gunshot rang out from his musket and one of the ghost-like beings dropped. Immediately, the others fell upon him; even as he still breathed his last, he was carved up by his former comrades who fed like voracious rats upon the raw human flesh. His screams ceased and one lucky cannibal won the battle for the dead man's head. (The brains were the tastiest part, it had been later explained to Takayama.) The gleeful winner scrambled away, the others, too weak to pursue, slumped back into the bloodied snow.

Takayama finished his report and a silence fell on the chamber. Even Nobunaga could not find a joke or barb to add. Yasuke felt the dark tenor fill the room. Even for a hated enemy, it was not a fitting end. Several of those in attendance excused themselves rather quickly.

The siege had ended after two hundred days. The commander offered to give his own life so that the remaining survivors might live. Hideyoshi, on Nobunaga's behalf, immediately accepted and Kikkawa cut his very empty belly open. Feeling sorry for the surviving defenders and full of respect for their long endurance, Hideyoshi ordered them to be fed forthwith. They gorged themselves and, tragically, more than half died on the spot from the effects of overeating. Kikkawa's head was couriered to Nobunaga in an ornate head box where Yasuke viewed it a few days later with awe and a little unease. If only the man had surrendered a little quicker. Hideyoshi, meanwhile, continued westward along the coast before returning south to finish whatever was left of the Mori.

Deciding the rest of his realm was finally under control, Nobunaga was ready to travel to Iga. Yasuke was again going to war.

CHAPTER SIXTEEN
The Dead are Rising

A few days after General Takayama's Tottori report, in early November, Yasuke and Nobunaga, and his group of samurai pages headed south for Iga. They rode at the head of Nobunaga's several-thousand-strong hand-picked corps of samurai horse guards and were also accompanied by Nobunaga's second son, Nobukatsu, and *his* vassals, who'd joined the original attack from his personal domain in the southern Ise Province just to the east of Iga.

If Yasuke or the others had hoped for some battle, they'd arrived too late. The enemy was already dead. Nobukatsu had clearly redeemed his crushing defeat of three years before. This time, only a week after the Oda troops had originally crossed the province's borders, the war was already won, most defenses had collapsed and the province was engulfed in flame. Only the castle at Kashiwara, with around sixteen hundred desperate locals, held out and their future looked equally dim. The defenders of Kashiwara were preparing a suicide attack when a mysterious Shinto priest appeared and somehow managed to organize a peace deal. Presents and hostages were exchanged and the garrison was allowed to leave. The war in Iga was already over; there'd been no ninja triumph nor Oda disgrace this time.

Nobunaga and his entourage had missed the action. The invasion of Iga became a tour of inspection, rather than a march into true battle. Still, Nobunaga could glory in his victory and gloat at the defeated ninja who'd so long been an irritation and assassination threat. The warlord, Yasuke among the riders at his side, toured the now-desolate land where every building in every village had been torched and only white snow-capped trees speared the tired grey sky. They were escorted and guided by the field commanders responsible for each of the six districts and put up in hastily constructed, but luxurious, residences. The lords each competed to see who could provide the best banqueting and entertainment for their liege and so it became less a military campaign and more of a leisurely victory tour through the ravaged mountains, complete with stops at noted sightseeing points.

They rode through the deep snow which blanketed the ruins of one small village, a sharp contrast with the conqueror's palatial, if temporary, accommodations. The remains of wooden house frames, charred black beneath the drifts, rose from the snow-mantled ruins. Headless bodies lay frozen solid, no doubt to stay until the spring thaw when they'd rot and bloat. Nobunaga would not waste his troops' time on the enemy dead. There were no kin left to bury them either. *Had these been the same beings who enjoyed such a fierce, almost superhuman reputation?*

Yasuke took a closer peek at a headless corpse, toed it. This ninja was definitely dead, and had definitely been human. Not some mysterious specter from legend. A woman too, by the looks of things, although it was hard to tell the sexes apart with no heads and in full, if tattered, battle ninja dress; a dirty, washed-out, shadowy blue from head to toe. She still gripped a short sword in her right hand, but her left was affixed to her belly, clearly trying to keep in the exposed, now frozen, innards.

The party of horsemen surrounding Nobunaga trotted on-ward, the straw-bound horses' hooves almost noiseless in the crackling frosty snow. Yasuke straightened, getting ready to mount his ride again.

Suddenly, the world exploded.

Where Nobunaga's party *had* been there was now only smoke drifting in the frosty air. The pong of gunpowder filled the area, burned Yasuke's eyes. The detonation had been more than gun-fire. Something else. An explosion like five cannon going off at the same time. But there'd been no cannon. *Was this the ninja magic people spoke of?* The snow was spattered crimson. Limbs of Oda soldiers lay scattered across the forest pathway. Dying men screamed.

And to add to the shock, the dead were rising.

Several Iga bodies between Yasuke and the black fog which now engulfed Nobunaga's party, had stood. Tossing smoking large-bore guns, they drew their swords. Iga men who'd been hiding among the headless villagers. These were not the walk-ing dead, they were ninja. Very much alive and with total sur-prise on their side.

The shadow-like killers ran into the spreading smoke to find and smite their arch enemy Nobunaga, who seemed to have somehow, again, survived the initial explosion and their ensu-ing sharpshooting.

Yasuke charged in after them to save his lord.

Within the acrid cloud, all was chaos. Men and horses stag-gered and fell. Oda soldiers had drawn their swords too, but could not tell friend from foe, thrusting out blindly at empty air, cutting at each other, slashing horses in the chaos. Meanwhile, the ninja danced through the Oda soldiers, from all different directions, spreading death as they went.

Above it all, Nobunaga's voice: "Kill them!" Loud and clear. *"Kire!"* And Yasuke struck at the darting shadows, heading toward where the voice had come from. The ninja had their backs to him as they tried to reach their target, Nobunaga, in the middle of the group. They had eyes for no one else, only Nobunaga, the arch devil who'd destroyed their world.

Yasuke swiped and missed one man who danced under a horse and onward. Another was not so lucky. Yasuke's sword took him in the head, cleaving it in two. The Iga ninja, his grey clothing hardly visible in the smog, dropped into the churned, blood-splashed snow.

Yasuke charged directly into the dissipating smoke, past the whirling bodies and swords, just managing to reach Nobunaga who'd dismounted and was engaging a ninja with his own blade. Nobukatsu fought beside his father; their swordsmanship made it look easy and the last would-be assassins fell under the father's and son's blades even as Yasuke took on a final enemy, hardly more than a boy.

Soon the dancing shadows, those seemingly risen from the dead, had returned to that state. The Iga boy had been no match for Yasuke's bulk and power, however much he feinted and twisted. He lay now at the African warrior's feet, head severed and hanging from a few sinews of flesh. The smoke had now fully cleared to reveal the carnage hidden beneath. Interspersed by the dead ninja lay seven Oda clansmen. Some had been blown apart by the initial explosion and volley of high-bore lead, their bodies shredded, and mangled limbs were all that remained. Others were cut down by the enemy after the explosion had thrown them from their horses. They lay still in their magnificent armor, whatever wounds that had killed them largely hidden under hard leather and cold metal.

"Hoooooo!" shouted Nobunaga, raising his sword with the customary Japanese battle cry.

The Oda survivors gulped in the mountain air, the tinge of gunpowder smoke on every breath. "Hooooo," they repeated. *It was good to be alive.* Yasuke had had his first Japanese battle. He was a blooded samurai of the Oda. Yet, no one but him seemed to notice.

They left Iga quickly, shaken but in one piece. It felt good to depart this land of fairy tales, death and snow. He could feel the eyes of the few starving, sorry survivors watching them from deep in the icy forest. They dared not reveal themselves. Many more would soon die of starvation and exposure. Yasuke wondered whether they'd attempt another attack, but they never did. And the surviving Oda pages and horse guards returned to the warmth of their homes in Azuchi Castle and the town nestled beneath the mount.

Despite slaughtering more than one third of the province's population in the war, Nobunaga approached the peace in a positive and magnanimous vein. The survivors who surrendered received immediate relief and came under the purportedly evenhanded governance of his son, Nobukatsu. Many were even granted the privilege of fighting *for* him or his vassals as samurai. Iga lived on, but the people's resistance was finished. Now they would use their special talents to serve their former foe.

Two more enemies vanquished. The list of remaining adversaries was growing ever shorter. Those areas Nobunaga hadn't yet addressed would fall all the more easily.

One day in winter, Nobunaga decided to pay the seminary a surprise visit so see what his resident Jesuits were up to. The warlord was eager to see the foreign conditions in their natural state, rather than having something prepared especially for him. Sudden tours of inspection like this were common with Nobunaga and they kept his people on their toes, sometimes costing their lives when things were not up to scratch. This time, he invited Yasuke along. Yasuke had passed by the seminary many times on trips to the Matsubara horse ground, but he'd not entered before and naturally was curious to revisit some of the world he'd left behind.

Father Organtino, acting school principal, was delighted and went about acting as tour guide of the thirty-four-room mission with gusto. It was a fine construction of wood and plaster, roofed with superior ceramic tiles, and the second floor, where Yasuke and Nobunaga now stood, boasted an internal verandah around the central courtyard. It was shady and cool in the hot months of summer, explained Organtino. Nobunaga was shown a clock as well as a harpsichord and a viola and he wanted to see them all demonstrated by the students straightaway. The instruments surprised and delighted him with their sound, and the bubbling of the brook flowing beside the mission below added the perfect accompaniment.

Nobunaga then spoke in a very affable, and informal, way with the Japanese students and their teachers. Trying to get *their* take on things, as opposed to Organtino's spin. When it was time for Mass, Nobunaga excused himself and returned to his castle on the mountain above. Yasuke, at Organtino's invitation and with Nobunaga's permission, stayed for the service.

As the New Year approached, tradition demanded that lords, great and small, flock to Azuchi to pay their respects. The castle was abuzz with activity, like snow caught in swirling winter winds. They made preparations for the three days after the New Year when the kitchens were, by tradition, not to be used. Even the women were allowed a rest and all food for the festival had to be prepared in advance.

At the same time, it was the season for a grand clean. The kitchens were scrubbed, the barns and pens all cleaned. Everyone had new clothes made. Yasuke's house—with all the others in town and running up the mountain—was swept inside and out; mats, blinds and damaged doors were mended. Yasuke's servants decorated the porch with two *kadomatsu* bamboo and pine sprig decorations as a sign for the New Year god to enter the house. They also hung an ensemble of pine, a small orange and a circular straw rope, a *shimekazari*, on his front door to keep misfortune and unclean spirits away.

Throughout the town below, people visited each other, exchanging greetings, gifts and blessings for the New Year. The more prosperous citizens kept a scribe at their front door to record who'd visited. Nobunaga had multiple secretaries for the supplicants who'd traveled for days to Azuchi pay their New Year's visit to him.

Yasuke stood to the side of Nobunaga in awe-inspiring attendance. His incredible size and foreign appearance one more adornment in a palace already bejeweled beyond compare. As he stood, the two swords of a samurai tucked into his belt, returning opens stares with calculated dead-eyed indifference—his instincts and training as a bodyguard remained firmly in place. He perused each person who stepped toward his liege; the parade of supplicants from all across Japan approached Nobunaga, prostrated themselves, offered sumptuous gifts through an Oda representative and then withdrew.

Prime among them was General Hideyoshi, flush from his northwestern victories against the Mori and clearly foremost in Nobunaga's favor. The monkey jokes had not, Yasuke admitted, been completely out of hand. Hideyoshi was a small, dark man with a squat, lean and hard face. But this in no way took away from his behavior; Yasuke could not believe the gifts the conquering hero had brought. Despite the grandeur of all the other gifts Nobunaga received—gold and silver, works of art from China, ornate garments, fine bolts of cloth—Hideyoshi managed to outdo them all, in quality and quantity. His gifts of two hundred silken kimono for his lord and a magnificent present for each of the senior ladies of the court, Nobunaga's mother, Dota Gozen; his wife, Nōhime; sister Oichi; and all the principle concubines. These were worth all the other supplicants' tributes combined, and then some. It was a generous and noteworthy tribute, especially from a man whose first role in Nobunaga's service had been nothing more than sandal bearer.

Then one of those curious moments occurred, when Yasuke *still* found it hard to understand the Japanese world. Nobunaga, with complete sincerity, honored Hideyoshi by ordering his favorite general to be given a few rough pots and a set of simple bamboo spoons to be used in the tea ceremony. The value attached to these simple and seemingly worthless items was clearly incalculable. Hideyoshi, a great practitioner of "the way of tea," cried with joy and reverenced the valuable boxed gifts with deep bows, touching them to his forehead to show he received them with respect and gratitude. In Japan, these were priceless. For him, these would be items he would treasure throughout his life. Yasuke would never understand this tea culture of *wabisabi*, the cherishing of imperfection, rusticity and intransigence; he'd rather have the money and weapons that his lord gifted him.

Nobunaga had planned his own celebration for the first day of the first month—he would have his guests, hundreds of them, guided around the castle. The most important step on the tour would be the golden audience chamber, a space built especially for a future visit by the emperor. The emperor had never visited anywhere outside Kyoto, and the fact that Nobunaga was preparing for this would have awed his visitors. This was again another unsubtle Nobunaga message about his unassailable power and glory, his national hegemony.

Yasuke was again part of this splendor and majesty, and took his place in Nobunaga's guard of honor. The vibrant winter day started well, with a ceremonial procession from the foot of the mountain up to the keep and the golden audience chamber.

Halfway up the mountain though, disaster struck.

THOMAS LOCKLEY & GEOFFREY GIRARD

The road fractured away from the mountain in a landslide; a whole section, and fifty-plus men tumbled down the steep face in a tangle of mud, ice and flesh, ripped by rocks and the sharp close-cropped bamboo which adorned the slopes.

At first Yasuke, at the front of the procession with Nobunaga, in bodyguard mode today, did not realize what was happening. *Could it be another attack? An earthquake?* The screams and the rumbling behind certainly sounded like one. But the ground under him remained firm, and the procession ahead continued as if nothing was happening. The only sign Nobunaga gave he'd noticed was a curt order to one of the other guards to go back and investigate.

The report was rendered, the butcher's bill high. Many of the warriors of the castle had been at the end of the procession and they'd borne the brunt of the destruction. Dozens were wounded, countless priceless weapons mangled, sumptuous robes reduced to filth and shreds, and several men even killed. Nobunaga listened to the account, ordered that the damage was to be seen to, the wounded treated, and then quickly got on with his day. No major personage had been injured, and a day as important as this could not be interrupted.

As the unhurt guests entered the main gate to the upper complex, Nobunaga stood at the stable entrance. Each guest handed their "voluntary" courtesy fee of one hundred copper coins directly to Nobunaga, a notable breach of protocol, which normally demanded an intermediary received presents before offering them up reverently to the warlord. But Nobunaga was Nobunaga, and could do as he liked. Each time, he'd throw the coins behind him into the stable, again a breach of behavioral protocol that only he could get away with, as receivers were supposed to show they valued their gift, not simply throw it away.

The guests carried on with their tour, oohing and ahhhing at every turn. A happy New Year's Day for all, topped off with feasts and drinks. The year had, all agreed publicly—carefully ignoring the road collapse and ensuing deaths—begun well.

Two weeks after the initial celebrations came the traditional New Year bonfire feast to mark the end of the festive season and the time to get back to business. All the decorative trimmings, the *shimekazari*, the great *kadomatsu*, which had adorned the castle and city houses were piled into a great pyramid-shaped pile and burned in a massive purifying conflagration. The pine needles and cones, the fern leaves, the felicitous calligraphy, wreaths of braided rice straw, the now long-dead flowers, and sprigs of early plum blossom, the huge bamboo decorations all burned fiercely. Nobunaga had started his own tradition for this bonfire event, combining it with another cavalcade of horsemanship and spectacle for the huge crowds of citizens who turned out to be both entertained and to revere their lord and guarantor of peace.

Yasuke and the other samurai pages were the first band of horsemen to enter the Matsubara horse ground near the Jesuit seminary for the festivities. As they rode by the large mission building, the bell pealed out to celebrate their passing and the boys waved in excitement. Following the pages were the senior lords, then Nobunaga's sons and finally, again in glorious isolation, Nobunaga himself, accompanied only by a groom to lead the three horses he'd chosen to ride that day. Once more, the warlord was dressed to astound the crowd—in scarlet, plum and sheer white—and he wore a four-cornered black hat nonchalantly, not a care in the world.

As the bonfires burned away the old year, Nobunaga and his samurai pages heralded in the new with wondrous displays of their equestrian skill. It is reasonable to imagine Yasuke, who'd been riding again for some time now, also joined them for part of the performance. It was near dusk when softly glowing embers were all that was left of the decorations, the horses stabled and the crowds dispersed, satisfied they were to be led once more into a prosperous and profitable new year.

As if to substantiate that belief, only weeks later, in early March 1582, treachery again provided Nobunaga with an excuse to address an old enemy.

The hated Takeda clan territories to the northeast in Shinano Province were no longer the existential threat to Nobunaga they'd once been in his youth. Takeda Katsuyori's father, Shingen, had once led tens of thousands of cavalry samurai, mounted on the local Kiso horses, terrorizing Nobunaga's home province of Owari in a generations-old battle for supremacy in the region. He'd even narrowly beaten an Oda/Tokugawa allied army once in 1573, one of Nobunaga's very few defeats in a life of victories. But shortly after vanquishing the Oda and Tokugawa, Shingen died of unknown causes.

His son Katsuyori took over the leadership, and although initially successful in battle, succumbed to Nobunaga's invention of the massed musket volley at the Battle of Nagashino in 1575. Katsuyori lost some 80 percent of the fifteen thousand samurai he'd brought to the field that day, including the cream of his commanders and advisors. The Takeda clan had never recovered and ceased to be a major threat. But having that strong and mountainous region near holy Mount Fuji in enemy hands, however impotent they'd become, was theoretically dangerous and—more frankly—an insult.

Fortuitously, rule of law in the domain had apparently disintegrated to such an extent that even the most loyal of Takeda retainers were considering defection to the Oda, and in the end one did: Lord Kiso. Nobutada, informed of Kiso's intentions, sent to his father for permission to invade and support Kiso's rebellion. Once more Nobunaga grabbed an unexpected opportunity with both hands. Nobutada, whose own domain bordered Takeda lands, was ordered to cross the border with as much force as he could muster and without delay. Nobunaga would join him as soon as possible and together they would pulverize the Takeda heartlands, rewarding those—like Lord Kiso—who welcomed Oda power, while burning any who resisted.

Once more, the Oda were at war.

CHAPTER SEVENTEEN
Collecting Heads

The heads—ten, then fifty, and then a hundred—were displayed in a viewing gallery below Azuchi Castle for everyone to see. Eventually *five* hundred severed heads were spiked on multiple rows of shelves in the open air hall. Out of curiosity, Yasuke came down to have a good look himself.

Each pate had been meticulously coiffed and face made-up: eyes and eyebrows lined in black, rouge for the blood-drained cheeks. And each was tagged with a label revealing the name and rank of its former owner, wrapped around its samurai topknot. Crowds milled about all week, inspecting each head individually, commenting on the reputations of each man, if known, and gazing in awe at the sight of so many vanquished enemies. Most of the town had come multiple times. A wonderful chance—for most, the only—to ever see famous lords and samurai up close.

Soldiers protected the display and shooed lingering dogs and

Townspeople in Kyoto view a severed head.

crows away. As the first heads spoiled, they were carted off and burned; it was too far to return them to their relatives as was the normal custom. But every few days, more arrived, couriered back to Azuchi as corporeal proof of Nobutada's victories against the Takeda clan.

This war with the Takeda was reaching its inevitable end—a decisive campaign against the defiant devils of the mountainous east. Nobunaga issued orders to his lords far and wide to muster their troops, gather provisions and, above all, first secure their own borders. There would be no repeat of the northern attacks which had occurred during the *umazoroe* Cavalcade of Horses last year when all of Nobunaga's major vassals were in Kyoto cavorting about in front of the emperor.

Only *after* their own domains were safe from attack were they to join him in Azuchi for a final attack on the Takeda. And while he waited for their arrival, Nobunaga followed his own advice and secured his own territory.

Over enormous maps laid out before his key vassals and advisors, Nobunaga worked out his next steps carefully. The Mori front, despite Tottori Castle falling, still remained hot, but General Hideyoshi clearly had it in hand. Nobunaga turned his attention elsewhere. He waved his hand over an area that had remained stalemated and uncertain for years—that of the Saika bandits, a group of renowned gunners, in the very south below Kyoto—and ordered an overwhelming surprise attack against an old, yet virtually dormant, enemy.

His strategy, again, proved successful. The Saika were overwhelmed and defeated in a matter of days. It was a fine start.

For the next weeks, however, the warlord waited atop Azuchi castle while he took in reports but still refrained from joining the Takeda front with his son. Frustration mounted within Yasuke and many of the other men, yet none dared let it show. Based on the ever-arriving heads, Nobutada and his men were clearly doing just fine in Takeda lands *without* his father's help, and Nobunaga's intended army had not yet fully assembled anyway.

New bands of men arrived each day, swelling the camps around the castle and Azuchi town. The merchants and madams of the town fattened their purses as well, as the Oda host grew larger and larger. And, where there are rough young men with weapons, there will be fights. Despite Nobunaga's attempts to clamp down, the streets after dark had become quite dangerous, some men were killed, a few local daughters raped and several brothels burned. Yasuke, who'd seen the same scene play out in half a dozen Indian towns, recognized they'd need to leave soon; it was supposed to be *Takeda* lands they devastated, not their own.

More reports arrived. Riding deep into the Takeda lands, Nobutada had evidently met little resistance, leaving only enemy bodies and smoking ruins in his wake. It was, as Nobunaga had planned, a fast advance with overwhelming firepower.

A hundred years before, battles had been a formulaic affair

where individuals, after stating their fine warrior lineages for hours, challenged each other to individual duals on the battlefield; conflicts, very efficiently and politely, were decided by only a few deaths. In the Mongol invasions of the thirteenth century, the defending Japanese samurai had stood aghast as the invading Mongols and their Korean levies engaged without etiquette, without "honor." The Japanese fell quickly to the foreign invaders and it was only the weather, the famous *kamikaze*, "divine winds," that wrecked the Mongols' fleet and stymied the invasion.

The Age of the Country at War had changed all that. War was no longer a rich man's undertaking. It was vicious, dirty, deadly, professional and, above all, driven by numbers. Thousands of trained Oda men had poured into Takeda's territory, their lightning advance aided by the many locals who'd despaired of their new lord, Takeda Katsuyori, and actually *welcomed* Nobunaga's coming.

Under Katsuyori, corruption, lawlessness and heavy punishment for minor offenses had all become commonplace. Crucifixion *à la Japonaise*—the victim tied to two cross sections attached to a vertical pole and poised spread-eagle in the air before being run through with spears—was now routine. Nobunaga's reputation for relative mercy (a few employees notwithstanding) and good governance, meanwhile, was well known across Japan. The locals may not have particularly welcomed outsiders (they were mountain people, insulated by slopes, forests and snow), but most were not unhappy to see the end of Katsuyori. For those few Takeda who couldn't bring themselves to surrender and swear allegiance to the old enemy, or to those the Oda would not *permit* to, there was no future. Many killed their wives and children and then put themselves to the sword before the Oda advance could reach them.

Lord Katsuyori himself, however, had not given up so easily. He'd first marched out from his own new, sumptuously built, capital, Shinpu, with all the force he could muster, some fifteen thousand men, and advanced to meet Nobutada straight on. This—everyone, especially Nobunaga, understood—was Katsuyori's best and only real hope. If the Takeda could inflict a speedy defeat on the Oda whelp as the Iga ninja had done to his brother in 1579, then Katsuyori might be able to halt the rest of the Oda campaign. But Katsuyori never got the chance to test this notion. As vassal after vassal defected or retreated, and Nobunaga's ally, Tokugawa Ieyasu, the powerful lord of the lands to the southeast of Mount Fuji, now attacked in a carefully planned pincer movement, the Takeda lord found himself in danger of being outflanked and retreated back to Shinpu. To shield his withdrawal, he'd left a garrison of three thousand men within the impregnable Takatō Castle, the key to the mountain passes through which the Oda advanced, and also, symbolically, his birthplace. These holdouts would harry Nobutada's advance while Katsuyori staged a tactical withdrawal back to the potential safety of his new capital.

Takatō Castle was a formidable barrier, built on a mountain with cliffs on three sides and raging rivers at its base. Its entrance was at the rear and only approachable via a long narrow winding cliff road. Its descriptions reminded Yasuke of the first castle he'd ever visited in Japan and the boy lord Arima, but Takatō was apparently many times more imposing.

Still, Nobutada had to take it. To leave the fortress intact was to allow an enemy garrison of three thousand to raid and cut Oda supply and communication lines. Worse, it would leave his own father vulnerable to attack when he finally advanced, following in their footsteps. Several leading samurai, including Ranmaru's older brother, Shozo, forded one of the rivers downstream in the dead of night and, after stealthily slashing the throats of guards in their way, sneaked unseen up to the main entrance. At daybreak, while the samurai of the night distracted the defenders within with an attack on the main gate, the rest of Nobutada's army made their way along the now-undefended narrow cliff road that lead to the castle gate. They entered the outer precincts easily and the fighting was fierce, but after many hours of hand-to-hand combat, the defenders could take no more and fled over the dead bodies of their comrades into the central fortifications, leaving the outer two baileys to Nobutada's men.

Nobunaga's son led from the front, fearlessly tearing down the palisade around the moat and then scaled the wall of the central bailey, cutting down the weary enemy as he went. His retainers, not wanting to suffer the dishonor of losing their lord, followed their liege over the wall and the fighting intensified. Nobutada's superior numbers, however, told in the end. The defenders first opened the throats of their own families and then made a final charge. They were led by Lady Suwa, whose husband, the castellan, and her children, had already perished. She was determined to seek as much vengeance as possible in the brief time left to them and led the suicidal defenders on attack, whirling like a dervish, slashing and stabbing at her foes. It was, Nobutada admitted with genuine respect, a glorious and fitting death for a samurai lady. After she was killed, another four hundred heads were taken and sent home to Nobunaga for viewing. Those were now displayed in Azuchi.

Further reports came back from Nobutada. The Oda heir had pushed on, laying waste to the countryside. Katsuyori sat panicking, in what he thought had been the safety of his new capital, Shinpu, only just finished.

No expense had been spared to create this architectural piece of art, but, carelessly, despite all the attention lavished on decorations, someone had forgotten the defenses. Katsuyori had mistakenly believed no enemy would ever reach this mountain complex of gold- and silver-gilt pavilions. The mountains had always been the Takeda's most powerful defensive weapon, but they'd finally failed. As the Takeda ladies—who'd been greeted by crowds of cheering citizens in their new home only four months before when they arrived by palanquin from the old capital of Kofū—fled barefoot into the mountains, Katsuyori himself set the first flames. Nobunaga would be denied his city, at least. The myriad hostages from families throughout Katsuyori's domains, held to secure their relatives' obedience, were left locked within the Takeda capital's finer buildings. The whole city became a gilt funeral pyre. Katsuyori knew all was lost, his vassals had deserted or failed him and this was his last act of petty cruelty and revenge. As the flames overwhelmed them, Nobunaga's biographer Ōta Gyūichi noted, his victims "wailed and howled all as one in their dying agony." The scene was "pitiful beyond words." Katusyori, however, escaped with his sixteen-year-old son, Nobukatsu (by his first wife, Nobunaga's adopted daughter, Lady Toyama—another failed political marriage), the six hundred men remaining to him and two hundred ladies of his court. By the time they reached a roughly fortified manor house in the mountains, Katsuyori had no more than forty fighting warriors left to his command. The others had melted away into the mountains and forests like snow in the spring thaw. Those who stayed were only there because they were his close relatives; there was no point in trying to escape anyway. As Katsuyori's relatives, they would be mercilessly hunted down and executed to avoid all likelihood of revenge seeking later. Such were the fortunes of war, and the Takeda clan themselves had dealt with others in this exact manner time and again.

The last handful of samurai and their ladies prepared for their imminent deaths by doing what generations had done before them, composing themselves for what was to come by writing their death poems. Katsuyori's second wife, Lady Hōjō, was the eighteen-year-old daughter of Hōjō Ujimasa, one of Nobunaga's key allies, but she would not demean herself by begging for mercy. Her final poem read:

The last stand of the Takeda, by Utagawa Kunitsuna.

My black hair is disheveled,
this world without end,
as fragile as a chain of dew drops

And, drawing inspiration from a skein of wild geese flying through the cold pale April sky, she beseeched the birds to deliver her poem home:

Returning goose,
won't you carry these few words,
to my old home of Sagami

Death was not long in coming. Having burned the new capital city, Nobutada proceeded to the old capital, Kofū, and established new temporary headquarters there. There he stayed put, dispatching hundreds of scouts to establish the whereabouts of the defeated foe. An old Oda vassal named Takikawa tracked them down and his troops surrounded the rugged makeshift palisade that Katsuyori's men had erected around their position. The Takeda men killed their wives, then burst out in a final suicidal attack. Katsuyori's male lover, Tsuchiya Heihachi, drew his bow and emptied his quiver, spreading death. When he could fight no more, he fell to his knees and cut his belly on the battlefield. Katsuyori and his son, Nobukatsu, charged into the midst of the surrounding enemy together, taking many Oda lives before kneeling in the churned and bloody snow of the battlefield to die by their own hands.

The last remnants of the Takeda nobility died that day or were rounded up and executed elsewhere by Nobutada's death squads. After the female *shigeshoshi*, the death beauticians— brought along to war especially for the task—had made up the enemy heads to honor their passing, Nobutada inspected them and then sent them by courier, with the captured poems, to his father, Nobunaga.

A full month after Nobutada was ordered across the Takeda border, preparations were finally ready for the main Oda force to join his attack. Yasuke's blood was up and his heartbeat strong as he rode proudly, clad in warm padded armor at his lord's side through the gate of Azuchi Castle at dawn on the fine spring day of March 28, 1582.

Valignano and the Jesuits had always brought gifts, manipulation and the Word of God. Nobunaga and the Oda now brought hot lead, the rule of law and the will of Nobunaga.

It had not been long since Yasuke had tasted battle, but the Iga ninja attack had been only a brief skirmish. This promised to be a whole different affair, as Nobunaga took the field with an army that now numbered sixty thousand. The snow-clad mountains ahead glimmered gold as the sun rose slowly behind them. The clouds moving swiftly in the brisk wind were a soft burnished pink. A fine day of riding beckoned.

Yasuke had been in Nobunaga's service for exactly a year. A formative year, too, clearly. His status had soared. He had his own residence for the first time, been back to war *and* his Japanese had improved considerably, having heard virtually no other tongue since leaving Kyoto. And now he was off to war again, where he'd most likely be able to prove his skills and worth to the Oda clan and its ruler.

Yasuke and the rest of Nobunaga's entourage with their warrior bands moved out. Generals Tsuda Nobutsumi, Takayama Ukon and Akechi Mitsuhide.

The town had grown fat and rich off the thousands of troops stationed there and the locals were, despite the fair share of unruliness and destruction, sorry to see them all go. It would be quiet again and business slow, but perhaps it was a good chance to take stock and rest while the samurai were off doing their business. Nice indeed to have a fight-free night and safe streets again.

Nobunaga's army made good time, crossing the flatlands outside Azuchi, then through the low mountain passes toward the plain of Gifu to the east. Nobunaga's old capital, and now Nobutada's fief. Thence to Inuyama Castle, deep in the Oda family's ancestral heartlands of Owari Province, and then north into the tree-clad mountain slopes. Friendly valleys at first, then rising ever higher from the lowlands into the mountains of the enemy Takeda's domain. Spring turned back to winter as they climbed.

Yasuke entered this part of Japan for the first time, and the biting late winter cold of the mountains was something new. The rising valley-bottom snows were just melting and the paths, well trodden from constant troop movements, were like rivers of mud and detritus. Nobutada had ensured there were levies of defeated locals to keep the way clear of snow for them; they could be seen working themselves to death in the never-ending battle against the elements as Nobunaga's cavalry, taking little heed of their defeated inferiors, trotted by.

As they entered the battle zone, they followed in the son's wake of destruction; Nobutada had left few buildings intact. Most were blackened ruins and the bodies lay awkwardly frozen in the snow, half picked clean by the huge crows who perched in trees waiting for their chance to gorge again on the unexpected late-winter feast.

The men camped late in the day and left before dawn. Servants raced ahead and prepared the rough field tents by the time the samurai arrived each night and there was only time for a quick meal and then sleep. Grooms saw to the horses during the night. The men were never dry and there were no baths remaining to ease their muscles unless they happened upon a piping hot *onsen* spring, common enough in the mountainous and volcanic Japanese islands. The army was high up, in a place called Iida, when Katsuyori's head arrived in an ornate box, couriered by a messenger with the news of his ultimate defeat.

The head of the former Takeda lord, along with the head of his son Nobukatsu (also technically Nobunaga's grandson, if only by adoption) were brought before Nobunaga. Each head was carried by *two* men, a special honor assigned only to *daimyō*. Normally, a head was held up only by a single samurai for his lord's inspection. Nobunaga had never met this grandson, had hardly had any time for his adoptive daughter, and the enmity with his old foe excluded any possible feelings of affection he may have held.

The viewing was meant as an opportunity to take stock in victory and pay last respects to a worthy foe. Victors often shared a drink with the set head and prayed for the departed's soul, addressed some words, perhaps even asking for its blessing. A ritual observed throughout samurai history. One thing the ceremony was not for was disrespecting the dead, however hated they'd been in life. That said, this was a unique moment.

As icy rain poured down, splashing off the three-day-old relic of the former Takeda lord and Yasuke looked on, Nobunaga gloated over the severed head. He taunted it verbally and even slapped Katsuyori's cheek. "I heard your father wanted you to go to Kyoto," Nobunaga jeered, "so, I will forward your head there to fulfill his wish. There, it will be exposed to the spectators, including all the women and children." Perhaps ominously, Katsuyori's right eye was closed and the left clearly scowling, an unlucky sign.

The dead eyes of Nobunaga's adoptive grandson also stared blankly upon this terrible spectacle and Nobunaga's entourage shifted awkwardly in their sodden straw capes, their sandaled feet frozen and soaked in the mud. They worked to hide their displeasure, looking on in shock, and what should have been a moment of sweet victory for them was diminished. It was a terrible breach of protocol, more reminiscent of Nobunaga's undignified youthful behavior than anything they'd seen in many years. They wished to look away; did not want to be there amid these ruins, watching their dear lord disgrace himself so. Yasuke, still learning the cultural norms, had little reaction but could sense the other men clearly enough. He leaned forward, kept quiet, tugging his thick straw cape closer against the rain.

Finally, Nobunaga gave Katsuyori's head a final hard slap and walked back to the relative warmth of his tent. There, the braziers were bright and steam rose gently from the sodden soldiers as they tried to force some jolliness into the sorry evening. Nobunaga was surely now a little lost *without* Katsuyori—a man who'd provided focus and purpose for years. The Oda warlord had other enemies, yes, but dealing with them would not possibly involve the personal emotional energy, the all-encompassing loathing, he'd engaged for the Takeda due to their generations-long family feud. He'd finally defended his family's legacy and avenged every defeat and slight suffered at the hands of the Takeda. The final stages on his journey to national domination would now be smooth and almost perfunctory, thought Yasuke. *And where*—perhaps the issue—*was the fun in that?*

By the time the eating and drinking were done, the day's events were a matter for the historians to debate. The Oda samurai and troops had a campaign to finish up.

The next day, another head arrived: Katsuyori's cousin, Tenkyū.

After Katsuyori's death, the only Takeda leader of any note still alive had been Tenkyū. Now he was dead, also. Per the messenger, Tenkyū had escaped north hoping to find a still-loyal clansman to aid his flight. The lord of Komoro Castle agreed to help, and gave him and his small corps of twenty samurai refuge for the night. Tenkyū should have guessed treachery was in the air; why would a minor lord risk Nobunaga's wrath to help a fugitive from a ruling house who'd clearly lost heaven's mandate? During the night, his temporary residence was torched, the weary refugees had no escape, they followed their training and burst from the flames in a half-hearted final attack. Tenkyū slipped to his knees, pulled out his short sword and prepared for seppuku. His lover acted as his second, and as Tenkyū stabbed down and pulled the sword across his abdomen, the young man decapitated him. The samurai lover then took his own life as did several other survivors from the flames. The people of Komoro Castle looked on with respect, giving space and dignity to this last act of their former rulers. The remainder of the exhausted men gave themselves up. (Not all samurai, as mentioned before, had the strength to perform seppuku.) These men were killed and the final Takeda heads were delivered, as promised, by express horse relay to Kyoto to be displayed in public. More proof of Nobunaga's victory and the Takeda's demise. It was expected the crowds in Kyoto would rejoice. Yet another violent step on the road to total peace and Japan's ascension.

Along with Tenkyū's head, the messenger had brought Katsuyori's personal sword. In return, the bearer was gifted a lined silk kimono for his trouble.

The days, and road, ahead were increasingly filled with fire, ferocity and blood—even by Nobunaga's standards. Farmers were rewarded with gold for turning in the heads of their former overlords and a regionally important temple, the Erin-ji, had been turned to cinders by Nobutada.

The fugitive son of Rokkaku Yoshikata, the man who'd commissioned the first sniper ninja attack on Nobunaga twelve years earlier, was reported to Nobutada as being under the protection of the temple's abbot. A bad decision for the temple. In retaliation, all the monks and inhabitants within were pushed into a second floor room, then the temple's first floor was packed with straw and oils, and ignited. The straw was wet from the winter damp and smoke billowed around the building hellishly as the clear, sonorous sound of the abbot praying rose above the crackle of flame. Soon, however, the prayer was overwhelmed by the screams of the priests, acolytes and children trapped within. One hundred and fifty people were burned to death.

Nobunaga, Yasuke and the army pushed north through the mountains, past the ruined castle of Takatō, and onward to the town of Suwa, an ancient holy site that Nobutada had also reduced to ash. There, dozens of smoldering fires were still banked with charred bodies. Sunlight broke through the snow-swollen clouds, slashing over the corpses and the company of horses and armored men picking their way through the burned stumps of shrines and homes.

Nobunaga commandeered the only building left standing, a temple, and made it his provisional headquarters. The god of Suwa was called "The Great Shining Deity," and devotees from across Japan had made pilgrimages to worship there. It seemed apt for Nobunaga to appropriate the divine aura. The first thing he ordered was disposal of all the dead. It was all very well leaving corpses strewn along the roadside, but he had no desire to share a town with them.

For the next two weeks, Yasuke and Nobunaga remained within the smoldering village. Nobunaga would go no farther for now, trusting Nobutada, still encamped in Kōfu to the south, to work independently. His first son had more than proven himself capable. In Suwa, Nobunaga made good use of his time and held court. The turncoat lords flocked to him in droves to pay obeisance and swear allegiance to their new ruler. Nobutada subsequently presented himself to his father where Nobunaga officially passed command of the theatre of battle to him.

Nobunaga then set to reorganizing the governance of the conquered territories by his terms. No longer would the peasants and townsmen be taxed to within an inch of their lives to raise a vainglorious ruler's new capital city—tax rates would be lowered; nor would customs be collected at the border as the provinces would now be unified with those of Nobunaga. People were to be treated fairly and with respect and rewarded appropriately. Infrastructure was to be well maintained and criminals punished in accordance with the local legal penalties for their infringements.

Yasuke listened carefully as Nobunaga proclaimed his edicts, setting new provincial regulations for the ruling samurai, a group of enemies turned allies, and newly appointed Oda men to replace those who had remained loyal to the Takeda and paid with their lives. He eliminated various tolls and taxes, established legal regulations, gave specific orders for establishing defenses and commercial infrastructure. His final pronouncement was: "There must be no animosity. In case of an unsatisfactory resolution of a matter beyond what has been determined above, bring the case directly before me." Yasuke thought on Nobunaga's words, and contemplated what made his lord a great ruler, not just a great soldier.

Battle and destruction were not everything Nobunaga was about, although it sometimes felt that way. The rule of law, and the welfare of his people were central to Nobunaga's concept for a New World. Farmers would farm in peace, merchants would create personal wealth, artisans could engage in industry and samurai could engage in cultural and martial pursuits while policing any wayward tendencies within the new society. Nobunaga was extending the economic policies that had been so successful in Azuchi, *rakuichi rakuza*, "free markets and open guilds," cutting red tape he believed restricted economic growth. And it made a difference. As farmers, merchants and townspeople who had been under Takeda rule understood just exactly what this "domination" by the Oda would mean, they calmed down, accepted their defeat and looked forward to a brighter future.

All of this security was being built on the boxed heads of Nobunaga's enemies. *If ever a ruler of people*, Yasuke thought, *would I follow my new lord's example?*

Yasuke's imagining of a lordship of his own was not idle daydreaming. It had been an offhand quip at first, his new comrades jesting, but as the months passed, the joke had spread into a rumor which soon transformed into genuine possibility. Seemingly already far beyond his wildest imagination, it was possible Nobunaga was only getting started with his plans for Yasuke.

The Jesuit Mexia reported that the rumor in Azuchi town was *"Nobunaga would make him 'Tono'"*—a lord in his own right.

A fief as reward for his services. Taxable lands. Troops to deploy. His own samurai to command. Anything was possible with Nobunaga, it seemed.

The conversation around the campfires even turned to which specific fief Yasuke might receive; there were plenty to go around in these vast new provinces where the old enemies had been dispossessed. Nobunaga had the right to go as far as he liked in doling out his spoils of war, and not waste opportunities to provide patronage and promotion to those who deserved it. How far would Nobunaga promote his new favorite? Such a thing had never happened before, after all. Foreigners rarely had need of real estate; they craved rather money or goods, and then went on their way. Never had a foreigner been given a lordship or fief and the implied permanence and responsibility to the lower orders who lived there which such a designation entailed. A normal samurai lord would be bound to fulfill his social obligations to high and low, ensuring harmony in the land he controlled. A long-term and resource-intensive undertaking.

Yasuke heard the rumors for weeks, dismissed them, but then started to seriously consider the implications as the gossip intensified. A fief meant glory, wealth and position, but it also meant the responsibility for hundreds or even thousands of lives, economic management, trade, swift responses to rebellion and playing at politics. A tingle went up his spine. Such a position was unimaginable. A highborn wife, concubines and heirs would also have their part to play. He imagined himself adjudicating legal disputes, dealing in vast sums and chastising wayward peasants for not seeing to their irrigation properly, and bowing as he entered the women's quarters to visit his wife and/or concubines. He grinned at the image of himself charging through the woods to hunt a boar, then frowned as he realized his own retainers would compete in New Year calligraphy contests, perform *noh* dramas and serve tea correctly to his visitors, but, lacking the deep classical training, he himself would be unable to. At least not for a long time. How would he gain their true respect? Being a samurai lord was more than fighting; it implied an apprenticeship in the house of another samurai, a lifetime's education including a thorough knowledge of the Chinese classics and a true appreciation of the arts. How would he gain the respect of his neighboring lords? It was a maelstrom of competing bittersweet ideas, fears and jubilation.

For several days, Yasuke held his expectations, dread and daydreams down. Still, he felt Nobunaga's eyes on him differently, more evaluating than even his first audience with the Japanese lord the year before; he told himself that was all in his imagination. And, as if the universe had heard his thoughts, over the next few days, the rumors died down. Replaced with other camp gossip and groans about the cold and harsh living conditions.

Nobunaga doled out fiefs, of varying sizes and values—there was no set measurement—to his new followers and promoted his old. The lands from which they were promoted went to new men. Nobunaga's teenage lover, Ranmaru, received two, even as Ranmaru's older brother was given four, with a castle, for his bravery and heroism in the taking of Takatō Castle. As a reward for his long years of alliance and service, Tokugawa Ieyasu received the whole province of Suruga, encompassing the south of Mount Fuji. Ieyasu, from an old blue-blooded family, had been allied to Nobunaga for over two decades, and although an independent *daimyō* in his own right, not an Oda vassal, had consistently and loyally supported his Oda overlord, including sacrificing his own wife and firstborn son to maintain the alliance. The rich and ancient province of Suruga was Nobunaga's gift in recognition of this.

The matter of Yasuke, however, never came up.

Yasuke felt disappointment in his heart, but also surely breathed a sigh of relief. A marriage first, perhaps. Then, Nobunaga might gift him a fief in the future to provide for his family; that was often the pattern.

But such ambitions would have to wait.

CHAPTER EIGHTEEN
Fuji-san

Nobunaga's work nearly done in the former Takeda lands, the warlord desired a proper holiday.

Despite all his triumphs and adventures over the years, there'd never been time for him to visit the holy mountain, Fuji. Now, with the old Takeda lords out of the way, this land was *his* to command, and he wished to see it. And Fuji lay directly to their south. At the same time he could grace his prime ally in the region, Tokugawa Ieyasu, with a visit, and confirm him in his new domain, Suruga. He could leave the final mopping up and enforcing details to his son.

But, an issue first: Nobunaga's men were hungry and tired. They'd marched hundreds of cold, wet miles up mountains and down valleys. They'd dispensed Nobunaga's justice on the local population, burned uncooperative villages and patrolled now-friendly towns. Furthermore, they'd carried almost everything on their backs. (There were few beasts of burden due to lack of land to grow fodder.) Several soldiers had even frozen to death in the bitter conditions. Yasuke—less used to the cold even than the rest—was fortunately part of Nobunaga's inner circle and afforded the luxury of each night being quartered indoors, in a temple, manor house or specially constructed lodging. Still, once inside, the Japanese seemed to have a flippant attitude to heating, perhaps because all buildings were wood and they preferred to lessen the fire risk. Instead, they smothered themselves with futons or woven straw mats rather than waste energy on heating large spaces with their small braziers. Yasuke struggled each night to stay warm under the futon covers, shivering and tossing and wondering how such cold could exist.

Nobunaga knew his soldiers would not appreciate this tour in the same way as he. Also, turning up at Ieyasu's border with only a small entourage would prove to the world he believed in his personal power alone to keep him safe in another man's land. For Nobunaga trusted in only one thing, his own supremacy. The warlord believed to his core that was enough. Thus—despite the inherent dangers of Ieyasu perhaps betraying him—Nobunaga dismissed his vassal lords *and* their armies, and tens of thousands of troops turned for home. In their place, Nobunaga took only a few hundred men with him, including Yasuke. Observed on this tour, Matsudaira Ietada, a prolific samurai diarist of the times, noted, "His skin was black like ink and he was around 6.2 *shaku* (over 6'2") tall. He was called Yasuke."

They proceeded south from Suwa to tour the remaining Takeda lands before exploring the environs of the holy mountain, Fuji. Once more Yasuke found himself part of a glorious procession that everyone, high and low, came out to witness. Yasuke rode proud and tall in the vanguard, feeling ever more at ease in his role as samurai and in his place within his group of comrades.

Mount Fuji viewed from the sea. Hokusai, 36 views of Mount Fuji.

As they rode through snow-dusted forest, he daydreamed about buying a razor-sharp long-handled *naginata* when he got home to Azuchi. He'd never used one in battle, but he loved the look and feel. With one of those on horseback, especially for a man his size, a warrior might easily scythe his way through a group of enemies. The daydream was interrupted abruptly by cheers and whoops from those around him.

As a mountain pass opened, the dazzling white snow-covered peak of the majestic Mount Fuji itself came into view, sparkling and haloed against the light blue sky in the brilliant winter sun. It was awe inspiring, and Nobunaga's party was in high spirits.

Mount Fuji was Japan's tallest mountain, some two-thousand feet taller than her closest rival, and an active volcano that had erupted throughout human history. The root of her name was lost to the mists of time, but it is thought she has been worshipped as a kind of mother goddess since the Japanese islands were first inhabited tens of thousands of years before. She rose gracefully far above any other peak, her silhouette viewable for hundreds of miles—a glowing, lingering visitor from the heavens.

Yasuke and Nobunaga approached Fuji from the north, and only her peak was visible, other lesser mountains obscuring her base. Still, their view encompassed the cone peak clad in the purity of icy-white snow. With Fuji gleaming in the distance, they rode contemplatively through the ruins of Katsuyori's new capital, and then on to the traditional Takeda capital of Kofū where they were delighted to find Nobutada had built a temporary palace for his father. There, they slept in warmth and comfort that night, toasting their victory and feasting on five hundred freshly hunted pheasants which arrived as a gift from Nobunaga's ally Lord Hōjō, Katsuyori's former brother-in-law. Nobunaga distributed them to his men, an unusual delicacy, fit for kings and conquerors.

The next morning, they moved into Suruga, the province which Nobunaga had just added to Ieyasu's growing collection of lands. Sunlight filtered through the swollen clouds above and they'd left the death and decay of the former Takeda battlefields behind. Now they were in Tokugawa lands, and Ieyasu—the Tokugawa lord—made sure Nobunaga's holiday was more of a luxurious pilgrimage.

New bridges were built to span rivers. Their paths cleared of stones and drizzled with water to reduce the nuisance of dust; now that they'd left the mountains, the first balmy heat of spring was upon them. Tokugawa soldiers lined the route and trees were cleared on both sides of the road to create a broader highway, for ease of travel and to reduce the risk of hidden assassins. (There would be no repeat of the Iga ninja attack while Nobunaga was Ieyasu's guest.)

One afternoon, Nobunaga desired to climb a particular mountain, so Ieyasu had his men clear a path, felling trees and removing boulders just so the Oda lord could ascend unimpeded. The journeying became slower as Ieyasu took Nobunaga aside for visits to sites of legend and historical interest. For each site, Ieyasu had constructed a teahouse so that Nobunaga might contemplate the surroundings with refreshment and in suitable comfort and his host could impress with his tea implements collection. And

THOMAS LOCKLEY & GEOFFREY GIRARD

all the while, the black-and-white goddess, Mount Fuji, grew bigger and then loomed over them as they approached and circled her lower slopes.

As each day's ride ended, Nobunaga arrived at a newly constructed pavilion, complete with three defensive palisades and a teahouse. As many as twenty had been built along their intended path weeks before by hundreds of men in anticipation of Nobunaga's arrival. Here, he and his entourage were treated to magnificent banquets. Ieyasu made sure both high and low were provided for—respecting Nobunaga's troops showed reverence to their commander; hundreds of huts were constructed for Nobunaga's soldiers each day.

The first act of loyalty from the locals, Ieyasu's new subjects, to their new lord was to provide the troops morning and evening meals, the best food and drink they possessed. Rice, dried and fresh fish, hunted beasts, wild fowl, forest roots, dried fruits from the autumn, fungi, spring leaves, tofu, eggs, rough grain wine and perhaps even distilled spirits kept the soldiers going— far better and varied than their normal marching diet.

For Nobunaga's banquets, Ieyasu provided nothing but the finest delicacies available in the land. Couriered sea fish and shellfish, plump freshwater fish and elvers from Fuji's streams; finely polished white rice; mushrooms; and wild fowl and game, euphemistically known as "mountain whale" to avoid religious scruples about consuming animal flesh. Only the finest of local foods, as well as those transported especially from distant Kyoto and Sakai graced Nobunaga's table.

Ieyasu himself was a noted gourmand, a rather large man, and had brought along master chefs, *houchonin*, from Kyoto to prepare the fine fare for his special Oda guests. Their performances of knife skill and delicate carving—one renowned *houchonin* was said to have been able to carve a carp a different way for one hundred days—were a highlight of the evening's entertainment as well as purifying and pacifying the spirits of the dead animals who formed the feast. Yasuke and the others marveled in respectful silence as the chefs' clever knife and chopstick work seemed to resurrect the lifeless fish so the light glinted off their silver skins as if they were darting through a stream. But the finale was a heron; the fowl had already been cooked and then reconstituted, feathers and all, and appeared to be alive once more. After the *houchonin* had finished their ceremonies and the eating had begun, Nobunaga asked his host, Ieyasu, a hundred questions about the land. There was always more to learn.

Yasuke, too big to fit inside the teahouses, often ate elsewhere, away from the tiny rustic tea huts. Not without good-natured ribbing about his size from Nobunaga and the others, and apologies from Ieyasu.

Coming down from the mountains, spring now budded around them in mountain cherry blossoms, birds feeding their young and melt-swollen rivers, but it still felt as cold. To counter the chill, Nobunaga and his band of pages held horse races. On mountain plateaus, they staged wild charges on their small but hardy Kiso horses, giving the whip liberally, seeing who could get the most from their mount. Yasuke and his comrades returned to camp laughing with the exhilaration of being alive at this time and in this place. Suddenly the cold did not seem so bad as Yasuke cooled down with a dip in a swift flowing ice-cold stream. That didn't stop him swiftly jumping into a hot pool when Ranmaru found one bubbling in the middle of the brook. *Hot, cold, hot, cold, was there ever such a feeling?* The mountain remained an ever-present backdrop. So huge, but also often wreathed in cloud. The lakes they passed reflected her majestic presence almost making it seem like there were two goddesses.

In the middle of the month of May, as they descended toward the coast and the weather got warmer, Ieyasu brought them to a place named Hitoana, a famous cave where the goddess of Mount Fuji herself was supposed to reside. There, another teahouse had been built specifically for their arrival and, once gathered properly around a bubbling kettle, Ieyasu treated them all to a terrifying legend regarding the cave and a samurai's descent into hell.

In the story, a young samurai accepted his lord's challenge of exploring the cave where many men had vanished or died. Within, the intrepid samurai found the realms of the Buddhist afterlife, and was treated to a terrifying guided tour of the various levels of hell and occasional fleeting views of heaven. He observed, with horror, the crimes of dead souls and their punishment in the afterlife. The Dante-esque story—a European tale Yasuke had heard of from the Jesuits—took them past giant talking snakes dripping with blood, beautiful maidens of death reeking of rotting fish, and flesh-peeling demons punishing souls for their crimes during life, from vanity and infidelity to those who "didn't like to farm" or "overburdened their ox with heavy loads."

The Oda warriors grinned thinly, glancing behind as the cave's opening gaped at their backs behind the teahouse. The story fully explored sin, death and karma; though his audience would surely have known the tale, Ieyasu judiciously avoided those moments addressing the fleeting nature of wealth, power and worldly fame. Yasuke stared up into the black canopy of night above, no less a cave than that behind them; no doubt recounting his own actions over the past ten years and the eternity that awaited them all.

The next morning, their leisurely pace continued toward Suruga Bay on the Pacific Ocean, stopping at various Tokugawa clan castles along the way. Ieyasu was demonstrating the strength of his realm to Nobunaga, but also emphasizing his obeisance to Nobunaga's hegemony through the cost, thought and details that had gone into hosting him.

Just before sunset, they reined weary horses along a rocky crest above a long strand of sandy beach on the expanse of Su-

ruga Bay. Gulls darted and wheeled across the shoreline looking for food. Behind them, the sea lolled in long meandering
ribbons of grey and blue flecked in gold-capped waves beneath
a setting sun which gazed hotly like the single eye of some pri-
mordial god fixed atop the farthest horizon.

The party led their horses single file down a narrow path to
the water as the surf grumbled softly, echoed in the wet drag of
golden sand. A crane rose at their approach, white wings flapping slowly, then settled in a patch of reeds a short distance away.
The men dismounted and walked the beach at the mouth of the
bay quietly. Fuji still loomed above, while all around the bay,
gentle hills extended to the waterline. Nobunaga reached the
surf first and leaned to wet his sleeves in the breaking waves.
He turned back to his men, grinning.

Yasuke and the others joined him at the water's edge and imi-
tated his gesture. The cold sea embraced Yasuke's arms, urging
him forward. He lifted the water to his face, closed his eyes as
it ran cool across his skin. He'd spent half his life on coastlines
and welcomed the familiar swish and rumble, the faint crinkle
of salt on his cheeks, on his tongue.

After a meal of sweet local shrimp and delicious whitefish
pulled straight from the water which Ieyasu had ordered prepared for them, they took the Tōkaidō main road, westward
along the coast through the Tokugawa provinces, back toward
home. Mighty rivers such as the Tenryū, which had never been
bridged before, now had gilt pontoon bridges thrown upon
them. Ieyasu conspicuously constructed the biggest ever bridge
in Japan, of boats with a highway laid over them, at massive expense. All for no other reason than Nobunaga should be seen to
ride across this river of roaring rapids, to conquer the river that
had its origins in the mountains of the former Takeda lands. One
more symbolic victory over his old foe *and* proof that Ieyasu's
reverence for Nobunaga was boundless. Yasuke thought he'd
understood the Nobunaga effect, but this was something else.

In thanks for this unparalleled Tokugawa hospitality, even Nobunaga failed to find the words to describe the joy he felt at what Ieyasu had done for him. Instead, he bestowed upon his host a prized short sword and a gorgeous black-spotted horse from his personal stable. Tokugawa's men were not forgotten either; Nobunaga rewarded them all from his own pocket, earning their love and affection as well as that of their lord's. A wise investment in the future.

In this manner, the journey continued back to Oda lands, where, in the latter stages of the return home, Nobunaga's own vassals and sons competed every night to outdo each other in the lavishness of their welcome for their lord, and Ieyasu's, hospitality. Truly it had been a homecoming of triumph.

The Takeda heads sent to Kyoto for display had been an effective publicity stunt. All of Japan now knew of Nobunaga's latest triumphs. Merchants, monks and travelers from throughout the land had returned home with word of the harvest of heads and imposition of peace. They cared little for the deaths of warriors; their one desire was to have a quiet land with ample food where business could flourish.

And so it was that on the final approach to Azuchi Castle, everyone who was anyone from Kyoto, Sakai and all the domains and fiefs in the region had gathered to congratulate Nobunaga on his victory, offer up gifts and welcome him home.

Yasuke rode at his side, basking in the reflected glory. "What a blessed reign," wrote Ōta, Nobunaga's biographer, equally versed as Ieyasu in sycophantic praise. "One with such a paragon of power and of glory."

Several days later, Ieyasu followed Nobunaga to Azuchi to formally recognize Nobunaga in his overlordship and express gratitude for the new territory he'd been awarded. Nobunaga wished to return the many favors that Ieyasu had done him, and ensured all the roads which he would travel were newly repaired, the accommodation was fit for a king and that the food was, as ever, the best available.

Nobunaga charged the always-capable Lord Akechi with the impending ceremonies and welcome in Azuchi Castle. Akechi sent his men far and wide gathering the delicacies for a three-day banquet. No effort was spared. But when Ieyasu had nearly arrived, Nobunaga inspected the offerings and declared them unfit for human consumption. In a rage, he tossed all the food to the floor and kicked it from the kitchen into the garden outside. The humiliated Akechi could only look on, head bowed deeply to the floor, while his careful preparations were trampled under his lord's feet. When Nobunaga left the kitchen, another meal with equally fine fare of white rice, delicate soups and fresh seafood was prepared and served to the visiting Ieyasu who, none the wiser, was well pleased by his warm and gracious welcome.

As soon as the feast was over, Nobunaga forgot about the whole incident and already needed Akechi for other things. For the warlord had just received word from General Hideyoshi in the west that the Mori were preparing a final counterattack. Hideyoshi requested reinforcements urgently. The general had just used thousands of laborers to quickly engineer the rerouting of a river to cut off one of Lord Mori's frontline castles at Takamatsu—turning it into an island! He'd also constructed floating gun boats to pound the suddenly stranded defenders day and night. Now, it was only a matter of time. The thousands of Mori soldiers trapped within would suffer a similar fate to the people of Tottori, and perish from starvation. But Lord Mori had assembled troops elsewhere and was advancing on Hideyoshi's position to attempt to lift the siege and rescue those stuck on Japan's newest island. If everything fell into place, it would be a perfect opportunity to end Mori resistance once and for all.

Akechi was ordered back to his home castle to prepare for all-out war against the Mori. Orders were also sent out to all the generals who'd been with Nobunaga in the Takeda campaign to muster for war. The Mori clan still had formidable armies in the field, not like the beardless boys Takeda Katsuyori had sent out to their untimely deaths. Akechi, still seething from the shame of the botched dinner, was one among those tasked with taking his personal troops westward, though Nobunaga himself would lead the line this time. No one else would take the glory from this decisive campaign.

Nobunaga treated Ieyasu and his team to dramatic entertainments as well as a final feast at which he himself served his guests. He then presented all of Ieyasu's men with kimonos and recommended to Ieyasu he should take a holiday to Kyoto, Osaka and Sakai. See the sights, do a little shopping. A subtle reminder that although he, Nobunaga, was going to war, he did not always need Ieyasu's support. Ieyasu took the hint and gladly agreed. He duly left the next day in the company of his small entourage with one of Nobunaga's own men to act as a guide.

That evening, Nobunaga grinned as he enjoyed a quiet drink with Yasuke and a small group away from the bustle of the almost-daily state banquets. Yasuke lifted his cup, grinned back and thanked Nobunaga for the honor of being allowed to travel with him.

The remaining Mori clan samurai would, for the first time, now face the full might of Nobunaga's power. And after the Mori were finished, his amassed troops could press on through the summer into Shikoku where the Chōsōkabe family still ruled. After, only the northern lands of eternal ice and snow and the balmy southern island of Kyushu would remain, but those remote regions would come next year. The realm reunited, *tenka fubu*. After that, farther afield, Korea perhaps? China? Or maybe that was not necessary; all things in their time. He would first travel to Takamatsu to personally make an end of the Mori.

Nobunaga and thirty comrades, including Yasuke, set out at

a gallop for Kyoto where they'd rest for the night before pressing on to the besieged Mori castle. They would stay, as always, at the Honnō-ji Temple.

Only one of them would live until morning.

CHAPTER NINETEEN
Battle Cry

Atop Mount Atago, in an ancient shrine dedicated to the spirits of the mountain, knelt Akechi Mitsuhide. The silence was broken only by the whisper of falling pine needles as trees braced in the stiff breeze. But for the lone priest in attendance who drifted within the temple's gloom like a shadow, Akechi had spent the whole night alone in deep meditation. In reflection and prayer, consulting Shogun Jizo, the Japanese Buddhist deity of victorious hosts. No common god of martial triumph, but one of the most divine and powerful in Japan, renowned for assuring victory and valor in war. Over several hours, he'd prayed to Shogun Jizo for divination three separate times and the sacred response, all three times, had been favorable. There seemed no need for a *fourth* consultation—the number four was unlucky anyway, due to it sounding the same as the word for death, *shi*. Refreshed and reassured by his meditation, Akechi hiked the five-hour descent with a bolstered stride, his brow unfurrowed for the first time in too long, a once-troubled heart restored and resolute.

Oda Nobunaga must die, and the time was now.

Commanded by Nobunaga to muster his troops and advance in support of Hideyoshi, Akechi had mustered his samurai as ordered. (No one would have expected anything less.) The order from Nobunaga had read: "You can be a more effective backup between Bizen and Bingo if you march directly from your own province in the next few days, so you'll get there before I do. When you arrive, wait for further orders from Hideyoshi."

Infantry, cavalry, craftsmen, laborers, armorers, grooms, cooks, carpenters, *shigeshoshi* head-dressers, and numerous camp followers set out on the road west to war. They now all waited for him in Kameyama Castle just below the shrine while he meditated and mulled over the biggest decision of his life.

It was June 18, 1582.

Akechi and his main force, typically based at Sakamoto on Lake Biwa, only a few short miles south of Azuchi, had skirted the mountains north of Kyoto and gotten as far west as his secondary fief at Kameyama where they'd remained to gather more men and supplies before the anticipated advance westward into Mori territory. His soldiers had seen to their weapons and the enormous support staff made ready for the long trek and the anticipated action waiting at its end. While his vassals organized, Akechi had climbed the mountain alone for his night of reflection.

The following night, he again ascended the steps to the holy enclosure atop. This time, however, he'd climbed in the company of eight gentlemen poets, among them Gyōyu, the head priest of the shrine, and Satomura Jōha, who'd risen from humble origins to be considered one of the most celebrated poets in all of Japan.

They'd gathered to hold a *renga* session—a centuries-old form of collaborative poetry where multiple poets took turns adding new stanzas after the poet proceeding them, competing in literary allusions and poetic skill. Each contribution had to conform to rules of structure, meter, stress and intention. The poem was to reach exactly one hundred stanzas in this case, and would take the poetic grouping all night to achieve. *Renga* were often employed as a powerful performance in the act of prayer, beseeching the dedicatory deity to grant victory in battle.

In the holy atmosphere of the old shrine, the stars shining through the open shutters, a servant fed the braziers to keep the participants' hands warm enough to perform their swift and light calligraphic brush strokes. While awaiting their turn, the poets quietly made tea from the bubbling kettle hung above the fire pit in the center of the room, ignorant of Akechi's treasonous intentions.

The first verse—the *hokku* (what some were now calling *haiku* as a standalone poem, rather than a poem's start)—was given by the most honored guest. And so, Akechi began:

Toki wa ima,
ame ga shitashiru,
satsuki kana.

The time is now,
the fifth month,
when the rain falls.

A seasonally appropriate opening, perhaps overly so. For there were also numerous homonyms in his opening lines, a *hokku* stuffed with double entendre. The word for time, *toki*, was identical to Akechi's ancestral clan's name, Toki, and the whole poem could just as easily be understood as:

It is the fifth month,
Now Toki shall reign,
Over the lands under heaven.

His celebrity guests continued on, carefully, building on an ode to beautiful brooks and summer blooms. Surely, Akechi hadn't meant anything sinister in his opening remarks. He was, after all, trusted and routinely honored by his lord, singled out for praise as an exemplary warrior; the first man to ever receive an entire castle from Nobunaga as a reward for his diligence, bravery and statesmanship. Akechi had faithfully served the Oda warlord for more than a decade.

This, despite Nobunaga's numerous insults and the betrayal of Akechi's mother whom he'd offered as a participant in a hostage exchange to a defeated clan. Akechi had guaranteed the clan lord's life, but Nobunaga countermanded his order and the vanquished warlord was crucified. In retaliation, Akechi's mother was executed. Nobunaga had shrugged off the whole incident and continued to rely on Akechi as one of his chief men. Akechi, also, seemed to have moved on, serving his lord another four years.

Which is precisely why it would be so easy to stage his coup.

Thirty miles away, Nobunaga conferred in one of the larger audience chambers of Azuchi Castle, a wide tatami-matted room on the third floor with glorious vistas of Lake Biwa visible through the slats of the windows. The warlord dispensed directions and orders to those he'd be leaving behind to defend his home, ladies and treasures while he went to battle. The orders were given out calmly and the soldiers went about their business with brisk efficiency. This had been done many times before and everyone knew their job. More harried were those soldiers arranging the hosting of the tired men who'd only just gotten back from the Takeda campaign, those returned to the countryside to see to their farms in the early summer. These men would take some time to marshal back to action, but the summons had gone out and they'd serve as ordered.

Yasuke informed his own small staff of his imminent departure. His servants cooked rice balls wrapped in bamboo leaves for the next day's journey, polished his armor, checked his straps and ties, and packed fresh supplies of sandals and underclothes and a change of formal clothing. He personally saw to his own swords and the *naginata* he'd purchased the previous week from the Azuchi armorers' showroom. He bade a cheery farewell to all, and as they knelt and bowed deeply to their African lord, he left the house to report to the stables at dawn on June 20.

Yasuke and another thirty pages were to accompany Nobunaga—with another thirty or so trusted servants in tow—through safe country to Hideyoshi's position in Takamatsu, some hundred miles away. There, they'd rendezvous with other forces—the armies of Takayama Ukon and Akechi—and then come to Hideyoshi's aid. All three armies would travel by similar routes, and Kyoto was the first stop on Nobunaga's journey.

Nobunaga planned to spend the night in his customary Kyoto residence, Honnō-ji Temple, where he'd use the time to pay his respects to some of the nobles of the imperial court prior to his planned conquests. He'd not visited the capital city for over a year, and to mark the occasion he brought along as gifts the most treasured tea implements from his personal collection. Among them a copper jar for waste water, a kettle named Otogoze, a tea leaf jar named Mikazuki, and several spoons, whisks and ladles, beautiful in their *wabisabi*, or contrived rusticity. With such souvenirs, the nobles would not soon forget their evening's entertainment with Nobunaga.

Yasuke was honored to be among those accompanying the

warlord to Kyoto. It was a clear sign of just how far he'd risen. Afterward, he knew, would come weeks or months of war. A perfect chance to truly prove himself, gain some spoils and maybe win that fief or further title the rumors hinted at.

The same day Nobunaga and Yasuke were heading toward Kyoto, Akechi traveled to his castle at Kameyama and made his final preparations. The one hundred stanzas of the *renga* were completed and there were more prosaic matters to attend to now. He summoned his four most trusted generals, swore them to secrecy and revealed his plans.

His men were astonished, frightened—having never heard their lord speak so—and treason was a heinous and dishonorable crime, perhaps the worst infraction of all. However, Akechi was their lord and they followed him, planning out their strategy for the initial coup and the subsequent political wranglings that would surely come afterward. It would not be easy. With the head removed, the monster that was the Oda clan would still writhe and thrash quite dangerously. Family members would need to go, and swift action was needed to secure the loyalty of the other chief Nobunaga vassals, especially Takayama Ukon (whose castle Yasuke had stayed at the week before meeting Nobunaga) and Niwa Nagahide (who'd led the *umazoroe* cavalcade the previous year). Both lords had field armies within easy march. Hideyoshi was far away, and could be dealt with later. Tokugawa Ieyasu, though in the area on ordered "holiday" after his visit to Nobunaga, had no army with him, and would be an easy and convenient pawn to control in the coming struggle. The plan was set.

First Nobunaga, then his sons, then Azuchi, then the realm.

By sunset, a hundred senior vassals and thirteen thousand unsuspecting samurai were marshaled for a highly unorthodox nighttime march. But the soldiers were eager to get moving and finally reinforce Hideyoshi, who'd been battling in Takamatsu while their lord played at poetry. If any of them had been thinking clearly, they'd have noticed that although they started out marching westward, partway through the march, their leaders seemed to have a change of heart about the route and turned them back east—the wrong way. They were told simply the route had been changed and that they were going via Osaka instead. It was slightly out of the way, but there must be a good reason. Akechi's trusted generals, those in the know, led the way.

That Kyoto was *also* now an option had not yet entered the troop's consideration.

Nobunaga's entourage arrived in Kyoto an hour ahead of dusk.

They'd made good time. Yasuke was returning to the city where he'd once almost been torn apart by the mob, but this time there were few crowds, and their swift and unannounced arrival went almost unheralded. They trotted over the Kamo River and into the city. As they passed, the citizens quickly got down on their knees and knelt in the dusty streets, not daring to show their faces until the Oda warriors passed. The familiar scents of incense and charcoal, and the sounds of hammers and hawkers filled the air.

As they approached their destination, the Honnō-ji Temple, servants rushed out to greet them. The same temple where Yasuke had first met the warlord fifteen months before. Both a lifetime ago and, also, seemingly only days before. Now, however, Yasuke was neither confused or worried, panicked or disoriented. He was a trusted and valued member of Nobunaga's vassals. The first African warrior to grace Japan's most powerful halls.

They rode through the main gate and the grooms took their horses. Bruise-colored shadows gathered in the corners of the courtyard, reaching out like a hand against the temple walls. Yasuke slid off his horse and passed the reins to a waiting groom as if he'd done it a thousand times, and by now, perhaps, he had.

He wasn't dressed as a Portuguese valet this time, or unsure of what was going to happen next. Rather, the two swords of a samurai were thrust through his sash and he marched, strutted, down the hallways behind the most powerful man in Japan.

Menials scurried out of their way. Nobunaga's full entourage escorted him to the door of his inner chamber, where two servants waited, bowing their heads to the floor. Nobunaga gave the group a curt nod and disappeared inside. One of the servants slid the door shut again and the group dispersed to their own quarters.

Yasuke was assigned a room with a number of other warriors. Various servants waited to show him and his roommates to the bathing facilities. There, he cleared his mind as he was scrubbed down by one of the bath attendants. She couldn't help stroking his smooth skin again and again; she'd never washed a black man before and she couldn't stop talking to him about it, wondering at his youth and beauty, the different shape of his face and size of his muscles. Yasuke couldn't help but recall Nobunaga doing the same a year earlier, though the girl's touch and talk was much softer and appealing than his lord's had been. Afterward, he entered into the deep steaming water and allowed it to soak off the aches of the day's ride. Several more samurai joined him in the water and the girls waited to massage them again when they were done. Nobunaga would entertain imperial nobles alone tonight with a reduced security team. Yasuke and most of the others had the evening off. And a massage before dinner sounded like a glorious start.

Mount Oino was a crossroads for Akechi in every sense.

The right trail led to Osaka and the left to Kyoto.

Another mountaintop, another decision. It was now time for the final call. Taking an army into Kyoto was a clear violation of orders and even if Akechi called it all off when they arrived, the original intention would be clear to all.

Behind him were battle-hardened, experienced men. The samurai who'd fought in Akechi, and Oda, war camps for many years. In the previous year, this same army had taken part in two major campaigns. In Tottori against the Mori, and against the Takeda in Shinano. They'd marched far more than a thousand miles in the process. They were sworn firstly to Akechi and many had been raised in his fiefs, doing only his bidding. Their second loyalty, however, was to Akechi's own lord, Nobunaga. In the main, they were typical Oda fighting men, lightly armored foot soldiers, armed with a mixture of guns and spears with a senior hereditary samurai archery corps sporting composite longbows. Almost all had one or two swords, the poorer men loaned them by their lord. Although the sword was no longer used as a weapon of first choice, it was a badge of warrior honor and a useful weapon when opposing samurai got up close. The introduction of guns had seen to that. Only the officers were mounted. (To maintain their vast size, Oda armies were assembled and maintained as cheaply as possible, and horses were costly to buy and prohibitively expensive to maintain. Mounts were available to the elite only.)

Akechi thought his men would follow him, but he was not entirely sure, therefore most of them were kept in the dark about their target. *Right or left?* Right to continue on the same path as Nobunaga's stalwart general, a solid life with guaranteed title and honor for his descendants forever. The hallowed name of Akechi would be counted among the highest in the land. To the left, another trail entirely; the one which led down to Kyoto. Left to where Nobunaga—and a few hundred soldiers at most— slept soundly for the night. Totally unprepared for what Akechi could throw at them.

Nobunaga's death was certain, Akechi thought; there was no escape for him. After that, however, the picture was fuzzier. He'd need Takayama Ukon and the other lords to recognize his leadership and combine their forces with him swiftly. Then they could stand against any other Nobunaga vassal or Oda son who chose to take the field against them. Takayama had a weakness for the Jesuits and would value their council highly, so Akechi would need to target the foreigners in a charm offensive if at all possible. And, if successful, he would be the highest in the land, the next shogun, and all Nobunaga's conquests to date would fall now to him. If not, he and every generation of his family would be hunted down and exterminated. The name Akechi would be shamed for eternity. The traitor's name.

Fifty years of life and it all came down to this most basic choice: Right or Left? Osaka or Kyoto?

His son-in-law, Hidemitsu, approached warily, bowing deeply. "Lord?"

"Left," Akechi said.

Nobunaga spent the early evening nearby in the Nijō Palace, a residence he'd had constructed for himself some years before, but later donated to Crown Prince Masahito, the heir to the imperial throne and eldest son of Emperor Ōgimachi. The prince, the governor of Kyoto Murai Sadakatsu, a rich merchant and old friend of Nobunaga's, Imai Sōkyū and his son Sōkun (the Imai family ran Oda's Sakai gun factory) and several other important guests graced the tea ceremony they held that night.

The guests swooned over the tea implements, treasures the warlord had spent a lifetime collecting and brought along to honor them, and show off. They were spread on the tatami floor for the viewing and, as was proper, the guests revered each one slowly, turning it in their hands and admiring the specific inimitabilities of handcrafted wares and the beauty of the priceless pots, bowls and bamboo utensils. Although it was not his residence, Nobunaga acted as the host, taking the prime position next to the kettle so he could symbolically serve "his guests." Prince Masahito, as the highest-ranking "guest," knelt beside him.

Shortly, the pre-tea meal was served on brand-new lacquered

trays in front of each guest, the food presented as an elegant feast for the eyes as well as the belly. The first tray held a *namasu* dish of lightly broiled fish and ginger; a dark miso soup with delicately cooked radish, tofu and miniature mushrooms; a selection of pickles and a bowl of pure white rice. The second tray held a lightly citron-flavored simmered fish with scrambled egg, a soup of crane in a light miso broth, and a grilled dish of salmon in a wasabi sauce. The third tray consisted of carp sashimi, another soup of grated yam and deep green laver, and a broiled duck on slices of tangerine. With the serving of each course the type of sake changed, but the sedate and reverent tone of the party did not alter, such was the "Way of Tea," as the ceremony they partook in is called. Having eaten the feast, the guests retired to wash and purify themselves before the main event, the preparation and drinking of the tea itself. Nobunaga used the time to symbolically clean all the utensils and sweep the tiny room in which they were to perform the ceremony.

At the sound of a gong, Nobunaga's guests reentered the room, sat in their places again and waited while Nobunaga prepared the tea, whisking it into a thick froth with his bamboo whisk. The Oda lord then bowed to Prince Masahito and offered up the bowl of bitter, almost luminous, green tea. The heir to the imperial throne bowed back deeply before accepting the bowl, then bowing to the second guest to his left, rotated the bowl in his hands three times, and as protocol dictated he passed the bowl to the second guest to his left who then repeated the pattern, as did the final two men in the room. The last guest returned the bowl to Nobunaga and the evening continued with another serving of tea, this time a whole bowl for each guest. Finally, the post-ceremony meal was served. Cold noodles, spicy radish soup, rice, steamed dumplings, a pickled plum, diced conger eel, mushrooms and dried tofu. The whole process took around three hours, after which the guests took their leave, and Nobunaga bid his farewells and returned to his

Honnō-ji Temple residence. The decorum and sophistication of his evening's entertainment served to prove the rightness of his mandate to rule the land.

Meanwhile, after a meal with the other pages, spent mainly in the silence of the post-bath stupor, Yasuke enjoyed a rather more boisterous few drinks to finish off the day before the off-duty pages all retired to their futons for sleep. The next few days would bring more dusty roads and forced gallops before they reached Hideyoshi at the battle front. Some rest would do them all good.

Akechi and his army entered Kyoto during the Hour of the Tiger, before dawn. As they crossed the Katsura River, just outside the city, the first flush of daybreak edged the mountaintops to the east.

"The enemy is at the Honnō-ji Temple," Akechi told his shocked men. They hesitated, but did not falter. Hidemitsu, his son-in-law and chief officer, led them onward. Akechi's intentions were now clear to all. And no one argued, it was too late for that. They followed where he led, and now there was no going back. They approached the temple, on the outskirts of the city, and took up strategic positions around it.

The citizens were just stirring, fires being kindled for the day and a few people were already hurrying through the predawn streets. At the sight of the soldiers, however, they made themselves scarce. Gates slammed, shutters slid swiftly shut again and sandal-clad feet clip-clopped speedily up the narrow alleyways. Kyoto had not been a battleground for a decade thanks to "Nobunaga's Peace," but the citizens still recognized war when they saw it and knew what to expect next.

Akechi's men stormed the temple without difficulty. Kyoto was supposed to be safe territory, long since pacified and policed. The token corps of guards on duty at the Honnō-ji Temple that dawn were both surprised and, subsequently, quickly eliminated.

The Akechi men climbed the temple's outer walls and entered through the few small side entrances like the night's wind. The sizable temple yard made it easy for the small army to move just as swiftly to surround the numerous buildings within. Those sleeping inside, unsuspecting in every way, had no escape.

Yasuke jerked awake, sitting up.

A peculiar sound had awoken him, though he couldn't remember what it had been, and he crouched in the dark, head tilted, holding his breath to hear it again. He'd already grabbed his sword.

The room was dark and remained silent except for shifting bodies and light snores from his roommates. Behind the pure white paper shutters came a pinkish glow, like dawn, but it was surely still early for that and it appeared different from other dawns. *Something wasn't right.* His whole body tingled with a terrible expectancy, but his head was still slightly cloudy from the drink the night before.

The African samurai stood fully and crossed the room silently and swiftly, gently sliding open his door. A narrow strip of the rising sun cut the hallway in half through a narrow opening in the shutters at the end of the corridor. He looked up and down the hall but the guard was not outside Nobunaga's room, which meant Nobunaga was not there; he must already be up. Yasuke clutched his sheathed blade tighter and, barefoot, stalked silently down the hardwood corridor. Nobunaga was probably outside at the spring washing his face. It was the time of day when he often cleaned out the cobwebs of the night and exercised his muscles for the day to come.

Suddenly, the piercing wail of a conch shell sounded outside. It was a battle cry; someone had sounded an attack.

Several Oda samurai grooms, Yashiro Shōsuke, Nobunaga's favorite horse boy, Tomo Shōrin and Murata Kitsugo, renowned sumo wrestlers, and Ban Taro, the son of a faithful old retainer who'd been getting the horses ready for the day's journey realized what was happening and charged the advancing Akechi men in an attempt to slow their advance. Ten, twenty Akechi attackers fell dead before them. But then these young men were cut down, surrounded on all sides by ever-growing numbers of enemy warriors. Their bodies drowned beneath the tide of the Akechi soldiers. The other twenty grooms, those who'd been inside with the horses, recognized they did not have the time or position to retreat to Nobunaga's quarters. Instead, they made their stand at the stables. The Akechi men ran onward, and took up firing positions against the main temple building where Yasuke and Nobunaga were.

Yasuke raced down the hallway toward where he thought Nobunaga would be. The fading night had erupted in the deep trumpeting of more conch-shell war calls, battle cries and the earthquake-like rumble of oncoming armored bodies. Then screams of men dying outside, and the clash of swords pierced the clamor. As Yasuke ran, the rest of the men within the Honnō-ji Temple stumbled out into the hallways, startled, grouchy, heads muddled by the sudden rude awakening.

Yasuke had finally reached the secluded courtyard behind the lord's quarters. He found Nobunaga and Ranmaru there, confused, but still both wielding their bows and clutching quivers full of arrows. Other men rapidly joined them on the run toward the front of the residence, grouping behind their leader and lord. Nobunaga cursed loudly, demanding an explanation.

"Some of the locals are brawling," someone suggested, but no one really believed this. Brawling locals would never dare approach Nobunaga's presence under arms.

Lord Nobunaga raged. "Yasuke, get my *naginata* from the room and—"

The whole world exploded with a thousand cracks of thunder.

The wood and paper walls around them shattered on both sides, vanished into thousands of fragments as Yasuke and the rest crouched down against the explosion of gunfire and flying ordnance. Pellets ripped through the walls. The men scrambled to find shelter behind columns; grabbing up bedding for protection, the covers still warm from the night's slumbers.

Splinters of wood and paper filled the corridor, drifting down on them like ghastly snow. The thumping of arrows and lead on wood and the soft thwack as a futon, sleeping mat or soft flesh was pierced. Another battle cry filled the early morning, the voices of thousands of warriors outside in the temple courtyard, unified in battle rage. Yasuke and the rest braced against the terrible sound which felt it would flatten the temple compound. Even as they chased that thought away, another noise swallowed the first.

Volley after volley, unceasing, burst through the dawn in the signature shooting pattern of three ranks shooting in turn to provide continuous fire, a method which Nobunaga himself had patented and passed on to his vassals. This was, without doubt, an Oda attack. An Oda army attacking the clan head? Treachery. *But who?*

Musket balls swept past again, slashing through the paper and wooden shutters. Between volleys, Yasuke and other Oda men moved, readying their weapons for the inevitable storming attack to come. Yasuke clutched his sword in its scabbard, and kept Nobunaga's *naginata* in his right hand ready to pass to his lord when the order came. Nobunaga was not going down without a fight. As screams of terror and pain echoed down the hallway from every direction, the warlord, Yasuke and the surviving few men ran to the front of his quarters and took cover behind stout wooden columns and the dead bullet-ridden bodies of their fallen comrades.

Another volley ripped through the temple.

Yasuke and other Oda men moved again, taking cover behind

Oda Nobunaga's last stand at the Honnō-ji Temple. Nobunaga can be seen on the right, and on the left Mori Ranmaru is running to try and save him.

anything they could find. Then another volley. The outer walls were in tatters. The gunfire ceased and the enemy charged in with hand weapons, spears stabbing and swords raised. The clan symbols on the flags were clear to the defenders now.

The five petal bellflower symbol of the Akechi.

It was stalwart Akechi Mitsuhide who'd betrayed his lord. Yasuke and the others couldn't fathom it. Akechi? Akechi had—

"What's done is done," grunted Nobunaga. The how and why of it mattered not to him. The next hour had already been decided by the actions of days and years before, and the vagaries of fate. He was resigned to the inevitable, and as the crane in the story he'd once shared, Nobunaga would surely meet his death with a calm and reconciled defiance.

He and his men burst out firing their arrows again and again at the horde of advancing traitors. These highly skilled men had trained in archery since they could stand. The proof came in Akechi deaths, men forming piles at the bottom of the stairs that led up to the main temple edifice. But there were simply too many.

Nobunaga's bow string snapped. With no spare on hand, he called for Yasuke to pass over his *naginata*. The enemy had reached the top steps to the residence, and the warlord slashed out with his new weapon and then stabbed, impaling an Akechi samurai on his wide, long blade.

Fifty more traitors started up the steps toward them.

Yasuke drew his sword and charged.

CHAPTER TWENTY
The Honnō-ji Incident

Bodies lay strewn across the gore-splashed steps, some still quivering in impending death, others crawling over each other to die away from the incessant slaughter. Yasuke had lost count of the men he'd killed.

Still, there was no escape here.

More and more Akechi soldiers wielding swords and spears fought their way steadily up the stairs, overwhelming the ever-dwindling number of defenders. The traitor lord had clearly brought several thousand men. With no other choice, Oda's men fell back, retreating into the main building. The shattered hallways behind the audience chamber were narrow, the only advantage against the numbers waiting outside.

Yasuke stood gasping beside Nobunaga. With them were Ranmaru, Ranmaru's two brothers and a dozen other surviving comrades. All now had the look of men accepting their fate—this morning, they *would* die—but before that fate was met they would exact as much vengeance as they could manage together.

For now, their retreat had worked. The attack *had* slowed, and the first six Akechi men to enter after them into the tight passage were cut down easily. More Akechi troops had filled the opposite end of the hall, but held their position. The hallway grew still within the chaos brewing outside as the two groups, on opposite sides of the hall, now studied each other with wide eyes, sizing up their opponents before the imminent bloodletting.

Nobunaga breathed deeply, looking away. What he was seeing against the shredded wall, Yasuke could only guess: A future that now would never be? His children? Some imagined hellish punishment for the traitor whom he'd trusted so? Or, maybe some simple memory from the past? The coy smile of a passing peasant girl, an accomplished courtesan's touch. Or, the dark silhouette of a wild hawk passing directly overhead. Nobunaga took only a moment. The warlord turned back to the others, glaring, his face alive and fierce, destiny was destiny. Yasuke could have sworn the man was smiling.

They pushed forward together to take on the next wave of attackers. Battle cries and combat erupted again in the hallway. It was now twenty against fifty, a hundred, pushing and screaming through the corridors and sleeping chambers that led off them, where the defenders had, only minutes before, lain peacefully asleep.

The fighting was steel-against-steel, hot spit and blood splashing as the two sides slammed together. Guns were useless in close quarters, and this battle concluded in a more time-honored way. Spears and samurai swords sliced through doors and stuck in crossbeams. Wooden pillars provided protection but also obstacles, massive splinters carved by downward sword strokes flung dangerously through the air.

Yasuke used his massive body to drive back several attackers, shouldering the smaller men, punching them with the sword hilt in his right hand, and then reversing the cut to dip his blade in their lurching bodies. They died quickly, which would not be Akechi's fate, should Nobunaga, or any of the Oda, ever get his hands on the usurper.

Having cut the heads off several spears, Yasuke nimbly danced between the splintered wooden poles and slashed his sword down into the attackers' faces before they could draw their own blades to counter his cut. Arrows flew past him from behind, the enemy dancing back and falling at the end of the hallway as Oda arrows struck home. The white walls were now spattered in hundreds of crimson blossoms.

In the courtyard, Yasuke knew, several hundred more soldiers waited to replace the first fifty who'd dared enter and commit the treacherous sacrilege of turning on their ultimate liege. However many foes were felled by Yasuke and Nobunaga's dwindling vassals, there were always more to take their place. Yasuke fought beside half as many Oda pages now as wave after wave of attackers crashed in, seemingly from all directions. Perhaps they'd burst through the wall at his back. He no longer knew. His forearms were gashed from multiple cuts, his hands swollen and sore, soaked in blood. Nobunaga had taken a spear to the shoulder, the warlord's left arm dangling, his white sleeping robe stained in dark, scarlet blood.

The samurai beside Yasuke wiped the gore from his eyes as the slash across his forehead continued to run hot down his face. Within the next surge of attackers arrived two Oda comrades—Yuasa Jinsuke and Ogura Matsuju—samurai who'd been staying in town but rushed to the temple when they heard the commotion, concealing themselves among the Akechi troops until they'd reached the dwindling Oda samurai.

The two men joined the defenders joyfully, to cheers from Yasuke and the others, but were soon cut down by the Akechi soldiers they'd pushed between to achieve their deaths. Loyalty had its price.

A new smell filled the air and a strange glow suddenly filled the gaps in the walls. Fire. Whether it was Akechi's men who'd set the temple on fire, the defenders who'd rather perish in the flames than surrender, or simply a stray spark from the guns, who could say. The result was the same. To the smell and smoke of lingering gunpowder, was added the sweet perfume of burning pine and cedar.

An Akechi arrow struck Nobunaga in his leg, sinking deep. The Oda warlord cursed, collapsing to one knee as more Akechi soldiers forced through the entrance. Yasuke and the others met them, fresh hot blood splashing against the shattered walls. "Form up on me, close ranks," Nobunaga commanded and the men clustered together, retreating again slowly down the main hallway. Toward the inner chambers, Nobunaga's hegemony was reduced from a nation to a few small rooms within a matter of minutes.

Within the doorways of other rooms, servant women crouched and wept. "Get out!" Nobunaga snarled, passing them. "Hurry! They won't harm women." Whether this proved true or not, the men did not know, but little could be worse than burning to death in the growing inferno. The last few men continued their tactical retreat toward Nobunaga's personal quarters as the women scurried down the hall to the courtyard and the approaching soldiers. This residence was an annexed place of worship, not some warlord's stronghold. The temple walls and most of the secondary buildings were made of wattle and daub, paper, and wood. They'd provided very little cover from the initial salvoes of bullets and arrows and had never been designed to. And now, they burned exceptionally well.

This morning's end was already written. Death. But Nobunaga could still decide where the final stand would be. It was his last command in a life of orders.

There were less than ten Oda men left standing now.

Nobunaga commanded all, except for Yasuke and Ranmaru, to hold out as long as possible. Those men left behind on no account were to perform seppuku, as Nobunaga himself needed the time to do that. Holding out to the last would be their final sacrifice for their lord. The teenager Takahashi Toramatsu rushed to the kitchen entrance at the rear of the building to stop a rear attack, and his quick thinking won his master precious minutes.

Nobunaga stepped into one of the last inner chambers and waved Yasuke and Ranmaru into the room.

"It is time," Nobunaga said and Yasuke followed.

A block away, at his residence in the Myōkaku-ji Temple, Nobutada woke to the sounds of battle, the strange glow to the south over the city, the acrid smoke. He'd heard the commotion and now gotten word of the reason. He had only two hundred men with him, but it was his duty to try and rescue his father. If he'd perished, Nobutada was now the clan head and must lead—whatever the circumstances, and for however long. Even if only another hour.

Nobutada started for the Honnō-ji, but several Oda faithful rushed into the courtyard with more news. "The temple has fallen," they related. "It's collapsed in flames. They'll come here next."

His father was surely dead. Betrayed.

As heir, Nobutada thought briefly of escape. Regroup, pull together his father's many allies, but it was already too late for escape really—his father's power, he knew, would end this morning. Whoever had turned on them—*Akechi? Takayama?*—would have already eliminated such options. He needed to make a stand somewhere and straightaway.

The new Nijō Palace next door, which Nobunaga had constructed at great cost and where he'd just spent his last evening on earth, had strong defenses. There, Nobutada would hold out as long as possible and procure what revenge he could. And, if it were destiny, there he and his men would die also.

As the final Oda survivors in the burning Honnō-ji Temple held the door to the chamber with Nobunaga, the Akechi troops withdrew to let the flames do their work for them.

Within the growing inferno, there was no time to stand on ceremony—no seppuku formality of white robe and death poem for posterity and no paper-wrapped blade for the stomach cut. Nobunaga must die by his own hand right away to avoid capture and humiliation.

The warlord knelt, took a short sword, and held it to his belly. He looked up at Yasuke. "My head and sword to Nobutada," he ordered. "Never let them fall to the enemy."

Yasuke bowed deeply, understanding his role. Neither could fall into the hands of Akechi as they would become a powerful symbol of Nobunaga's downfall and Akechi's legitimacy as his conqueror. With the head, Oda vassals would more easily accept Akechi as their new overlord.

Nobunaga had a chance to stop that, striking one final blow of revenge at the man who'd betrayed him while giving his own son, Nobutada, a shot at legitimate succession. Nobunaga chose Yasuke for this extraordinary task.

Yasuke swore he'd deliver the head and the power of legitimacy it would provide. He bowed again, his eyes never leaving the warlord's.

It was the last order Nobunaga ever gave.

Throughout the city, people gathered at windows and doorways arguing and shushing each other and crying over what might be happening. Kyoto had been at peace for a decade, but memories were long and much of the inhabited city was still surrounded by burned-out blocks, crumbled walls and a waste-strewn no-man's-land from the preceding hundred years of war and strife. They'd gone to bed thinking the nearest enemy was a hundred miles away, that their protector and benefactor Lord Nobunaga was untouchable here. But they had been wrong.

Nobunaga breathed in deeply as smoke snaked through the joints in the room's panelling. He rammed the short sword fully into his belly, doubling over before starting to draw it across horizontally.

Ranmaru, acting as Nobunaga's second in this ritual suicide (his *kaishakunin*), stepped forward with his sword and neatly severed his lord's head, skillfully leaving the skin at the very front intact to avoid the head bouncing to the floor in an un-dignified manner. The beautiful samurai knelt to the floor and prostrated himself before Nobunaga's remains, then deftly cut the final lappet of skin. He picked up the head, his right hand grasping Nobunaga's topknot, his left the bloody remains of the neck as he'd been taught, careful not to get any more blood on his lord's already paling face. With great gentleness, an almost caressing touch, he wrapped the head deftly into his jacket, tying the package at the top.

Bowing once more to the floor, Ranmaru offered up the bundle to Yasuke who, kneeling also, bowed back more deeply. There were no tears in the young samurai's eyes as the teen asked Yasuke to spare another minute for one last request before he escaped, the ultimate honor of acting as *his* second. Wasting no more time, Ranmaru quickly knelt and stabbed himself in the stomach with his own short sword.

Yasuke did as asked, and as *kaishakunin*, sliced the boy's neck to sever it, copying the technique he'd just witnessed, leaving it hanging by a narrow strip of muscle and skin. His cut had not been as good as Ranmaru's, but it had done the job; keeping the boy's head off the floor, the face now hung upside down fac-ing Ranmaru's chest. Yasuke made the last cut tenderly, lifted the head, and placed it reverentially next to Nobunaga's body. Master, mentor, hero, lover.

Long grey fingers of smoke ghosted across the ceiling as flames licked up the walls and the first timbers cracked. The temple was collapsing. It was time to go.

Yasuke cinched the cloth package with Nobunaga's head to the sash at his waist. Nobutada, his new liege, was only five minutes to the north in his own temple residence, but those would not be an easy five minutes. Yasuke knew Kyoto's streets well enough, but he also knew it was crawling with the enemy.

However, if he could reach Nobutada, he could fulfil his lord's last order and strike a blow at the traitor who had so vilely brought down the house of Oda. He could not bear to leave the corpse to simply burn; Nobunaga deserved better. Yasuke hoisted his lord's body and sword aloft and set about escaping the hell-like inferno.

Suddenly, the last ten Oda warriors burst into the room. They immediately took in the situation. It was clear Yasuke had a greater chance of fulfilling Nobunaga's last orders if he had less to carry, so they relieved him of the body, and sent him on his way to Nobutada with only the head at his waist and the sheathed sword in his hand. Both were needed by Nobutada, but the mauled and lifeless body could be left for them to deal with. At least it would be treated deferentially and given the proper rites. They could then, maybe, perform seppuku themselves afterward, knowing the most important job, defending their lord's final dignity, was in good hands. If the Akechi troops did not find them first.

Yasuke nodded in appreciation and farewell and then dodged through a side hallway, jumping over the slaughtered bodies which were spread everywhere. His own face was smeared with fresh blood and soot. He slashed open a side wall, saw several shapes moving in the wall of smoke ahead. A few soldiers.

Yasuke emerged from the inferno to face them. With the

fire raging behind, he hoped to cut through them quickly. To somehow flee before another three, or even a hundred, blocked his escape. The warriors stopped, frozen, then openly cowered back. Yasuke stepped forward, looming over them, focused, undaunted. Wrathful. A living nightmare. One of the soldiers glanced at the spear in his own trembling hand and his look revealed all: *It was not weapon enough.*

Yasuke smirked grimly. Fear was a much-needed ally this night. These were mere flies, he reassured himself, flies to swat on this, the last mission for his dear lord. The three soldiers remained spellbound, unable to move. Even words failed them. *"Yasuke de gozaru,"* Yasuke said as he stepped forward into attack position, ripped his sword from the scabbard and performed the attacking stroke simultaneously, a technique Nobunaga had personally taught him. He'd knocked their blades aside. With the return stroke he swiped the tip of his sword across the startled faces of his three foes, and as they fell back, he delivered the death stroke to each man.

Outside, the city was still mostly in shadow, the sun was only just reaching over the Kyoto mountains, but the temple already glowed brightly. Flames engulfed its roof and walls and outbuildings, mingling the sweet aroma of burning wood with the acrid stench of burned gunpowder.

The Akechi troops had now largely vacated the burning residential structure, edging back into the main temple courtyard to avoid their own fiery deaths; they believed everyone inside was either already dead or about to burn. The rear areas, occupied mainly by smoldering residential buildings were left unguarded, and only the neighboring streets were being patrolled as the rebels fanned out to seek survivors and the main force headed north to kill Nobutada.

In the confusion of the smoke, working around the burning temple buildings, Yasuke swiftly jumped the wall at the back of the compound and found himself in a deserted street. He knew it, for it was the one that led to the Jesuit mission, visible in the distance due to its abnormal height. The smoke and the early morning mist curled together, choking each breath. Ranmaru's jacket cinched at his hip slapped heavily against Yasuke's upper thigh as if urging him onward. He'd vowed to see Nobunaga's mortal remains to his lord's heir, and he hadn't come seven thousand miles to break such vows; no time for a break despite his heaving lungs, he ran for Nobutada, his new liege, whom he thought was in his customary residence at the Myōkaku-ji Temple.

Yasuke did not take the roads; there was too high a chance of being intercepted. Instead, he vaulted over fences, ran through houses and gardens dodging patrols, even climbed over roofs, surprising the shocked, nervous, rudely awoken householders of the capital city. When he found his way blocked by Akechi troops, he relied on shock value, jumping out at them, wielding his weapon above his head. As before, the Akechi troops he encountered cowered before this seemingly giant dark god. They were quickly overwhelmed by Yasuke's size and strength.

Nobutada's men had taken possession of the Nijō Palace, barred the gates and taken up positions on the wall. Nobutada breathlessly prostrated himself and advised Crown Prince Masahito, whose residence it was, that it was preferable for his safety if he left. It was not *his* day to die, and the prince and his entourage were escorted quickly from the scene by the poet Satomura Jōha. The man who'd composed verse beside Akechi two nights before had been sent ahead *by* the traitor to protect the crown prince. Akechi would need imperial allies in the days to come and killing the crown prince would be unwise. They hurried out without the usual pomp and ceremony that typically attended imperial travel around the capital. In the meantime, to the din of a thousand approaching soldiers, Nobutada and his men prepared for their ends.

Just ahead of the enemy ran Yasuke, dodging gunshots and arrows as he sprinted the last fifty yards toward the Myōkaku-ji Temple. Nobutada on the palace walls saw him, shouted his new position next door, and Yasuke swerved to the right as another arrow sped by. On Nobutada's orders, the palace doors were opened enough and slammed quickly shut again as Yasuke gained the temporary safety of the walls.

Yasuke dropped to his knees before his new lord. Gasping, lungs heaving. Head bowed to the floor. Finding the right words in Japanese, he respectfully reached up and offered Nobutada the bundle which contained the head.

Nobutada, eyes welling, accepted it formally and opened Ranmaru's tattered jacket to reveal his father's head. He touched the mortal remains to his own head in a sign which both reverenced his father and showed deep gratitude to Yasuke for having completed his mission. Yasuke then offered up the sword, which Nobutada grasped with his free hand by the scabbard and nodded again at Yasuke. There was no time for anything else. The ground already rumbled with the running feet of the enemy troops and the first musket balls were already buzzing over or striking the wall.

Nobutada passed the head to a vassal with orders to guard it with his life, then turned, lifted the sword and shouted, "To the walls!" A largely unnecessary order as his men were already standing to arms at the meager fortifications, but one which filled his small corps of Oda samurai with heart. Their new clan head was now in secure and confident command. *Tenka no tame.* "For the realm!"

Defending the palace, alas, was a hopeless cause.

There were too few of them, again. And, although better than the temple, Nijō Palace was still low-lying and built largely for ceremonial rather than military purposes. While it was smaller than the Honnō-ji Temple and Nobutada had marginally more men than Nobunaga, it was still too large for this handful of men to defend effectively. Once the enemy gained the wall, the battle became a series of sallies and retreats. The Oda forces burst at the invaders with swords whirling, killing or being killed, then withdrew in ever smaller numbers to group together around Nobutada again before attacking once more.

The deciding factor, however, was the guns. Akechi's forces had commandeered the roof of a neighboring residence which overlooked the Nijō Palace's courtyard, and set up their firearms *above* the defenders. Gunning them down as they attempted their last brave, but futile, stand was a simple matter. Yasuke charged out again and again beside his comrades, his energy and life-blood seeping with every attack and new wound. They simply could not stop the enemy as they fell, one by one, to bullets, arrows or blades.

Nijō Palace, as before, was set aflame and Nobutada, as his father had, realized it was all over. He tasked one of his retainers with hiding his own remains and those of his father and then— as Nobunaga had done less than an hour before—cut his belly, losing his head in turn to the samurai who performed as his second. The short-lived clan lord's head was put reverentially next to his father's and then both were buried under a walkway, which was then covered up to be later devoured in the flames.

As Nobutada performed his last act, Yasuke and the final score of Oda men held the courtyard and prepared to fight to the death. Despite his battles and the wild running flight through the yards and backstreets of Kyoto, the African warrior had clearly only delayed his own death.

They were now completely surrounded by Akechi warriors. The circle tightened with each round of attacks, and then the final defenders somehow became separated, each one surrounded by a vortex of spears and swords. As before, the enemy soldiers were not quite sure how to react to the warrior, the likes of whom they'd never fought before.

They circled Yasuke nervously, prodding with their spears

and sallying, only to fall back before the exhausted African samurai could swipe at his tormentors. The final moments of the
combat had descended to something akin to bear baiting. And
as with bear baiting, many of the attacking dogs died. As the
other Oda warriors around him fell, Yasuke cut down another
six Akechi men. Ten.

The circle continued to tighten. He could hardly move. Spears
and swords cut cautiously at him. Blood ran down into his eyes
from cuts. The world began to blur.

In time, Yasuke dropped to one knee, but kept swinging his
remaining blade. Countless men swarmed him, kicking and hitting with the hilt of their swords, and the world turned black
for an instant.

He awoke seconds, minutes, later to find he could not move.
Five men held him down, but he couldn't have moved had there
been no man holding him. Every muscle was completely spent.

An Akechi samurai came forward and screamed angrily at Yasuke, made him promise to properly surrender his sword. And
he reluctantly complied.

Dawn had come only two hours before. There was nothing
left to fight for. Yasuke offered up his sword to the officer in
charge, and he knelt on the ground, neck outstretched to await
his coming execution. He hoped it would be quick and painless.

But the stroke never came.

Instead, the Akechi samurai gestured and Yasuke was dragged
from Nijō Palace by four men into the streets of Kyoto. It didn't
take long for him to learn why.

Lord Akechi had set up his field headquarters in the house
next door to the large residence from which his troops had
gunned down Yasuke's comrades. The warlord sat on a raised
dais, his legs crossed with his helmet beside him. His long sword
rested in his lap.

As Yasuke waited to be seen, he could hear Akechi barking

out orders as men bowed and rushed out on urgent commissions as others entered to report before just as swiftly departing. Akechi was fuming when he saw Yasuke, having little time to spare for niceties. He demanded Yasuke tell him where Nobunaga's head and sword were. The African samurai claimed he didn't know—not a complete untruth, for the temple and palace were now both consumed in flames.

Akechi's face grew flush with anger and Yasuke held his breath, waiting for his death sentence.

To Yasuke's surprise, however, the usurper did not order his execution or seppuku. Akechi said wryly to the samurai beside him, "*This* man is not Japanese and has no honor; otherwise, he would already be dead." Then, Akechi gestured with his fingers and shouted for his retainers to get the "black beast" out of his sight, to take him to the Temple of the Southern Barbarians, the Jesuit church.

It was the first time in all his years in Japan Yasuke had ever heard a Japanese use his skin color in denigration. His eyes narrowed at Akechi, challenging. He understood the insult had come because the traitor would not, could not, kill him today.

Akechi knew well where Yasuke had originally come from— he'd even been present at the original audience with Nobunaga— and he needed no more enemies tonight. Yasuke was with the Jesuits. And the Jesuits, Akechi clearly hoped, could be a useful conduit to Takayama Ukon, whose ear they had. It was best to return Yasuke to the priests and perhaps gain their favor.

Several soldiers led Yasuke from the house and escorted him on the five-minute walk to the same Catholic church he'd left the year before.

Yasuke had no fight left in him, but there was nothing left to fight for anyway. Akechi was already carrying out his next moves, ordering searches for Oda survivors, and sending troops on to new battlefields beyond Kyoto. On the way to the mission,

Yasuke heard shrieks and screams from behind walls as house-holders were "encouraged" to reveal any hiding Oda fugitives. The searches proceeded with a brutal efficiency; scores of people mercilessly tortured and killed in the streets of the capital. As he walked with his escort, Yasuke bore witness as a handful of Nobutada's men who'd escaped the battle lost their heads like common criminals, without the honor of seppuku. Yasuke looked away; he could do no more for them.

At the Jesuit church, Yasuke was surprised to hear his captors speaking respectfully to the priest who opened the door.

It was Father Fróis.

Akechi, whom Fróis had known for years, had never once shown any particular affection for the Europeans' foreign ways. The Jesuits within had been following events as best they could with their doors barred, wondering how this change of circumstance would affect them. A score of Catholics had taken refuge with them and they were all praying together, kneeling in deep concentration before the crucifix in the chamber of worship.

Father Fróis and the other priests gave thanks in prayer for Yasuke's deliverance and then gently pressed him for as much information as he could manage. It was clear Japan had just changed in astonishing fashion, and forever. As in the past during times of crisis, they should evacuate Kyoto to take refuge in some other, quieter place, such as Takayama's castle or Sakai.

Yasuke collapsed in the chapel foyer. Brothers and priests rushed to help him as blood spread beneath the enormous warrior.

The battle, like the reign of Nobunaga, was over.

PART THREE

Legend

CHAPTER TWENTY-ONE
Japan, Tomorrow

All that day, Akechi's men kicked down doors throughout Kyoto, seeking Oda soldiers or loyalists who'd escaped the two battles and were still hiding within the city. Such men, and their allies, were dragged into the streets for threats, interrogation and summary executions. The Akechi troops killed hundreds and those corpses not quickly claimed by relatives were tossed onto makeshift funeral pyres, adding to the deathly pall of smoke and rank smell still permeating the city from the morning's battles.

Yasuke mostly slept, fading in and out all that day. His many cuts and burns were cleaned and wrapped by the Jesuit attendants. The missionaries had barricaded the compound gates and, like most of the townsfolk, kept their heads down, following events as best they could from behind the barred doors as they debated how the sudden changes would affect them.

Akechi's men now stood guard, unchallenged, at strategic points throughout the city. By nightfall on the first day, the streets of Kyoto were silent again. As if, except for lingering wafts of smoke, nothing at all had happened. Yasuke's wounds and memories of the morning battle seemed like figments of his imagination. It made feverish sleep even more confused.

The head of the mission was still Father Organtino, but he'd remained in Azuchi as principal of the new seminary there. It was Father Fróis, the translator who'd first welcomed Yasuke and Valignano to Japan, who took command in Kyoto in the absence of the superior. He was a brilliant and kindly man, but lacked the authority to make key decisions as the Jesuits fretted over next steps and the fate of their brethren in Azuchi.

From the word on Kyoto's streets, the bulk of Akechi's troops were already halfway to Azuchi to seize Nobunaga's capital and greatest castle. They'd gone east, spending the night first in Akechi's fief in Sakamoto before, at first light, boarding fast sculling skiffs across Lake Biwa toward Azuchi. *What would happen? When would word of the Jesuits' safety arrive?* And then there was the matter of "their" man Takayama, the powerful converted Japanese lord who'd also been mobilizing to aid Hideyoshi and had his army ready in the field. *Under these new circumstances, who would he now back?*

Lord Akechi, in contrast to the agnostic Nobunaga, was a devoted Buddhist and the Catholics feared he might take a far dimmer view of the foreigners and their alien religion. The Jesuits made moves for years assuming Nobunaga and his heirs would rule Japan for the foreseeable future. All the years of hard work and graft were now lost and who they backed in the next few days and weeks could determine their place in Japan for centuries.

Two days later, Kyoto received word that the entire town of Azuchi had burned. And its extraordinary castle—Nobunaga's pinnacle of power and Yasuke's home—had been razed from the earth, its treasures and gold looted by Akechi's troops.

Yasuke had assumed the castle would hold as it was well garrisoned and virtually unassailable, but something had clearly gone wrong. Perhaps the defenders had panicked or the warden of the castle, Gamō Katahide, a senior, long-standing and trusted vassal of Nobunaga's, had judged the numbers insufficient to mount a proper defense. Yasuke and the Jesuits had imagined Akechi would simply seize one of the most glorious castles on earth. Instead, destruction. No one knew who'd kindled the flames—some said Akechi's men, others claimed that the garrison had done it to deny Akechi the satisfaction of its capture. Either way, the grand castle was no more. A pile of smoldering ruins was all that remained. Along with the castle, the Akechi soldiers burned, pillaged and raped their way through the surrounding estates and the town at the foot of the mountain. Yasuke's house, he assumed, was among the many destroyed, his servants murdered or fled.

There was no word yet on the seminary or the Jesuits in Azuchi. If they'd been killed also, it was only a matter of time until the Akechi rebels came for them in Kyoto. All they could do now was wait and pray.

Several days later, a ragged bunch of footsore refugees from the Azuchi seminary staggered into the Kyoto mission. They were headed by Father Organtino himself but escorted by a dozen Akechi soldiers. Organtino brought a more definitive narrative of the events at Azuchi, far more detailed than the ever-contradictory rumors which swirled around Kyoto.

When word of Akechi's coup came, the students and staff, around thirty in all, had narrowly escaped the attack on Azuchi by fleeing just ahead of the advancing enemy by boat to hide on tiny Oki Island in the middle of Lake Biwa. They'd not been alone; the garrison and townspeople had also fled wherever they could. Lord Gamō, the warden of the castle, had, it seemed, not given up after all. He'd rather weighed up his duty to fight to the last in defense of the castle, and his duty to protect Nobunaga's family (mother, wife, sister, kids *and* concubines). Father Organtino didn't know for sure, but thought it likely they'd escaped to Gamō's smaller and more easily defended Hino Castle fifteen miles away.

Although the Jesuits had first managed to save many of the irreplaceable articles in the seminary—their candelabras, a gilt crucifix, several devotional images, some textbooks and the more portable of the musical instruments—the boatman who took the Jesuit party to safety on Oki Island had then, cruelly, robbed them. They'd been left with nothing, starving and stranded on the small island with no way of escape.

Soon after, however, Akechi's men arrived on the island, "rescuing" the Jesuits and taking them back to shore farther down Lake Biwa. The Jesuits expected to be executed, crucified or set afire, the latest Christian martyrs. Instead, to Organtino's astonishment, he was brought before Akechi himself at Sakamoto Castle and solicited to use Jesuit influence with Takayama Ukon to persuade the powerful warlord to support Akechi's coup. For without Takayama's backing—military, pecuniary *and* political—the overthrow wouldn't last long. Akechi, alone, did not have enough samurai or regular troops, and he needed support from other senior Oda vassals to solidify his position and swell his ranks.

Organtino had little choice. His and his students' lives were unmistakably in jeopardy. At once, he sent a formal letter in Japanese to Takayama through Akechi's messenger, recommending that—for the sake of the church—Takayama support Akechi's revolution. However, Organtino also managed to slip in another message, *in Portuguese*, which only Takayama could read. The secret message suggested Lord Takayama follow his conscience, supplanting any recommendation received in the Japanese missive.

In return for his perceived cooperation, Father Organtino and the Jesuits were given safe conduct back to Kyoto by Akechi's men.

For days, Yasuke had suspected why he'd been spared when all his comrades were killed. Now he'd confirmed it. Akechi's mercy had been one of several gestures of goodwill to the Jesuits to secure their support and influence with Takayama.

But Takayama and the Jesuits were only a small part of a much bigger picture. Akechi—it was whispered throughout Kyoto—had made little progress with the emperor, despite a generous donation of silver. Nor with the other warlords throughout Japan. Even his own son-in-law, Hosokawa, refused to join him; the stain of his treachery was too great. Nobunaga had indeed been loved, if also feared, by most who knew and served him. Yasuke was not alone.

And, Tokugawa Ieyasu (who'd traveled to Fuji with Yasuke and Nobunaga), perhaps the most powerful player still on the board and the man whose military and moral support could make or break Akechi, was on a jaunt to Sakai where Nobunaga had sent him to express gratitude for his support and hospitality. When the coup occurred, Tokugawa was isolated from his fief, *and* armies, in the east and vulnerable to coercion should Akechi catch him. Tokugawa knew only one way to make it home alive and retain his independence. He renewed an old acquaintance by hiring the infamous ninja leader, Hattori Hanzō—a man who only the year before had led the resistance *against* Nobunaga in Iga—to smuggle him through the Akechi forces who sought him. The long route they took passed through the ruined province of Iga but the ninja held his side of the bargain and managed to keep Tokugawa safe. At one point, the boat Tokugawa was hiding in under a cargo of rice was speared by an Akechi patrol hunting for him. The spear caught him in the leg, but Ieyasu was (or, so it's told) quick thinking enough to wipe his blood from the steel before the Akechi man retrieved his blade. Within a few days of the coup, Tokugawa had made it back to his home province of Mikawa alive but, one of his senior vassals, Anayama Baisetsu, who had been ordered to take a different route so as to confuse the enemy, was not so lucky. He was caught and executed.

Nobunaga's iron grip had brought peace to the capital and its surroundings, the first real peace for one hundred years. Across Japan, Akechi's coup was not embraced or popular. Instead, everyone simply waited for what came next.

Just over a week later, in early July, Yasuke and the Jesuits received word that Akechi Mitsuhide was dead.

General Hideyoshi—more than one hundred miles away from Kyoto and still embroiled in a fight against the Mori clan—had been assumed to be unable to take part in toppling the usurper; Akechi figured he could deal with him later. But, a messenger from Akechi to Lord Mori proposing an alliance—sent *before* Nobunaga's death—had been intercepted by Hideyoshi's men only a day after the coup. Hideyoshi reacted quickly, first offering the commander of the besieged Mori castle, Shimizu Muneharu, generous surrender terms, which allowed his family and the defenders to keep their lives in exchange for Shimizu's immediate seppuku. Shimizu, who had no idea yet about the coup in Kyoto, accepted, and duly sacrificed himself in full view of both armies on a boat in the middle of the artificial lake Hideyoshi had built to surround the castle. The castle now taken, Hideyoshi marched his army—some thirty thousand men—covering an impressive twenty-five miles per day, to meet Akechi, before most of Japan had even gotten word of the coup.

Meanwhile, Takayama had followed Father Organtino's Portuguese message, and his conscience, and brought his own three thousand men to join the anti-Akechi forces forming under

The siege of Takamatsu, where Hideyoshi used extensive earthworks to change the course of a river and turn the castle into an island.

Hideyoshi's leadership. The two lords marched together on Yamazaki, a village just outside of Kyoto. Here, Akechi, realizing his attempt at a truce had failed, prepared to make his last stand against the combined Oda forces, taking up a defensive position behind the Enmyoji River. Akechi's forces had declined to fewer than ten thousand, dwindling by the day as his position weakened. Those lords he'd counted on to join his efforts had declined the invitation and most were now formed up *against* him. Hideyoshi and his allies arrived on the field with up to forty thousand men. Akechi knew he was doomed.

The very first night, Hideyoshi sent hired mercenary ninja across the river to set fire to the Akechi camp, spreading chaos, confusion and dread among the defenders. He, also under the cover of night, sent a force of musketeers to control the high ground before the battle had even begun. The next morning, Hideyoshi and his allies' troops routed the Akechi army in under two hours. Akechi's troops died in the thousands along the river and in Shōryūji Castle, Akechi's headquarters, where many had retreated. Others fled. Akechi himself managed to escape the battle too, no honorable suicide for him, and attempted to flee to Sakamoto Castle, his main fief and final stronghold. He didn't get far.

As Akechi sneaked through a bamboo grove on the outskirts

of Ogurusu, a tiny village, he was speared by a common bandit with a simple bamboo spear. He died in a muddy ditch. A mighty comedown for one who'd gambled that he could take the entire realm, and a fitting end for the traitor.

Akechi had "ruled" Japan for thirteen days.

Now that General Hideyoshi had broken the coup, Yasuke waited in vain for a message from any of the remaining Oda sons or Nobunaga's former vassals. None came. Weeks passed, then months. The others were all too preoccupied with the chaos that had engulfed their own lives. For Yasuke was not the only one who now found himself in a precarious position, in fear for his life. Every man who'd served Nobunaga was now unsure of his future.

In July of 1582, a conference of leading Oda retainers was called in Kiyosu, the original stronghold of the clan, to discuss the succession. The course was not clear. Nobunaga's assumed heir, Nobutada, was also dead. No consensus could be reached over which of the several surviving Oda brothers should succeed. Yasuke assumed that Nobunaga's second son, Nobukatsu, would automatically take his father's place, but his rash nature did not endear him to some of the senior advisors present, most of whom preferred the *third* son Nobutaka.

Instead, Hideyoshi in a masterstroke which nobody else had thought of, proposed Nobutada's infant son (Nobunaga's grandson!) as the immediate heir. Who would dare argue with such direct and legitimate lineage—the firstborn heir's heir? Hidenobu, all of two years old, became Oda clan head, thereby ensuring Hideyoshi was the real power in the clan.

Technically, Yasuke remained a samurai.

But, one without a master, a *ronin*. He needed to find a new lord to serve or find a different path. He was eager to serve under Hideyoshi, a general he'd always respected. However, Hideyoshi had his own loyal men to promote, and most of the Oda samurai did not figure in his plans; even the new infant clan head was only a puppet to give initial legitimacy to his rule.

Hideyoshi never did call for Yasuke. Nor did anyone else.

The Jesuits didn't really know what to do with him either.

If anything, with Hideyoshi's new men in ascendance, Yasuke's close association with the Oda became something of a liability. Yasuke was a potential embarrassment to Jesuit relations with the new regime and needed to be moved out of the way so as not draw unwanted attention during the uncertain times.

For the next few months of 1582, Yasuke remained a virtual prisoner in the church compound. The curious in Kyoto still tried getting a peek at Nobunaga's "black man" but the Jesuits tried to play it down. Yasuke spent most of his days inside. Sweeping, doing odd jobs, helping to prepare food. It was maddening; after being at the center of the realm's affairs, he was confined within the four tall mission walls.

The answer, finally, was to remove Yasuke entirely from Kyoto, from the theatre of war and politics. It's likely he accompanied one of the regular Jesuit groups on their travels between their capital mission and the Jesuit stronghold in Nagasaki. With them, Yasuke would have retraced the steps of years before: through Osaka, Sakai and the pirate-infested Seto Inland Sea to find himself back in a further-developed Nagasaki, now surrounded with formidable stone walls, bristling with cannon and manned by the citizen militia.

A year passed. News from central Japan revealed that Hideyoshi was cementing his grip on power. Those Oda clan members who'd been dissatisfied with Hideyoshi's power grab, were picked off one by one, killed in battle or forced to perform seppuku. Nobutaka, the third son, cut his own belly, leaving behind a vengeful death poem:

You have caused the downfall of your rightful lord,
May you get your just deserts, Hideyoshi.

Yasuke was back to military life, spending his time in Naga-saki on garrison duty, cleaning and maintaining the guns, train-ing new gunners, wearing down the whetstone as he continually sharpened his weapons, and watching for enemies who hadn't come yet, but were getting closer every day.

War, as always, brewed in all the regions bordering Nagasaki.

From the south, the Satsuma clan had managed to domi-nate nearly half the island, and to the north, Ryūzōji Takanobu (Arima and Ōmura's old Catholic-hating foe) regularly made incursions into the Christian lands that formed Nagasaki's hin-terland. (Ōmura had essentially submitted to Ryūzōji's domina-tion and most of Arima was also now in his hands.) Lord Arima, the teen warlord, had barely been holding out against Ryūzōji's offensives ever since he'd welcomed Valignano and Yasuke to Japan three years before. Even with Jesuit aid—in the form of funds, weaponry, lead and gunpowder—it was never enough. Arima was no closer to reclaiming the lost portions of his do-main than when Yasuke had first arrived. And the young lord was desperate. Ryūzōji, it seemed, only grew stronger with each passing year.

In 1582, with Nobunaga dead and Japan appearing to be heading into another hundred years of civil war, Arima had decided to play a lesser evil against the greater and submitted to Satsuma overlordship in exchange for an expeditionary force to help reclaim his lands from Ryūzōji. At least then, he and his people would know peace and be allowed to practice their Catholic faith. (Though the Satsuma were not known for their pro-Christian sentiments, they, at least, *tolerated* the religion, whereas Ryūzōji did not.) The Satsuma clan took their time arriving, more than a year, but eventually brought more than eight thousand men to fight beside Arima's remaining forces.

Together, in December 1583, the new allies advanced to the town of Shimabara, one of the most important settlements in the Arima domain, long since taken by Ryūzōji. There, they attempted a siege of the garrison. The entire operation, however, was a mess. Ryūzōji's men held out against overwhelming numbers and dispatched to their lord for help. In response, Ryūzōji assembled a huge force of twenty-five thousand and set out overland to relieve his beleaguered garrison. By now, it was the spring of 1584. No one had expected a force of that size to be sent by a minor regional lord. And they were not only superior in numbers, but also had managed to acquire five hundred muskets, a massive number for regional warfare.

According to Fróis, Ryūzōji vowed, "the first thing he would do for fun after returning victoriously from the battlefield, was to order the crucifixion of the new mission superior, Gaspar Coelho," who had taken over from the hapless Cabral in 1581, and "give the port of Nagasaki to his soldiers to be sacked and destroyed as a reward for their trouble." The direct threat to destroy Nagasaki and the Jesuits was no idle banter. Ryūzōji had threatened this for years and his shadow had long loomed large over the mission. Now was his best shot at its destruction once and for all. No doubt, Yasuke himself would not fare much better than Coelho when Ryūzōji's forces eventually arrived. Ryūzōji's hatred for Coelho was well earned.

Coelho was a thin, small, weak and cunning man in his mid-fifties, and his provocative behavior would soon lead the Jesuits to many of their problems in the years ahead. The new mission superior had for long been the biggest advocate of destroying temples and shrines, to the extent that he even involved children in his plans, inciting whole families to go on Buddhist-destroying day trips, setting fires wherever they found signs of the infidel. Those who would not convert were expelled from Jesuit-influenced lands and monks who refused to become Christian forfeited their lives. Ryūzōji was not impressed.

As far as the Jesuits were concerned, they were staring the end of their mission in the face. Without Nagasaki, and with all their key personnel dead, the mission would find it hard to even survive, let alone spread. So when Arima once again turned to the Jesuits for help against Ryūzōji, they were only too happy to oblige. The alternative was likely the end of Catholicism in Japan. Their involvement effectively meant they would *also* have to bend their knee and accept Satsuma vassalage, but even that was better than extinction. (And, perhaps God, in his greatness, would intervene and soon convert the heathen Satsuma clan to the one true faith.)

The Jesuits put all of their limited military resources into supporting one of their oldest allies. They sent to Arima their private war galley, with three hundred, almost all, of their Catholic militia aboard, crucial supplies of gunpowder and lead, food and more guns.

And Yasuke.

CHAPTER TWENTY-TWO
The Guns of Okitanawate

Yasuke was sent to the Arima camp specifically to handle the two new cannon which the Jesuits had delivered to their allies. These guns were the very same *kunizukushi*, "nation destroyers," Valignano had offered to Arima all those years ago to help him find his faith.

The order had taken more than a year to get to Goa, several more to process, and another year to get back. Arima was just now receiving the cannon and no one in the Arima camp yet knew how to use them. These two bronze cannon, nearly nine feet long and of 3.7 inch caliber were the breech-loading type, relatively rare in an age where muzzle loaders were more common, and virtually unknown in Japan. Breech loaders were quicker to reload (at the rear of the cannon) and several charges could be prepared in advance, giving a near-continuous rate of fire of cannonballs or grapeshot if handled correctly.

Despite the high rate of fire, the relatively primitive nature of foundry technology meant that a tight fit for the breech was impossible, and the weapons lost a lot of power when fired. Hence they were little use against fortifications and ships, but quite deadly at close ranges against humans, and could reduce unprotected armies to bloody piles of cadavers far quicker than the slower, but more powerful, muzzle loaders. The two new guns were also mounted on swivels and could be aimed very quickly. Muzzle-loading cannon, meanwhile, normally had to be mounted on wheels or immovable breastworks and took a lot of effort, men and time to move.

Yasuke had been around cannon often in India and on ships, and worked with the smaller ones during his time in Nagasaki. He'd been taking *kunizukushi*-type weapons out for test runs during his year there. There was no one for a hundred miles who knew more about these weapons.

The Jesuit expedition headed inland through the mountains forming the backdrop to the port of Nagasaki. Yasuke traveled with this force of Christian samurai, porters, carpenters, smiths, camp followers and packhorses with the cannon strapped fast to their backs. The company of several hundred strode gladly together through the rough mountain passes and along the Arima domain's coast, before traversing the savagely beautiful Unzen volcanic mountains which would lead them to the battlefield on the other side of the peninsula. Here, they passed bubbling volcanic pools hissing steam, and endured the hellish stench of the acidic sulfurous waters for hours before emerging again on the mountains behind the normally small, sleepy fishing town of Shimabara, to support Arima and his Satsuma allies. The town, located in a wide bay and enclosed by two forested promontories, did not look particularly well defended. Some rough stone walls and a flimsy-looking wall of bamboo stood atop earthworks.

The allied army was camped in the narrow space between the green mountains and a sea of deep blue, a short distance from the town's wall. Sandy spits twisted out from the beach around which numerous small sampan boats bobbed. Men who'd stripped down to their loincloths continuously unloaded supplies, and now waded through the shimmering shallows bent under the weight of bamboo frames filled with stores on their backs. As the boats were relieved of their supplies, they cruised again across the narrow Ariake Sea to the Kyushu mainland or headed south to Kuchinotsu to pick up more ordinance and provisions.

As the Jesuit reinforcements first arrived, they were welcomed with cheers of *banzai* by the Arima–Shimazu troops. Earlier, when the Christian Arima men first mixed with the more numerous Buddhist Satsuma troops, they'd endured a fair share of mostly good-natured anti-Catholic baiting. Although they were now allies, Satsuma forces had fought Catholic armies many times before and old habits die hard.

Middle-ranking samurai commanders led the way to the ground that had been prepared for them. Yasuke was looking forward to this action, an exciting change from garrison duty which had become rather monotonous; he could finally be useful again. He was also keen to see what these new guns could do against a real army.

Yasuke had, as his second in command, a Japanese Christian samurai with the baptismal name of Martinho. Martinho had fled Ryūzōji's anti-Catholic purges in Ōmura and escaped to fight against his persecutor in any way he could, a typical member of the Nagasaki militia. Martinho also had some experience of these breech-loading guns, from the walls of Nagasaki where he'd been stationed with Yasuke.

They arrived at the Arima camp during Easter week, in early April of 1584. A mere three years before, another Easter week had forever changed Yasuke's life. One way or another, it seemed it was to be again.

★ ★ ★

Additional reinforcements from Nagasaki arrived a few days later by sea on the Jesuits' war galley, a sixty-oared and double lateen–sailed craft called a *fusta*, which dwarfed the tiny sampan boats. The ship was captained by the former-soldier-turned-Jesuit-brother Ambrosio Fernandes, the very man who'd met Valignano and Yasuke at sea prior to their arrival in Japan. The Jesuit *fusta* was probably the fastest ship in Japan, and crewed by Japanese-Christian members of the Nagasaki militia. For armament, the warship had two small cannon at the front and stored huge numbers of muskets for the entire crew, which could run to three hundred men when full. After offloading their supplies at Arima's camp, Brother Fernandes rowed the *fusta* galley to rendezvous with Arima's own naval forces, the many sampan which were still gliding in and out of the bay, farther down the coast at Kuchinotsu.

The arrival of Ryūzōji's force was imminent, although nobody was sure how large it would be.

The joint Arima-Satsuma allied forces of around nine thousand samurai were under the command of Arima Harunobu (now seventeen) and Shimazu Iehisa, the third son of the Satsuma clan lord, Shimazu Yoshihisa.

The news from the scouts came in, Ryūzōji's army was estimated to number more than twenty-five thousand, three times the size of the Arima-Satsuma alliance and far beyond what anyone could have predicted.

Iehisa, by virtue of commanding the most men, held the higher command and immediately decided upon a strategically advantageous frontline, extending from the coast to the foothills of the massive Unzen Mountains that lay above Shimabara at a place slightly up the coast called Okitanawate. There, they made their camp and dug in to prepare for the assault. They couldn't match the enemy numbers and a battlefield of their own choosing was their only real hope.

To Yasuke's eyes, this was an amateur affair compared with the professional armies of the Oda, more akin to the eager, excited chaos of an Indian army than the well-drilled, armored and disciplined troops of Nobunaga. Apart from a core of leading samurai, the lower Satsuma ranks were decidedly part-timers, farmers mainly, and only lightly armored, though clearly more than handy with their weapons. All had spears, and most *also* carried the two swords of a samurai and considered themselves, rightly or wrongly, the best swordsmen in the land. They stood proud under their white banners with the distinctive black cross within a circle, the *marujuji*, which signified their southern clan. The Satsuma men—despite nearly four decades before having been the first clan to ever use guns in battle in Japan—had relatively few muskets and still relied heavily, instead, on the longbow, spears and their famed swords.

Arima's men, meanwhile, had Jesuit supplied European-made guns, less precise than the Japanese ones as each weapon had a unique bore and could only use ammunition made specially for that one gun in a tool that the gunner carried at his waist. The Japanese ones were made to standard bore sizes and hence ammunition could also be mass produced and shared. This made them more effective because they ran out of projectiles less often. The European weapons were also slightly more unwieldy than the Japanese guns as they generally had to be placed on rods (firing sticks) before firing. The Japanese firearms had been especially modified with shorter butts to fire *without* the rod, although aim improved when they were mounted.

The majority of the Arima forces were not dug in on the hillside but remained on ships in Kuchinotsu awaiting the call, their guns ready mounted on the sides of their vessels. When it came, they'd have a special surprise for Ryūzōji.

Yasuke and Martinho busied themselves with readying the two large guns. These formed the bulk of the firepower available to the land-based Arima-Satsuma forces and would need to perform if they were to stand any hope of repelling the coming attack. Yasuke and Martinho had been assigned a team to help, and quickly went about the task of preparing positions which allowed both swivel guns to have maximized clear ranges of fire and some protection against enemy bullets and arrows.

The troops didn't seem to even blink at Yasuke's presence. They knew he was a professional, knew he was there to help them save their homeland from destruction. Besides, they were relatively familiar with black men from the docks and their lord Arima now had at least one in his entourage whom he used as a messenger and interpreter when dealing with the Portuguese. The Jesuits' close involvement in the area meant these otherwise-rural peasant samurai were far more cosmopolitan in some areas than the sophisticates of the capital city who'd mobbed Yasuke three years previously, almost to the day.

When all was in readiness, Yasuke inspected his two beauties slowly, knowing how many watched his every move. Any chance they had of holding off the impending invasion, everyone understood, rested within these two cannon. He pronounced himself satisfied with the positions and then he and his team got to work.

Throughout the next few days, they set about getting to know their new weapons better and molding lead ammunition. Yasuke assigned men to the jobs required to run a cannon: aimers, loaders, mallet men to ram in the chocks and an ordnance-preparation team. He taught them how to prepare the charge and keep a production line going so the rate of fire was uninterrupted. But there is never any substitute for actual experience, and with Lord Arima's permission, Yasuke directed his team in limited live-fire practices. Thanks to the Jesuits, there was a vast stock of powder and shot, and the ground below their bastion soon became churned and pocked with small craters. Under Yasuke's direction—trying a method he'd seen used to devastating effect—they lowered the cannon barrels so the lethal shot grazed the ground, bouncing on a murderous path for hundreds of feet before stopping.

Each display of deadly firepower was watched by Lord Arima himself, and accompanied by loud cheers from the defenders on the mountainside. As the rate of fire grew quicker, the sound of the guns gave them a newfound strength and confidence.

The end of April approached, and the scouts' updates on the proximity of Ryūzōji's massive army became more frequent. Lord Ryūzōji Takanobu himself was so fat he could not walk, nor could any single horse bear his weight, so he traveled in a palanquin borne by six men. This rigmarole slowed down their progress considerably, and also provided a degree of mirth for the defenders, increasing their confidence even further.

On May 1, the enemy army's advance units finally came into view and settled around the town of Shimabara. As Yasuke and his team watched, troop after troop arrived, spread out and made camp, never sparing more than a glance for their heavily outnumbered foe dug in above. And they kept coming: the twenty-five thousand warriors, and thousands of support personnel, porters for the most part, but also men and women to carry out every other service the samurai needed. The vast host was still arriving the next day as the enemy swarmed around the plain like angry hornets, in preparation for their impending assault.

From Yasuke's position, it was clear there would be three fronts: one along the beach, one along the main road leading directly to his guns and the last which would attempt to outflank them on the steep and densely wooded hillside. The third front, through the steep forest, would have the hardest going and take the longest to arrive, but might also be the most dangerous to Yasuke's side if Ryūzōji's men succeeded in making their attack from the flanking position and managed to cut off the defenders.

That night, Yasuke and his team prayed together.

His men were fervent Catholics and desperately believed that

God was on their side. As their leader, he gladly led them in the Lord's Prayer time and again. When a Jesuit brother came on his prebattle rounds, the men said their confessions and received his blessings.

Two days later, in the morning, the attack came.

As the first men formed their battle lines, Ryūzōji's *taiko* drums roared their challenge, and the conch shells sounded the age-old Japanese signal to advance. The African soldier found a grim smile. Whatever the numbers against them, it was remarkably better odds than the *last* battle he'd fought. And he was a lot better prepared this time. Ryūzōji's men advanced from all three positions steadily under their banner of two apricot leaves. Ryūzōji himself was in the middle-road column, his palanquin curtain opened so that he could direct the battle personally, wielding his war fan with gusto, urging his men onward. Ryūzōji was accompanied by his young lover, a boy of no more than sixteen, gloriously dressed in the finest armor by his lord and mounted on a sleek black horse beside the lacquered palanquin.

The gigantic columns of soldiers ran over a mile long, but the gunners had been concentrated at the front. As they got within range, the five hundred large caliber muskets fired at once in a massive crack of thunder that hovered as one long hedge of thick smoke in the tree line. The dug-in defenders braced themselves and kept their heads down beneath the defenses as the hot balls of lead buzzed past them. Men fell as round shot from the more powerful guns pierced the thin wooden walls or ricocheted off stones in the earthen ramparts; others spun about wounded and screaming. Hundreds fell.

Yasuke expected there to be a near immediate second volley of fire, as Nobunaga's troops had always kept the bombardments coming continuously, but it never came. All the guns, instead, had fired at once, forming a fierce and deadly hail of fire, but then they'd all paused to reload at the same time. It took several minutes, and he knew that when the next shots came, they would be ragged, the more inexperienced gunners taking longer to reload, aim and fire. The devastation of that first massive salvo would not be repeated. Nobunaga's trick for rapid continuous fire had not been learned by Ryūzōji, it seemed.

It was their chance. Yasuke jumped up, ordering his men to start their own bombardment in the momentary lull. Their well-practiced routine went smoothly into action: load, hit chocks into place, aim, take cover and pray and then Yasuke himself placed the match. As the team saw to the loading, Yasuke was left free to aim and fire both guns himself; he just needed to cover the few paces between them. Arima's two cannon roared, one at the Ryūzōji men on the hill and one at the enemy troops advancing up the road. A light sea breeze wind wafted the cannon smoke away and Yasuke could see clear gaps in Ryūzōji's ranks. A massive cheer went up from the defenders on the hill. *Could they get another shot in before the enemy gunners had reloaded?* They did, but soon the surviving musketeers in the attacking ranks had managed to reload and they let off their weak and untimed volley, still defenders fell all around, including some of the men in Yasuke's team, who'd been concentrating so hard upon their job that they had forgotten to take cover.

Thus went war. Yasuke shouted encouragement and reminders in Japanese against wasting powder and the remaining men (and the new replacements) went back to their deadly work. Already, hundreds of Ryūzōji's men lay scattered in heaps of splintered bone, entrails and shattered weapons.

Below, Shimazu—feigning retreat, a ruse his clan had mastered over the years—had successfully lured the bulk of Takanobu's road-column troops into a marshland directly between the river and Yasuke's hilltop. Here, their numbers meant little, thousands of men trapped in sluggish terrain.

Meanwhile, approaching the beach column was the flotilla of

Arima ships bristling with muskets and bows, led by the massive Jesuit *fusta* galley spitting lead and man-mauling grapeshot from the two small bow-mounted cannon similar, in all but size, to the two giants Yasuke was operating. With seeming impunity, the Arima seamen and the Jesuits' Nagasaki men blasted away at the beach column. With each volley, blood soaked the sand. Screams of terror and rage filled the air, but Fróis tells us above it all could be heard an unlikely sound. The gunners were "piously kneeling down with their hands turned toward heaven, they began reciting 'Our Father, which art in Heaven, hallowed be they name...' The first phase of the strategy having thus been accomplished, turning impatiently to load cannon with balls, they fired with such force against the enemy that with one sole shot the whole sky could be seen to be filled with limbs."

The beach was littered with fallen warriors, bodies bobbed beside the boats, but the Nagasaki gunners continued their deadly fire, and the waves that lapped the shoreline were edged pink with bloody froth.

Atop the hill, sweat and smoke stung Yasuke's eyes and he cuffed it away. The beach column may have been near collapse by now, but the other two were coming ever closer. Yasuke's team kept at their grim task knowing their lives depended on it. Ryūzōji's men were so close now, it was a massacre, but they kept on coming.

Suddenly the beach column disappeared, they'd had enough. The Ryūzōji men at the back turned and fled inland, anywhere to escape the deadly fire from the sea. The remains of the front of the column ran to the road and joined that attack, now even more men were bogged down in the marshland as the Satsuma charged from their positions and down the hillside. One flying column of samurai headed straight to the palanquin from which Ryūzōji himself was directing the sluggish assault. It was a glorious sight, the lacquered black armor shone in the spring sunshine as they charged forward, cutting down the weakened enemy with their drawn swords and leveled spears. This was the break Yasuke had waited for, and he silenced his guns; his men slumped down around the bastion and took deep drafts of water from their bamboo drinking flasks.

Directly below, Ryūzōji, atop his palanquin was run through with a spear by one of the Satsuma soldiers. He'd been so confident of his own safety, that he had assumed the approaching soldiers *must* be his own men. Moments before he was killed, he even chastised "his men" with threats, berating them for their indiscipline. Then, his litter's carriers were cut down and the Satsuma soldiers surrounded the enemy lord. As Ryūzōji felt the steel, he heard his killer say, "It is *you* we all came to get." The warlord fell from the palanquin to the ground and, as the Satsuma warrior took his head, Ryūzōji prayed, *"namu amida butsu"* ("I take refuge in the Amida Buddha") which he fervently believed would ensure his entry into Buddhist Nirvana. His young lover embracing the fallen lord, *also* lost his head, the two united in death as they'd been in life.

And suddenly all of Ryūzōji's men were running, the diminished host fleeing over the fields, some back into the ships' field of fire and others up the mountain and through the fields that surrounded Shimabara. The force that had so confidently advanced until only minutes before was now gone, replaced by chaos and killing.

Yasuke ordered his men back to their work. Aiming long, they felled as many of the swiftly retreating Ryūzōji soldiers as they could, before once again the guns of Okitanawate fell silent. Ryūzōji's dead lay in skewed piles of bodies; it was impossible to know the butcher's bill but it must have run to over ten thousand.

In the aftermath of the battle, the people of Arima and Nagasaki celebrated and gave thanks to God for their deliverance. Masses were held, bells pealed out the happy news. The threat of Ryūzōji, so long hanging over them all, was gone. In the villages and towns of Arima and in the city of Nagasaki, people filled the streets to dance, the brothels did a brisk trade, teahouses ran out of food, sake casks swiftly emptied, the first bounty of spring was enjoyed and the world seemed a different place. It seemed a huge victory for Catholic Japan; surely the rest of the country must soon follow the calling to Christ.

Honors and material prizes of war were given to all. The Jesuits, for their role in the victory, were bestowed a huge tract of land bordering Nagasaki to the north, including the large village of Urakami. Their foothold in Japan had grown considerably.

And Yasuke, as one of the main figures in the battle, must also have been well rewarded with a handsome bounty of treasure. He was again celebrated, valued. And, if those accolades would soon pass—as he well understood—he now had several pouches full of silver and long strings of copper coins. And with them, the power to bankroll more options and a world of future possibilities for the first time in many years. A man or woman with money was their own master.

In an age when few had options, the African samurai now had many.

CHAPTER TWENTY-THREE
Possible Paths

There is no verified record of Yasuke after the autumn of 1582. It's possible Yasuke was killed in the Battle of Okitanawate. It's also possible that the African warrior sent by the Jesuits may have been another man entirely.

Historic conjecture suggests otherwise. It's highly likely Yasuke *was* the African gunner at the Battle of Okitanawate and, as likely, that he survived.

But, we must officially now concede to the spoils and muddle of time and delve into the world of historic detective work to find possible paths for the African samurai. The search begins with verifiable accounts of men who *may* be him: African men not simply visiting temporarily (the sailors), but residents in Japan, carrying out various roles in the years and decades following Nobunaga's fall. Our story continues with them.

Yasuke may or may not be one of these men, but there were so few Africans in Japan at this time, that at least one of them is likely to have been him, and their lives and stories do mirror the most likely avenues he would have taken.

The Battle of Okitanawate put the island of Kyushu firmly in the hands of the Satsuma clan, but the real struggle for Japan's second largest island was only just beginning.

As part of the spoils of the battle, the Jesuits had gotten their hands upon new lands and were able to strengthen their hold over the original territory they held around Nagasaki. The local populations within, however, weren't so quick to accept new management. While many of the people of the new Jesuit lands were already Christian, they'd never actually been subject to Jesuit governance. Regardless of the gifting Arima may have done, there were still plenty of local power brokers, minor samurai and village headmen, who were *not* Catholic and bridled under alien rule.

Coelho, the Jesuit mission superior, was a soldier at heart, and headstrong, and thought he knew exactly how to deal with unruly peasants. While the Jesuits tried their best to win over their new wards' hearts and minds by fair means, some military backing went a long way. However, teaching Japanese peasants a lesson was not always an easy job. They could be violent, and there is a rich history of overbearing samurai getting their come-uppance at the end of various sharp farming and fishing implements or rusty weapons scrounged from the dead on battlefields. (Akechi Mitsuhide hadn't been the first samurai lord to end his life on the tip of a peasant's bamboo spear.)

Coelho sent his increasingly powerful and well-armed military to pay insubordinates a visit, to ensure their obedience and submission. This military force would, were he still alive, likely have included Yasuke. To these new territories, the Jesuits brought along their deadly *fusta* warship, manned by the militia, as backing, and spent the months after the Battle of Okita-nawate demonstrating who now was in charge.

Yasuke and the other militia men would have accompanied priests, brothers or acolytes ashore to various villages. While the Jesuits politely explained the new governmental arrangements, the soldiers stood nonchalantly in the background. The message was clear. And, whether the locals liked it or not, one or more of their young folk were taken to Nagasaki to live with the Jesuits and taught to be good Catholics. The Japanese villagers called them "hostages." The children were taken from more than twenty villages.

Meanwhile, the Shimabara peninsula to Nagasaki's south had fallen back under the full control of Arima, but in name only. All of Arima's decisions were now subject to Satsuma approval and the Satsuma clan left a garrison of hundreds in Arima's lands to make sure that remained the case.

And Arima's neighbor, Lord Ōmura (who'd given Naga-saki to the Jesuits four years before), had been under Ryūzōji's overlordship, but with Ryūzōji's death, he *also* now recognized the Satsuma clan as his new liege. This had major consequences for the Jesuits. From the Japanese political perspective, the Jesuits (residents in Ōmura land), were perceived *as* Ōmura's vassals and hence vassals of whomsoever *he* recognized as *his* overlord. In 1584, after the Battle of Okitanawate, the Jesuits, therefore, were also forced to submit to Satsuma power.

In evidence of this new arrangement, within a month of the decisive battle, the Satsuma had installed an overseer of Nagasaki, Uwai Kakken, and insisted the Jesuit military *and* diplomatic activities cease forthwith. The Satsuma were not at all happy with the Jesuits exerting any kind of military force, nor trying to build Christian coalition armies which could potentially one day threaten their new domination of northwest Kyushu. Priests, the Satsuma argued, should stick to priestly endeavors—especially *foreign* priests. Each week, the Satsuma troops arrived in Nagasaki in ever-increasing numbers. For the first time since its founding, the Catholic city was now under the suzerainty of non-Christians.

Meanwhile in Japan's heartland, the Kyoto region, competition to succeed Nobunaga as national hegemon was coming to a head. Hideyoshi's stroke of genius in taking over control of the Oda had still left Tokugawa Ieyasu as a powerful independent lord and the disgruntled anti-Hideyoshi Oda forces appealed to Tokugawa to combat Hideyoshi's rise. The resulting series of battles in 1584 led to an eventual stalemate, but Tokugawa realized he'd come out worse in the end and—in an exchange for hostages, which included Hideyoshi's own mother, in a gesture of goodwill—Tokugawa accepted the inevitable and agreed to become a vassal of Hideyoshi.

Nobunaga's legacy now belonged to Hideyoshi alone and those former Oda samurai who were not dead were now Hideyoshi's vassals, even Nobunaga's senior surviving son, Nobukatsu.

Father Coelho hoped to travel to Kyoto and petition Hideyoshi to intervene against the Satsuma, but Uwai outright refused to let the Jesuit travel. As far as the Satsuma were concerned, the days of the Jesuits stirring the Japanese pot toward their own ends were over. And as an anti-Catholic domain, they were adamant that the Jesuits would not extend their power and influence any further. The priest seethed under this Satsuma yoke, but until specifically forbidden from doing so, he spent the rest of the summer pondering how to reach Hideyoshi to obtain his support to balance out the Satsuma yoke, and trying to assemble an alliance of Christian lords to work toward enforcing Catholic domination on Kyushu.

Although Uwai kept his interference in the running of the city to a minimum, the Satsuma were concerned the Jesuits might appeal to the then seemingly pro-Christian Hideyoshi for support. If Hideyoshi came to the island of Kyushu, the Satsuma would meet him with blood and fire, but it would be best if Hideyoshi left the island alone entirely.

Had Coelho managed to leave Nagasaki for Kyoto in 1585, he'd likely have taken Yasuke, for the African samurai had met Hideyoshi several times at Azuchi during his residence there and would have been the perfect go-between. It was not to be, and Yasuke—grounded now with all the other Jesuit militiamen—would have remained on uneasy garrison duty in Nagasaki.

But where, before, Yasuke was trapped in the employ of the Jesuits, he now had more options. He'd been well rewarded for his role in Okitanawate by Arima and the Jesuits, and this led to the chance of a new life beyond the walls of Nagasaki. Sometime in the 1580s, suggested by the fact that no missionary source mentions him by name again, Yasuke seems to have left the service of the Jesuits.

With personal skills such as multiple foreign languages, extensive knowledge of foreign lands, peoples, cultures and of course soldiering, Yasuke would have been a very valuable man for any Japanese lord attempting to engage in the sphere of foreign trade for the first time or a ship's captain needing extra hands and protection from pirates.

He was also a relatively rich man. He would have had reward money from Okitanawate and perhaps retained some of the fortune that Nobunaga gave him. The dead lord of the Oda was known to be a generous master. Could Yasuke have reclaimed this fortune from the ashes of Nobunaga's death? Could it have survived the looting of Azuchi? The answer is that some of it probably *did* survive. In that day and age it was common to take much of your fortune with you, if it was portable. Yasuke probably carried at least the cash portion of his fortune with him at all times. After all, coins were specifically made with holes in them so that they could be hung on strings around the waist or on another part of the body. Samurai also commonly carried purses at their belts, and if the thirty-seven kilos of copper coins

Yasuke received as a meeting gift had been converted into silver or gold, they would have weighed considerably less and been even more portable. Akechi's rebel forces had thoroughly destroyed Azuchi Castle, but it's also possible Yasuke successfully managed to bury and hide some of his money there. We also have Yasuke's continued career to take into account. He accumulated pay, gifts and rewards in the course of service to the Jesuits and to Arima, and perhaps other Japanese lords as well. In any case, Yasuke would have been well bankrolled for the next chapter in his story.

That next chapter most likely, as with almost everyone in Japan in the 1590s, involved Lord Hideyoshi.

By 1586, Hideyoshi had become undisputably the most powerful man in the land, Nobunaga's true successor and along with a host of honors, he was formally bestowed with the new clan name Toyotomi by the imperial court. The Oda returned to their rural roots, and the "Toyotomi" ruled the nation from their new castle in Osaka, just south of Kyoto. Hideyoshi built it to surpass Azuchi Castle in every way, with eight stories, one more than Nobunaga had, the walls twenty feet thick and constructed of huge granite stones as large as ten feet wide and forty feet long. Three moats had to be crossed before reaching the center of the fortress. The decoration of course also surpassed that of Azuchi, and it is said that many of the embellishments were of pure gold, not simply painted with gold leaf. It quickly became one of Japan's most important political centers.

And it was Osaka Castle that Coelho finally managed to reach in 1586 where he had an audience with Hideyoshi that included a personal three-hour guided tour of the palace. Hideyoshi was buttering up the Jesuits because he wanted two Portuguese ships, and crews to sail them, and Coelho made a rash promise of compliance. The Jesuit was accompanied by the Jesuits' old friend Ōtomo Sōrin from Bungo, now rid of his former wife, "Jezebel," and living out his days in domestic peace. Ōtomo essentially remained the only other power on the island of Kyushu besides the Satsuma. The Japanese lord was quite rightly convinced that the Satsuma forces would turn upon him next, therefore, Ōtomo had considered the post-Nobunaga political winds and supplicated himself to Hideyoshi to ally himself with the most powerful force in Japan.

Hideyoshi was delighted; Kyushu was his last step to national domination anyway, and to be actually invited by a local lord would make things so much easier. Firstly, he sent a polite letter to the Satsuma asking them to return to their own lands and cease their thirty-year campaign to dominate Kyushu. The reply was scornful, and chose to emphasize Hideyoshi's low birth in comparison to the ancient samurai lineage that the Satsuma lords possessed. Predictably, Hideyoshi was not amused, and as the Satsuma troops crossed the border into Ōtomo's territory in November 1586, Hideyoshi's vanguard landed on the island of Kyushu to support their new ally.

Lord Hideyoshi's main force would take longer to prepare, but by the time it arrived in early 1587, it comprised the largest army *ever* seen on Japanese soil. The force comprised well over two hundred thousand men, provisioned by twenty thousand packhorses and supported by a vast fleet.

Hideyoshi's attack came in two main prongs. First, around a third of the army under the command of senior vassals "liberated" Ōtomo's lands and then headed south chasing the retreating Satsuma samurai. They met stiff resistance at several points along the way. Satsuma samurai had a well-deserved reputation for being among the fiercest fighters in all of Japan. The second prong headed down the western side of the island, past Nagasaki and Arima and into Higo, bordering on the Satsuma fiefs. This army was commanded by Hideyoshi in person, and supported by his fleet. Here too, the Satsuma had no option but to retreat, and Nagasaki came back under Jesuit control after its three years of non-Catholic occupation.

Coelho made the trip to Hideyoshi's field headquarters accompanied by some Portuguese merchants, the first time the warlord met non-Jesuit Europeans. It is quite possible that Yasuke went with him. Hideyoshi, like Nobunaga before him was fascinated by exotic things. In fact, in the next decade he even specifically requested African dancers perform for him on at least two occasions.

Hideyoshi was fascinated by the Portuguese mode of dress and weapons and acted extremely graciously toward the foreign merchants (shortly afterward, he requested that his tailors provide him with similar Portuguese attire, starting off another craze for European-style clothing among the elite). But there was little time for civilities; Hideyoshi had a war to win, and his army pressed southward. After a huge battle on the banks of the Sendai River, where Hideyoshi's one hundred seventy thousand men were held for a whole day by sixty thousand Satsuma samurai, the Satsuma retreated to their capital city of Kagoshima. The clan heads deliberated staging a glorious fight to the last man but, in the end, the Satsuma clan surrendered to reality and the Kyushu campaign was over. The entire island of Kyushu now belonged to Hideyoshi, who'd nearly completed the life's work of his former master, Nobunaga—the reunification of Japan.

Hideyoshi headed for home, but first requested another visit from Coelho to get an update on the ships he'd ordered at their earlier meeting. They met on the Jesuit's galley rather than on Hideyoshi's ground, a deep sign of respect to the Fathers, and Hideyoshi hinted that he desired it for himself. Coelho foolishly tried to bargain with the warlord, and outwardly Hideyoshi seemed to let the matter go. He also asked that a visiting Portuguese ship be brought to him so that he could see one for the first time in person.

But there was a problem—the ship was loaded with hundreds

of Japanese slaves and the captain judged, rightly, that Hideyoshi would not be impressed to see his countrymen tethered in the depths of the foreign ship. He apologized profusely and made his excuses, but Hideyoshi knew he was being lied to. That year he had seen conditions in Kyushu for himself, the power of the Church, the enforced destruction of native religious belief, places of worship. He had, for the first time, become cognizant of the vast enslavement of his countrymen as well. He was outraged, and the fact that the foreigners would lie to him or bargain with him like a merchant in the market, only made matters worse.

That same night, Hideyoshi drafted a promulgation that both banned the slave trade and ordered the missionaries to leave the country. Nagasaki, he took for himself.

The missionaries managed to avoid immediate expulsion, having disguised some Portuguese merchants in priests' robes to make it look as if the Jesuits were following Hideyoshi's orders to leave Japan. For the next decade, the Jesuits kept their heads down.

With the new leadership on Kyushu, a man named Katō Kiyomasa—a distant kinsman of Hideyoshi, and a young and ambitious lord with likely links to Yasuke—came to the island. Katō, twenty-five, had successfully risen, like Hideyoshi, from the lowest levels of peasant society, and just been granted by Hideyoshi part of Higo Province (modern Kumamoto) directly east of Nagasaki and directly north of Satsuma territory. Here, he was to keep a close eye on the defeated enemy and, no doubt, the wayward Christians too. He was already known across Japan as a soldier's soldier and, uncommonly for his time, disdained all the normal artistic pursuits of a samurai such as poetry and tea, choosing instead to lead a Spartan military life, practicing only the martial arts.

After claiming Kyushu, Hideyoshi completed his conquest of all of Japan in 1590 by defeating the last holdouts to his rule,

Nobunaga's erstwhile allies the Hōjō clan in Odawara (near modern-day Tokyo). Again, he attacked with overwhelming force, two hundred and twenty thousand to the Hōjō's remaining eighty thousand and although Hideyoshi's final foe was no pushover, after three months of siege and sporadic fighting, the Hōjō could see the outcome was inevitable and surrendered. Throughout the siege, Hideyoshi had engaged in somewhat unorthodox tactics for his last battle in Japan. While still ensuring the fighting fitness of his forces, he had treated them to lavish entertainments right under the eyes of the besieged foe. While the defenders dozed uncomfortably in their armor on the walls, Hideyoshi's men enjoyed their favored concubines from home, prostitutes, musicians, acrobats, fire-eaters and jugglers. All making as much noise as possible. For his part in the victorious siege, Tokugawa Ieyasu, Hideyoshi's most important vassal, was awarded control over the largely undeveloped lands around a large bay in the east of Japan (what would later become Tokyo).

Having claimed Japan, the warlord soon set his sights higher, deciding he deserved to sit on the Imperial Throne of China, and was even said to be aiming so far as the land of the Buddha's birth, India.

He fancied himself a modern-day Alexander the Great, heading *west* across the Eurasian landmass, and sent missives to neighboring states, Korea, Ryukyu (modern-day Okinawa), Taiwan, and the Spanish colonial authorities in Manila, demanding they support his ambitions to invade China. The Europeans were polite but firm in declining to help, and the Asian kingdoms either failed to answer or delayed again and again. The Koreans wondered who Hideyoshi even was, as they generally ignored goings-on to their east, considering the Japanese to be barbarians little worthy of attention. As Hideyoshi became more and more frustrated at the limit of his ability to project his power overseas, he decided to move unilaterally in his invasion of China. First though, to secure supply routes and a beachhead from which to attack the Chinese mainland, he needed to control the Korean peninsula. If the Koreans would not become his allies, then they'd become his subjects.

The ensuing Imjin War of 1592–1598 involved virtually the whole nation of Korea where, ultimately, three hundred thousand Japanese warriors faced the Koreans *and* more than one hundred thousand Chinese soldiers supporting their Korean allies. Hideyoshi's well-equipped, battle-hardened professional soldiers landed in the southeast, at the point nearest to the Japanese archipelago in 1592. The force consisted of one hundred sixty thousand samurai, twenty-four thousand muskets, seven hundred transport ships, and more than three hundred warships, and all the support staff and sailors that went with that. Hideyoshi himself commanded proceedings from his specially built castle in north Kyushu, where another eighty thousand-plus samurai waited in reserve.

The Japanese force met surprised, and badly prepared, Korean forces who largely lacked any experience of actual warfare and, crucially, had no muskets. Battles raged up and down the peninsula, but eventually led to a stalemate in 1593, after the decisive Chinese intervention with Japanese forces holding on to strongholds on the southeast coast and successfully, but only just, retaining their continental foothold.

Katō Kiyomasa led this invasion with his army of more than twenty thousand vassal samurai and traversed Korea from south to absolute north, laying waste to vast swathes of the peninsula. In 1592, his troops crossed the frontier into Manchuria—now part of China, but then the realm of the fearsome and warlike horsemen of the Jurchen people who turned them back. Then Katō's army fought their way back south again in the face of an onslaught by Korea's Chinese allies, and an increased Korean guerilla insurgency. The distance Katō's army crossed was roughly equivalent to that of Napoleon's famous campaign from Paris to Moscow, at around seventeen hundred miles. When the war was at a stalemate in 1593, Katō built himself a formidable castle at Sosaengpo near Ulsan in southeast Korea (now head-quarters of Hyundai and the site of the world's largest shipyard), where he was holed up while negotiations between the Chinese and Hideyoshi seemed to continue without end.

Katō and his troops were in a desperate situation in the So-saengpo Castle. His diseased and depleted army was hungry and running short on funds and munitions for the anticipated push back into the interior when the peace talks eventually failed and they'd enter the field again. Katō was a hawkish general, dedi-cated to the war and to the service of his master Hideyoshi. He turned now to his leading retainers at home in his fief Higo to deliver the materials that were so badly needed.

One of those retainers was African, quite possibly our heroic African samurai.

Just how Yasuke would have entered Katō's service is unclear, but Katō's proximity to, and trade with, Nagasaki would cer-tainly have meant that it would not have been difficult. Yasuke would have been a highly useful hire, and his warrior back-ground and Nobunaga connection would clearly have appealed to the martial-minded lord.

In a letter regarding orders for an overseas trading mission dated December 6, 1593, Lord Katō gives specific instructions as to the leaders of the mission. One of them is called *Kurobo*, or "Black man."

The word *Kurobo* was originally a corruption of the name of the Sri Lankan city Colombo (*Kuro* was the Japanese name for Colombo, in modern-day Sri Lanka, and Sri Lanka and India were often confused with Africa), and meant "a native of," but had become a catchall term in Japan for people with very dark skin, especially Africans.

It may be that Yasuke, away from Nobunaga's court, simply

became known by this name. It was common in Japan to be named for your place of origin, profession, rank or position in society and it seems this African man simply became known as "Black Man." This was an entirely logical name for the time and place in which he found himself and not the offensive epithet that it might be seen as today. Indeed, Elizabethan English references to Africans also use similar nomenclature, *Blackamoor*, *Blackman*, or simply *Black*. These were not meant to be offensive either and were simply an extension of common naming practices which described a person's looks (*Fair, Strong*) profession (*Barber, Gardner*) or location (*River, Castle*).

Kurobo had not initially accompanied Katō to Korea, but remained behind in the port of Ikura in Higo; the letter of 1593 suggests he was a retainer with responsibility for aspects of international trade and foreign affairs. In 1594, this man called Kurobo sailed aboard a Chinese-style junk, built by immigrant Chinese shipwrights in Japan and owned by Katō, with a cargo of silver and one hundred twenty tons of wheat bound for Manila in the Philippines. The Spanish in Manila were very keen to trade for wheat as it was in short supply locally and for them it was essential to make their staples of bread and ship's biscuit. The wheat was traded for munitions and then taken to Korea to supply the desperate garrison. The junk's next moves are uncertain, but it was again ordered on a supply mission in 1596. Katō had complained that supplies of saltpeter to make gunpowder had been insufficient in previous voyages, dressed down his retainers and demanded over three tons this time. A later letter revealed that lead had been sent, but he again demanded more.

We do not know how long Kurobo had been living in Ikura and serving Lord Katō, but another highly interesting piece of information revealed by Katō's letter is that it specifically mentions that care be taken of the African man's wives and children while he was away on this voyage. Kurobo—whether he was Yasuke or not—was clearly well established in Ikura and had been living there for some time. He was a man of wealth, rank and responsibility in Higo. After all, to support such a household, you needed to be very well off. To specifically mention his family, Katō also clearly valued his service, and was fond of him. It also indicates the African was given preferential treatment as a foreigner, allowed multiple wives—polygamy was not common in Japan, although most men of rank had concubines.

Geographically, professionally and chronologically, there is every possibility that the man was Yasuke in the service of another Japanese lord.

Remember the illustration on the writing box (in Chapter 8) created by the Rin School, the one of a very tall black man dressed in expensive Portuguese clothing and the two dark-skinned boys. The cloak carrier is clearly a servant. But the boy musician? Could he be the giant's child Katō writes of? The man in the illustration is likely Yasuke. The artists of the Rin School, who created this beautiful artifact, were based in Kyoto during Yasuke's visits there, and this is probably the nearest we will ever come to gazing upon the African samurai's genuine likeness. It may be Yasuke and his son captured in this image.

Another potential record of Yasuke's post-Nobunaga years appears in an anonymous Japanese document from the 1670s. It mentions an African man alive during the 1590s, who shared the physical description of Yasuke, "seven *shaku* (feet) tall," and was "black as an ink stone" *and* shared a name with Katō's retainer, for he was *also* called "Kurobo." This source gives us the additional information that this man was from a country called Kuro. (Again, the Japanese name for Colombo, in modern-day Sri Lanka, often confused with Africa by the people of the time.) The unknown author undoubtedly meant this man was from Africa.

This man from Kuro was recorded as being present in the town of Shikano in the domain of Inaba (now eastern Tottori Prefecture in western Japan). This has to mean he was associated with a local lord by the name of Kamei Korenori because, although Kamei is not mentioned by name in the source referring to Kurobo, he was the lord of Shikano at the time and did indeed lead troops in Korea. He was a naval commander, and had contributed five ships to the war, all of which had been lost at the disastrous naval Battle of Dangpo in July 1592, when he himself only narrowly escaped with his life. Most of the Shikano men who had left so confidently for war the year before would have drowned on that terrible day.

Sadly, nothing is recorded about what activities the dark-skinned giant undertook while in Shikano, but a look at Kamei's profile gives a good idea and establishes a key link with Higo and Katō's African retainer of the same name. Kamei had been granted the fief of Shikano by Nobunaga at the age of twenty-four for his contributions to the Oda campaigns against the Mori clan in Tottori. Yasuke was very likely with Nobunaga when Kamei was granted his new land in audience at Azuchi, so Kamei would have met Yasuke.

Katō's ship, with Yasuke onboard, and laden with weapons and food from Manila would have docked in Korea to supply his desperate men, then may well have undertaken a further voyage to Tottori to procure more supplies, or perhaps pick up silver from the local mines which Kamei controlled, for the embattled Katō troops. It's tempting to suggest Yasuke was perhaps renewing an old acquaintance when he met Kamei again in Korea while working for Katō and then carried out a further task for him.

Due to the very small number of Africans recorded as being resident in Japan at the time, and even fewer who would have matched Yasuke's profile, the man, or men, proposed above are highly likely to have been Yasuke, but what if they were not? What other possible avenues could his life have taken after the death of Nobunaga?

In all of the scenarios, Yasuke essentially disappears into the mists of time. Few people recorded detailed information about servants or farmers, sailors, soldiers or craftsman, in these times when the written word was still normally reserved for weighty and "respectable" matters. These were times when writing was a rare skill, something very few people could do and even fewer records were ever published or have managed to survive until the modern day for historians to analyze. Such a contrast to today when almost anybody can share their life or post a lasting statement online in minutes. Until very recently, published work covered only a miniscule fraction of the human experience and what remains for us today was necessarily weighted to the interests of those who wrote, almost invariably members of the ruling classes.

Some historical detective work is again necessary to uncover the stories of those who did not fall into this bracket, of whom Yasuke was one.

Here are the most likely possibilities:

Jesuit Muscle

Upon Nobunaga's and Nobutada's death, Yasuke was returned to the Jesuit church by Akechi troops and, as we have seen, probably returned to Nagasaki to provide some support for their mission. Conceivably, he could have left Japan then, returning with one of the Portuguese ships to Macao or even Goa. Jesuits regularly traveled these routes, for trade, ordination (there was no bishop in Japan at this time), promotion and a variety of other reasons. They would have needed Yasuke for the same reasons Valignano did, protection and intimidation in an unstable and dangerous world. Any long voyage would have run the risk of meeting pirates and the European controlled cities in Asia such as Macao, Manila and Goa could scarcely be called safe as many of the settlers had specifically ended up there because they were criminals or were fleeing persecution in Europe. India and the

Far East were, in effect, frontier worlds for Europeans, far from the writ of the monarchs who nominally made, but were rarely able to enforce, the colonial laws. Yasuke could have provided much-needed protection for the priests in these frontier towns, or on missions to the interior as he had done in Japan.

Hideyoshi's expulsion edict of 1587 was not seriously enforced for a decade as long as the Jesuit missionaries did not attempt new conversion work and kept a low profile. Hideyoshi had even hosted Valignano on his *second* visit to Japan in 1591, which had largely been to try and smooth over Coelho's huge mistakes and overly militant attitude.

All seemed well again until 1597 when a Spanish treasure galleon, *San Felipe,* wrecked on the Japanese coast. Hideyoshi's coffers were depleted by the Korean wars, and he saw an easy way to refill them by confiscating the large cargo of silver, worth many millions of dollars in today's money. The Spanish, mediated by Franciscan missionaries, tried to get it back. In the process, an indiscreet Spanish pilot, Francisco de Olandia, showed one of Hideyoshi's advisors a world map on which the Spanish Empire was portrayed and informed him that Spain's main strategy for colonial conquest was to send in missionaries to soften up local populations and create a fifth column of Christians who would fight for them, not their local and "rightful" lord or king.

That went too far. These foreigners were potentially plotting treason. They needed to be taught a lesson and it came on February 5, 1597. Twenty-six Catholics—four Spaniards, one Mexican, one Indian and seventeen Japanese Franciscans (including three boys), and three Japanese Jesuits (arrested in error), were executed by crucifixion in Nagasaki on the orders of Hideyoshi. One of them, Brother Paulo Miki, had been a student at the Azuchi seminary and hence an acquaintance of Yasuke.

After having their ears and noses cut off, they'd been marched to their deaths, all the way from Kyoto, twenty-eight days and well over six hundred miles away, where they had been apprehended in defiance of the nonpropagation law. The burghers of the Christian city of Nagasaki watched in defiance, praying and shouting out the names of Jesus and Mary. These first Martyrs of Japan would later be canonized by the Roman Catholic Church on June 8, 1862. In addition to the executions, 137 churches were demolished. The Jesuit missionaries limped along with increasing difficulties until 1614, when the new Tokugawa government finally issued a definitive expulsion order that was properly enforced. Circumstantial evidence suggests that this was directly after hearing the news from protestant English merchants that the Jesuits had been expelled from their country and were executed if apprehended.

Many Jesuits did leave Japan at this time, as did large numbers of prominent Japanese Catholics, including Takayama Ukon, the man who all those years ago had hosted Yasuke and Valignano at Easter and been present at Yasuke's first audience with Nobunaga. He steadfastly refused to surrender his faith, despite having lost his lordship, lands, influence and everything else a lord possessed. He had been demoted to the ranks, left dependent on others' mercy and charity. Most other prominent Catholics stayed, and apostatized or kept their faith secret. Later more would die for their beliefs.

Despite continued Jesuit attempts at infiltration, smuggling themselves into the country aboard Portuguese and Chinese ships, and continuing evangelism in secret, the Japanese authorities became better at weeding out what they saw as purveyors of a criminal code and a stability-threatening ideology. At first, Catholicism among the Japanese converts itself was tolerated in the belief that given time everyone would see the error of their "new" beliefs, and return to the natural way of things. But as the Japanese government under the Tokugawa shoguns, the dynasty that Ieyasu had founded in 1603, became more frustrated at criminal Jesuit infiltrations and continued public displays of religious passion from a population who very much clung to

A mass execution of Catholics in Nagasaki. The 26 Martyrs of Japan. *Painted by an unknown Japanese Jesuit exiled to Macau.*

their faith, policy turned to strong pressure which eventually by the 1620s turned into severe persecution.

Mass executions and extreme torture, designed to enforce recantation, including being hung in pits of snakes or held over burning hot sulfur lakes until nearly dead, then revived to go through it all again, became institutionalized.

Under this intense pressure, most Japanese Catholics in Nagasaki—and even some foreign Jesuits—recanted. Catholicism went underground or remained only in isolated rural areas, far from the eyes of the central government.

Among all this persecution was an African named Ventura, who was one of the Catholics to renounce his erstwhile religion. He worked for the local magistrate Suetsugu Heizo, searching for Christians and fugitive priests in Nagasaki between 1625 and 1632. Although few details about this man are recorded, Ventura was much feared among the local Catholic population. Yasuke would almost definitely have been dead by this time, few people reached their seventies in this age and so it is unlikely to have been him.

If Yasuke had remained with the Jesuits into the seventeenth century, he would likely have moved with them away from Japan back to one of the Portuguese bases in Macao or India.

Sailor

If Yasuke had been cast aside by the Jesuits or decided to go it alone, there was one avenue that would have always been open to him. Sailor. The regular Portuguese ships in Nagasaki constantly looked for new and healthy hands. They were always undermanned due to deaths and desertions and Yasuke would have been a prime applicant. He knew his way around the ships, spoke several languages and was very strong. He had ample experience of traveling by sea and could likely make himself useful on a boat of any nation, in a variety of situations and jobs. Africans, Indians, Chinese and Japanese men often made up a very large proportion of "Portuguese" crews in Asia. This would have been a last resort for Yasuke, however, as there were far more attractive avenues open to him.

Portuguese Buccaneer

A team of African bodyguards was a very fashionable accoutrement among the elite of the East Asian seaways at this time. Rich merchants competed to see who could have the grandest and most ostentatious entourage, and what could be grander than a six-foot-plus African warrior? Valignano and Nobunaga had certainly agreed with this sentiment, as clearly did other Japanese lords. A good example of a Portuguese man who also gloried in an African warrior corps was Bartolomeu Vaz Landeiro—a powerful ship's captain, pirate, trader and early Macao pioneer—who kept a bodyguard of around eighty Africans with him. The men were armed with halberds and shields and accompanied him everywhere. Landeiro, a Portuguese man of Jewish origin and hence fleeing anti-Semitism and pogroms in his homeland, left his background behind him after rounding the Cape of Good Hope and reaching the Indian Ocean. As many did, he faked his ancestry and claimed Portuguese aristocratic descent on arrival in Goa in 1560, and nobody seems to have had the gumption to contest it. Perhaps the African bodyguard corps, which he seems to have recruited in India, ensured he was taken seriously.

Landeiro was active in Asian waters between Macao, Manila and Japan, where he did much of his business of slaving, smuggling, piracy and occasionally more savory trading pursuits until his death sometime after 1585. His regular and generous financial support for the Jesuits ensured his ships were welcome in Nagasaki and he also paid calls to other Kyushu ports such as Kuchinotsu, especially to obtain Japanese slaves. His notoriety and success earned him the epithet "The King of the Portuguese from Macao." His ships' crews were notably multiethnic—employing Chinese, Japanese, European, Filipino, Indian and African sailors and buccaneers. It is entirely possible that Yasuke could have become one of them.

Chinese Pirate

Chinese pirate crews in the South China Seas, an area which no state power adequately controlled and where it was often in minor rulers' interests to turn a blind eye for their own financial benefit, often employed Africans who had escaped from slavery or gone it alone. An example, though shortly after Yasuke's time in the 1620s, was the Chinese pirate, smuggler and merchant, Zheng Zhilong.

Zheng had a large African bodyguard corps, more than three hundred men at its peak. The bodyguards were recruited from various places, but most entered his service via Macao, the Portuguese enclave in southern China, and many were escaped slaves.

They could also have been men freed in reward for their part in the successful defense of Macao against the Dutch in 1622.

In this battle, an attempt by the Dutch to wrest control of the inter-Asian trade from the Portuguese, Macao found itself virtually defenseless as the Dutch attacked when most of the Portuguese merchant militia were away on trading missions in China. In a desperate bid to defend the outpost, *all* African slaves—a large group who did most of the manual labor in the colony—were granted their freedom, *and* as much alcohol as they could drink, in exchange for fighting in the city's defense. These drunken, newly freed men and women were wildly successful in destroying the Dutch, and their mercenary Japanese and Thai troops, despite being heavily outnumbered. The Africans charged the Dutch musket fire fearlessly and gave no quarter; and as it was the feast of John the Baptist, allegedly celebrated by removing heretic Protestant heads from their bodies. The former slaves, having been released from their bondage, would have been searching for better employment (and quickly), and pirates such as Zheng Zhilong could provide this.

Zheng had lived much of his life in Japan, where he was safe from Chinese government authority and could take advantage of Japanese and European trade and smuggling opportunities. At the height of his power, his fleet was estimated at up to a thousand ships and controlled almost all interactions in the South China Sea.

Zheng's era coincided with the conquest of China by the Manchu tribes from the nomadic lands of the northeast. The Manchus usurped the Ming dynasty and formed their own new Qing dynasty that lasted from 1644 to 1912. The famous movie, *The Last Emperor*, is based on the life of the final Qing emperor, Pu Yi.

In 1628, Zheng turned respectable, to serve his country by resisting the invaders, who most Chinese people despised. He and his pirates became the legitimate naval forces of the retreating Ming dynasty. Zheng fought hard for his new masters (and old enemies), but eventually realized that further resistance to the Manchus was useless. In 1646, he surrendered.

The fate of Zheng Zhilong's three hundred African warriors is unclear. Upon his surrender to the Qing imperial authorities, his Chinese troops abandoned him and the Africans were the only men to stick by him. It is said they were massacred in a last ditch stand after it became clear their general had been double-crossed by his captors. Zheng was taken to Beijing as a hostage against his son's good behavior. His son continued the resistance and the father was executed like a common criminal in 1661.

Zheng was not unique in having an African bodyguard corps, even if his was the largest recorded. Although Yasuke's time, in the 1580s, would have been early in this historical trend, Yasuke could easily have joined one of these international mercenary groups in Chinese employ that would no doubt have enjoyed plenty of plunder when used in raids and outright war.

Other Possibilities

There are several other interesting African men recorded in Japan at this time, but the likelihood any are Yasuke diminishes as the years go by. The businesses that Yasuke was involved in—war and seafaring—were the most dangerous on earth and hardly conducive to longevity.

In 1615, Lord Arima Harunobu's son, Naozumi, is recorded as having employed an unnamed African messenger as a retainer who routinely traveled the considerable distance from Kyoto to Nagasaki to deliver messages and represent his lord. The English merchant Richard Cocks, resident in Japan from 1613–1623 "gave lodging [to this African man] in the English house with meat and drink, because he was servant to such a master." Yasuke would have been in his fifties at this time, probably a little old to be plying the cold waves of the Seto Inland Sea on such a regular basis.

Another unnamed African is recorded in the early seventeenth century in Mexico City, as the friend of a Japanese slave named Tomé. Although the details of why are unclear, this Mozambican *had* lived several years in Nagasaki before being sent to Manila and, subsequently, to Mexico. He was perhaps a sailor on the Spanish galleons which plied the route between Manila and Acapulco or a manservant to a merchant.

Fast forward two centuries, and this face appears.

Tamaki Mitsuya, twenty-three years old and a retainer of Kawada, Lord of Sagami (modern-day Kanagawa, near Tokyo) and senior vassal of the shogunal house, the Tokugawa. Both his lord, Kawada, and Tamaki were members of an 1864 Japanese mission to Paris seeking French support to renege on previous treaty commitments and close down the fast-developing international port of Yokohama near Edo (soon to be renamed Tokyo). Tamaki's photograph was taken in a Paris salon by the famous photographer Antonio Beato. Tamaki and his colleagues had traveled there by way of Africa, even posing for a photograph in front of the Sphinx on the banks of the same river, the Nile, by which Yasuke had been born approximately three hundred years before. There is no more information about this fascinating man.

However, if not for the samurai garb and swords he wears, you might easily mistake him for an African. There is absolutely no proof Yasuke was Tamaki's ancestor; it could just be one of those coincidences or even a trick of the camera lights which suggest such mixed features. Still, several clues make Tamaki worth considering.

First, the second Chinese character in his given name is the same as the *Ya* in Yasuke's name and it is quite a common Japanese practice to carry on the use of a character through the generations; such customs have sometimes lasted hundreds of years and continue to this day.

Tamaki Mitsuya, photographed in Paris, 1864.

Second, the house he served, the Kawada, had originally served Nobunaga's bitter enemies, the Uesugi, but around Yasuke's time, entered the service of the Oda-allied Tokugawa clan. It is entirely within the realm of reason that, after leaving the Jesuits, Yasuke was offered a position by one of Tokugawa Ieyasu's followers; after all, he had traveled extensively in their lands and one of their leading retainers, Matsudaira Ietada, recorded the African samurai in his diary. Perhaps it was at the behest of Ieyasu himself who later employed and promoted foreigners such as the Englishman William Adams (inspiration for the novel and hit 1980 TV series *Shogun*) extensively.

Running against this possibility, we have two hundred years of Yasuke's descendants marrying people with Asian features in Japan. There would have been no other Africans to marry, and thereby keep the facial features "African." In similar circumstances, ancestors of Africans who lived in England during Yasuke's time soon became indistinguishable from the local population as they and their descendants intermarried with indigenous people. This would have been the case in Japan too.

And could Tamaki have been the ancestor of another African or even black American? It is estimated there were around three hundred Africans resident in Japan in the sixteenth and seventeenth centuries, and there were many more temporary visitors as sailors on foreign ships. After the 1630s, foreigners who were allowed to trade in Japan were severely restricted, among non-Asian nations only the Dutch were permitted. They were restricted to trading at Nagasaki and their movement was limited to a tiny man-made island called Dejima. Contacts with the local people were highly restricted, so any Dutch-African visitor, even if they successfully impregnated one of the low-ranking local sex workers who were permitted to them, would not have had children who became hereditary samurai to a prominent family in the Tokyo area, hundreds of miles away.

The first African Americans probably arrived in Nagasaki in 1797 on the US ship *Eliza* which was flying the Dutch flag as it had been contracted to take care of Dutch trade at a time when the Netherlands were unable to sail due to the Napoleonic wars. The head of the Dutch station was approached by the Nagasaki authorities over the "unfamiliar kind of black people" on board. The locals were again fascinated by a racial type that they were unacquainted with or had forgotten about over the centuries. These sailors would have been confined to Dejima and unable to interact with local people properly.

Furthermore, shortly after Yasuke's time, the samurai became a hereditary caste. If you were not a samurai in the early seventeenth century, your descendants were not samurai in 1864 (there were some minor exceptions to this rule, but generally it held).

As far as we know, Yasuke was the only African samurai, ever.

It is almost impossible that any samurai with African looks would have been the descendant of another African-looking warrior. Thus, if Tamaki is of African descent, then Yasuke is highly likely to have been his ancestor. There is not enough information to draw definite conclusions, but the possibility that Yasuke's family retained his looks and position as senior samurai in service to a Tokugawa family vassal is there.

Yasuke's final years may have ended in any one of the historic paths above. But his legacy, as we shall see in the final chapter, still flourishes today.

CHAPTER TWENTY-FOUR

Yasuke Through the Ages

The story of Yasuke was first published in 1598 in the second volume of a compilation of letters and reports from the Jesuit mission in Japan entitled *Cartas que os padres e irmãos da Companhia de Jesus escreverão dos reynos de Japão e China II* (Letters written by the fathers and brothers of the Society of Jesus from the kingdoms of Japan and China—Volume II). The collection of letters—normally known simply as *Cartas*—covers the period from 1580–1598.

Yasuke's 1581 visits to Sakai, Kyoto, his audience with Nobunaga and his subsequent part in Nobunaga's final stand and the Battle of Okitanawate were revealed in letters from Luis Fróis, the prolific chronicler of the mission. Yasuke's time in Azuchi is also recorded in a letter by Lourenço Mexia, who was Valignano's right-hand man and accompanied him to Nobunaga's capital, Azuchi, in 1581.

The compilation was published in Europe as publicity/ propaganda to promulgate the huge success the Jesuits were enjoying in the Far East. The two volumes were disseminated among the literate classes in parts of Catholic Europe—Spain, Portugal, France and the German and Italian states—and then made their way clandestinely to Protestant Europe, mainly England and the Netherlands. For most readers, these were the first reliable eyewitness accounts they'd ever read about the Far East. Prior to this, much of what had been written of Japan was of questionable scholarship and often contained outright fabrications. Still, the Jesuits, mindful of their control over this valuable fount of information, made certain that only information they approved of was included, and edited to paint their mission in a very positive, even triumphal, light.

The final published versions were substantially different from the editions actually sent from the Far East. The letters within went through three or more rounds of editing and censorship— in Nagasaki, in India and finally in Rome—before being released to the public. This became abundantly clear when a hoard of the original letters was curiously discovered in a miserable state of repair behind a painted Japanese screen in Lisbon in the early twentieth century. They'd been recycled and used as packaging for the painting on its long trip to Europe by sea from Japan. When rediscovered as the painting was being restored, they proved a treasure trove for historians as they revealed many details that had been edited out in the final published versions.

Europe was crying out for reliable news of the Far East in the sixteenth century. It was an era of turmoil in "Christendom," as the continent was loosely known at the time. The Reformation had torn any religious unity apart and vicious sectarian warring continued unabated. Catholics saw the Asian missions—and Japan in particular, due to its noted success—as ways to revitalize the Church and make up for the loss of devotees who'd converted to Protestantism in northern Europe. The Protestants, in turn, saw access to the legendary wealth of the Far East as a means to strengthen their relatively weak economic standing and to establish allies against Catholic enemies, particularly against Spain. The Dutch and English put huge efforts into their attempts to navigate the unknown seas to China and Japan. Both had good reason to hate Spain; the Dutch had only recently declared independence from the Iberian monarchy and would be at war to validate that state until 1648, and the English had narrowly escaped invasion at the hands of the Spanish Armada in 1588, more by luck than military prowess. The threat of Spanish domination continued to hang over both countries for decades.

As the *Cartas* were going to press in Evora, Portugal, the first Protestant ship to reach Japan, *de Liefde*, was just setting sail from Rotterdam. Her voyage took nearly two years and she suffered the loss of most of her crew, but arrived in April 1600, a virtual wreck, off Bungo, in the territory of the Ōtomo (the very place Yasuke had begun his fateful journey to meet Nobunaga in the spring of 1581). Only twenty-four of the original hundred-man crew survived and many of those died shortly after their arrival in Japan. Among those who survived was William Adams, a man who'd rise quickly in the service of Tokugawa Ieyasu, Hideyoshi's political heir to leadership of the realm. Adams, an Englishman from Gillingham in Kent, was granted high samurai rank, *hatamoto*, generous trading rights and a small fief by Ieyasu. His notable achievements include becoming court interpreter, foreign affairs advisor, scientific and mathematics tutor to Ieyasu and the building of two English-style ships, one of which became the first Japanese vessel to reach North America in 1610 when it repatriated the crew of a Spanish treasure galleon, which had wrecked off what is now Chiba Prefecture near Tokyo, to Mexico (then called "New Spain" and Spain's most important colony).

The Spanish were so alarmed and threatened by the prospect

of a Japanese ship navigating the Pacific, that they confiscated it upon arrival. The Japanese crew and the officials onboard were sent back to Japan on a Spanish ship. To compensate the Japanese for the theft of their ship, the Spanish paid a generous compensation, agreed to trading privileges in Mexico and to share precious gold and silver mining technology.

Of the remainder of *de Liefde*'s surviving crew, the Dutchman Jan Joosten van Lodensteyn was *also* granted samurai rank and the few others who survived made small fortunes as teachers of subjects such as gunnery and mathematics. Both Adams and van Lodensteyn have areas of Tokyo named after them, and Adams's story, in particular, has been repeatedly told on screen, in theatres and in print. Adams provided the first information about Japan to be penned in English. He also facilitated the first direct diplomatic contact between England and Japan. (A suit of armor sent by the second Tokugawa shogun, Tokugawa Hidetada, to James I in 1613 can still be seen among the British royal treasures in the Tower of London today.)

Yasuke's story—derived only from *Cartas*—was repeated several times in works published in Europe during the seventeenth and early eighteenth centuries.

The first mention was in the French Jesuit educator François Solier's *Histoire ecclésiastique des îles et royaume du Japon* (The Ecclesiastical History of the Islands and Kingdom of Japan) of 1627. The second record was in the rather cumbersomely titled, but no doubt sincere, *Elogios, e ramalhete de flores borrifado com o sangue dos religiosos da Companhia de Iesu, a quem os tyrannos do imperio do Iappão tiraraõ as vidas por odio da fé catholica* (Praises, and Bouquets of Flowers Sprinkled with Jesuit Blood, to Those Whom the Japanese Tyrants Kill Through Hatred of the Catholic Faith) by the Portuguese Jesuit and historian Antonio Francisco Cardim in 1650. The third and final rendering was by the French Jesuit scholar, Jean Crasset, in his *Historie de l'eglise du Japon* (History of the Japanese Church) in 1669.

Other histories of the Jesuits in Japan such as *Histoire de l'établissement, des progrès et de la décadence du Christianisme dans l'empire du Japon* (History of the Establishment, Progress and Decadence of Christianity in the Empire of Japan) by Pierre-François-Xavier de Charlevoix, first published in 1715 and in print for over a century, fail to mention Yasuke's story at all.

Perhaps due to the increased climate of racism against Africans that rose with the Atlantic slave trade, the African samurai ceased to remain of interest to the eighteenth-century European audience.

By Crasset's day, 1669, there'd been little or no Jesuit activity in Japan for 30 years. There would be no more until the late nineteenth century, with the one exception of a lone priest, Giovanni Battista Sidotti, who, craving martyrdom, attempted to infiltrate the country in 1708. He was captured almost immediately and spent the final seven years of his life in captivity before dying, as he had wished, a martyr's death of starvation in a prison pit in Edo.

In Japan, where Yasuke's story was always destined to be a topic of fascination, Ōta Gyūichi's laudatory biography, *The Chronicle of Nobunaga*, was published posthumously in the decade after Ōta died in 1613. Notably for its time, the biography was printed using a movable type printing press—one of the first of its kind in Japan, the first to represent Japanese characters, and also a product of Hideyoshi's Korean war (the technique and the printers were originally abducted to Japan in 1593). Ōta's finished work was based on a lifetime of diaries and personal notes he'd kept.

In the chronicle, Yasuke is mentioned specifically in relation to his initial audience with Nobunaga, and in an early draft of Ōta's work (one which did not make the final edit and was never published, but is held in the archives of the prominent Oda vassal clan, the Maeda), we can today read of Yasuke's name and

An African man wrestles as Nobunaga and his court look on. This is almost definitely Yasuke, although painted several decades after his time with Nobunaga.

his elevation to samurai status: "This black man called Yasuke was given a stipend, a private residence, etc., and was given a short sword with a decorative sheath. He is sometimes seen in the role of weapon bearer."

Ōta's work is considered to be a reliable historical source and is the fount of much information about Nobunaga and his times. Ōta also later wrote lives of Hideyoshi and Tokugawa Ieyasu as well as a record of the wars in Korea.

After Ōta's *The Chronicle of Nobunaga*, as far as recorded history in Japan goes, the African samurai's story seems to have been largely forgotten for some two centuries.

But the fact that word of Yasuke's legend lived on into at least the seventeenth century in Japan is supported by the picture of a black man sumo wrestling from 1640, *Sumo yurakuzu byobu*. The picture shows a sumo tournament between an African and a Japanese man. Nobunaga is keenly adjudicating as members of his court look on and other wrestlers await their turn to fight. The location is unclear, but it is probably meant to be Azuchi. Somebody knew Yasuke's story sixty years later and was able to depict it.

That Yasuke's story was passed on as a popular legend in Japan is also shown by the 1670s reference, mentioned in the previous chapter, from Tottori, referring to the visit of a black man to the local region nearly one hundred years previously. Why the unknown chronicler was suddenly inspired to write about "Kurobo," or Yasuke, in 1670, we shall never truly know. But it is clear that some folk memory had remained of an African visitor to a remote corner of Japan.

A final eyewitness source, a line about Yasuke which informs us of his stipend, height and again his name, is from the diary of the senior Tokugawa vassal samurai Matsudaira Ietada, who met Yasuke just after Nobunaga had conquered the Takeda in 1582. Matsudaira's diary was not published until 1898, nearly three hundred years after the death of its author in battle in 1600. There is no reason to think its description of Yasuke was widely known until then:

> *Lord Nobunaga gave the black man who the missionaries presented to him, a stipend. His skin was black like ink and he was around 6.2 shaku (over 6'2") tall. He was called Yasuke.*

It was Yasuke's first published mention since the 1670s, and came about as part of the movement to shed a more scholastic light on Japanese history by publishing ancient tracts which had remained in dusty storerooms or family book chests.

Nearly fifty years later, in 1943 and 1944, came the first Japanese translation of the Jesuit letters, *Cartas*, by the eminent historian Murakami Naojiro. This was not his first translation of an early European account of Japan—he'd been doing it for forty years plus—but it was his masterpiece. Murakami opened up these valuable historical documents to the reading public in Japan for the first time. The few surviving copies of the original *Cartas* print runs of 1598 had been largely forgotten, and were slowly wasting away in a few European libraries, accessible to only the most dedicated scholars who could read the arcane Portuguese in which they are written. There is still no translation in English.

Yasuke's story was poised to be rediscovered in all its glory, at least by a Japanese reading audience. But times were not good in Japan. In the early 1940s, the country was at war and in imminent danger of absolute destruction. The Japanese had other things on their minds: survival and then national reconstruction. It would take another twenty years for Yasuke's story to receive published attention and for a second age of Yasuke to dawn after three hundred years of neglect.

In 1968, author Kurusu Yoshio, a pioneer of historical fiction for children, wrote a Japanese language children's book called *Kurosuke*. The story was illustrated by Mita Genjiro, a famous illustrator whose beautiful book covers are familiar to all Japanese children, even today. This was the first story ever to be dedicated in its entirety to Yasuke and his life.

The book is highly sympathetic to Yasuke's character and tells the African samurai's tale loosely based on the Jesuit sources. It seems neither of the Japanese sources, Ōta nor Matsudaira, were consulted. Perhaps Kurusu was unaware of them. Though the author does not say how she found out about Yasuke, in her afterword, Kurusu talks of being fascinated by a map of Africa which was on the wall of her elementary school when she was a child in the mid-1920s. She talks in scornful words about "European empires, fat merchants and bearded generals" who "chewed up" the African continent. Then links Yasuke with the optimistic new Africa that in 1968 was only then emerging, the independent Africa of nations. In doing so she was the first of many people to draw inspiration from his story as opposed to simply recording his existence.

In the decades following WWII, European nations could no longer justify colonialism, morally or economically. In 1950, only four African countries—Egypt, Ethiopia, Liberia and South Africa—were independent self-governing states. By 1968, however, the tide of independence had swept the continent and it was this optimistic spirit of liberty and change which inspired Kurusu's writing. It is clear her message in *Kurosuke* is that Yasuke is a story of hope for Africa's future, and Yasuke's story was being liberated at last even as Africa itself seemed to be on the verge of freedom from centuries of domination and plunder by outsiders. *Kurosuke* is a work of fiction aimed at a young reader, so does not dwell on the bloody wars, but looks at geographical variations between Africa and Japan, exotic animals and cultural differences such as meat-eating and weaponry. The mind and emotions of a young man in a foreign country far from home is one of its most poignant themes.

Yasuke is painted as a hero, an affable and cooperative guy, who makes friends with Nobunaga's pages and entertains his women, as well as being scrubbed to check the veracity of his skin pigment. The climax is the battle at Honnō-ji Temple, where Yasuke fights to the end before escaping. He is not present at Nobunaga's death in this story, but escapes with another page to fight with the Oda heir, Nobutada, playing a key part in the final Oda stand before being wrestled to the ground and taken before Akechi.

Having been escorted back to the Jesuit church, Yasuke is exhausted and demoralized and the last chapter is called "A Dream of Africa," in which he dreams forlornly of his family and childhood in Africa and cries silently.

For its revolutionary contribution to the, at the time, new genre of children's historical fiction, *Kurosuke* won the prestigious Japanese Association of Writers for Children Prize in 1969.

But the times were not exclusively sympathetic to Yasuke.

The next time he appears as a fictional protagonist is in Endō

Yasuke and Nobunaga as portrayed in Kurosuke, *1968.*

Shūsaku's 1971 novel *Kuronbo*. Endō was one of the best-known writers in postwar Japan, often tackling deep subjects, especially to do with spirituality; he was a practicing Catholic. The title of the novel is the same "Black Man" that Katō's African retainer was known by, the difference in spelling is due to Endō writing his title in Nagasaki dialect which changes the sound of the word slightly.

However, while "Kurobo" was a common name to call an African person in the sixteenth century, in the modern age, the word has taken on an offensive and discriminatory meaning for black people, making it a rather unfortunate book title. Endō may have intended it provocatively. Further, in Endō's novel, Yasuke is portrayed in a very different light from the positive African freedom icon in Kurusu's book *Kurosuke*. In *Kuronbo*,

Yasuke as portrayed on the cover of Kurosuke, *1968.*

Yasuke becomes a buffoon, perhaps even a simpleton, an object of fun for Nobunaga's inner crowd, and indeed the wider public entertaining his audiences with obscene shows such as farting in time to drum beats. Worryingly for some of the male Japanese characters, Yasuke is also an object of sexual interest to the ladies. He attaches himself to a girl called Yuki who looks after him like a mother while he performs tricks to make ends meet. They go through hard times, but also happy times, together, living in a hut.

The story seems, at least superficially, to be somewhat reminiscent of the belittling 1930s representations of Africans in European comic strip books like *Tintin* but can be read in several lights. Firstly, as a product of the racist times in which it was written; although ways of describing race were in the process of change in 1971, the contemporary Japanese dictionary *Daigen-kai* simply gives the modern definition of *Kurobo* as "Indian, or African American," and does not mention that it is derogatory. Judging by the contemporary definition, the Japanese public still saw no problem with this word which, although it had originally had no disparaging connotations, being the word for the inhabitant of a respected fellow Buddhist kingdom (Colombo in Sri Lanka), had come by the twentieth century to infer something similar to the English words "black boy." The modern word for people of African descent in Japanese is *"kokujin"*—literally "black person."

Second, *Kuronbo* can be seen as a story about the trials of living and the difficulties of mere survival in a strange and foreign land. Yasuke struggles with everything, depends on Yuki as if he were a child, and has to do humiliating tasks to earn money. The subject of alienation was a subject which Endō was himself familiar with as he spent his childhood in China and many years in France as a student. He visited this theme in numerous other works also. Finally, in *Kuronbo*, Yasuke's character can also perhaps be seen as an allegory for Africa, or the general plight of black people around the world in 1971. Endō had spent several months in the United States in 1969 and saw the American Civil Rights movement firsthand. He also attended the first Afro–Asian Writers' Conference in Tashkent in 1958, a gathering held to denounce imperialism and to establish better cultural contacts among delegates' countries. Yasuke, in *Kuronbo*, has his ups and downs, but the general theme of infantilization could be read as a further political comment, less optimistic than Kurusu's, on decolonization, the global treatment of people of African descent and the inevitable difficulties and humiliations that were emerging in the fight for civil rights and independence in 1971.

Japanese views of Africans since Yasuke's time have changed over the centuries and decades. During the long period of isolation from most non-Asian nations, the 1630s to the 1850s, there were few Africans in Japan to have any opinion of, and hence views became necessarily filtered through European lenses. That perspective, alas, was not positive and by the 1850s, any knowledge of Africa, minimal as it was, was bigoted and deleterious. It should be remembered that Japanese attitudes toward *Europeans* during this period were *also* overwhelmingly negative. (There were even rumors in Japan that the Dutch merchants wore high heels to accommodate a horn in their heels, somewhat akin to the paws of a dog.)

The international imperial power politics of the late nineteenth century changed Japanese people's view of Europeans and Americans, but did little to improve their view of people of African descent. On the contrary, the vast majority of blacks that the Japanese came into contact with were slaves or servants, and this clouded their views.

Times changed with the coming of the twentieth century, and Japan became only the second non-European nation to defeat a serious European military in war since the Middle Ages during the 1904/1905 Russo-Japanese War. People of color around the world celebrated, and Europeans and Americans took stock of this revolutionary threat. The recently coined term *Yellow Peril* became a buzzword and the United States started to restrict immigration from Japan to its west coast states. This conflict eventually led to the Russian Revolution and also became one of the "warm-up" wars for World War I.

However, the *first* non-European nation to defeat a modern European army in the field was Ethiopia at the Battle of Adwa in 1896, ironically with some support from Russia. The Italian invaders were soundly thrashed and beaten back to their colony in Eritrea to the north. This caught widespread global attention too, and cemented Ethiopia's independence until a more successful five-year Italian fascist occupation in the 1930s and '40s.

Japan was no exception to the rise in esteem for Ethiopia in

Front row, right to left: Lij Araya Abeba, His Excellency Heruy, Lij Tafari, and the interpreter, Daba Birru. On the back row are Mr. and Mrs. Sumioka.

角岡代部に於けるエテオビヤ特使一行
（前列右より）、デアラ・アベバ公、ヘルイ特使、リデ・ダフア・随員

The Ethiopian royal party in Japan dressed in kimonos.

the early twentieth century; in fact it was seen as a potential ally to the extent that in 1933 it was proposed that an Ethiopian prince, Lij Araya, marry a suitably aristocratic Japanese wife to cement strategic and commercial ties. A candidate was found, Kuroda Masako, a distant descendant of a samurai who fought for Nobunaga and undoubtedly knew Yasuke. However, the wedding never took place. The Italians put huge diplomatic pressure on both Japan and Ethiopia to cancel the plan, and other Western powers, highly uncomfortable with Japanese influence further expanding worldwide, agreed with the Italians.

In the early twentieth century, many African American intellectuals such as W.E.B. Du Bois lauded Japan for its role in supporting other non-white peoples in their struggle against colonialism and imperialism. Others were not so convinced, pointing out that Japanese imperialism was equally subjugating the Koreans and Chinese. Although writer and social activist Langston Hughes was warmly welcomed in 1933, he nonetheless went out of his way to condemn Japanese hypocrisy in extensively supporting colonial independence movements while fiercely maintaining its own right to colonies in Korea, Taiwan and the Pacific. Hughes stated that the Koreans and Chinese "were in somewhat the same position as Negroes in the United States." After these statements, he was promptly deported.

Then came 1945. The Japanese defeat in WWII brought a huge occupation force of American GIs (and a few thousand representatives from the other Allied nations), including many African American servicemen. The first major instance of black people on Japanese soil since the early seventeenth century. Of course there were far fewer black GIs than white, and the old prejudices from home were imported along with all the other baggage the US Army brought with it. Segregated communities of whites and blacks formed outside the newly established US military bases all around the country. And any local Japanese person could easily see where the power resided. Perceptions of people of color suffered greatly and took decades to repair.

Among Japanese intellectuals, the 1960s brought the dawn of a new world of social activism, revolution from below, and an age of peace and love. Japan and its views of black people, largely those of African Americans who were often portrayed, fallaciously, as cannon fodder for the US forces in Vietnam, became more pitying. These black men, although stationed in Japan on their way to Vietnam, were no longer imperialist occupiers to be reviled, but comrades in a larger war against the ruling and warring classes.

Fast forward to the twenty-first century and tens of thousands of black people reside in, or are citizens of, Japan. Racism? Yes, Japan does have major issues with accepting those who have originally come from elsewhere or are mixed heritage as normal citizens, but that does not mean that a foreigner cannot make their way. Someone who commands the language, marries into a local family and plays a full community role is respected for that, and becomes a *kokumin*, a member of the community. Their mixed heritage offspring are Japanese citizens.

In the past decade people of mixed black/Japanese heritage have played an increasingly prominent role in society. The best known is probably Miyamoto Ariana, born to a Japanese mother and an African American father, who in 2015 won the Miss Universe Japan pageant (achieving top ten in the global tournament) and subsequently decided to use her fame to help combat racial prejudice at home. Miyamoto said, "I want to start a revolution. I can't change things overnight but in one hundred to two hundred years there will be very few pure Japanese left, so we have to start changing the way we think." Other famous black Japanese people include the international sprinter Asuka Cambridge, and comedian and television presenter Ike Nwala who appears almost daily on children's TV shows as well as on more adult-orientated content.

During the 1990s and early twenty-first century, the Japanese public came to have a reasonably good concept of Yasuke's existence, even though few knew his full story. This was due to his increasingly regular appearances in a massive variety of media formats.

In the 1990s, Yasuke appeared briefly twice in the wildly popular year-long historical *Taiga* dramas from the national television broadcaster NHK which, since 1963, have told the life of a different historical character in a dramatic way each year. His first appearance was a nonspeaking role when Valignano meets Nobunaga in 1992's *Nobunaga: King of Zipangu*, and his second was, following the thread of history, *Hideyoshi*, a 1996 drama in which Yasuke is killed by Akechi's men during the Honnō-ji battle.

The fictionalized version of the African samurai's story then appeared extensively in the long-running manga (manga is the globally popular Japanese-style of comic book art which can trace its roots back to the twelfth century and became adapted for modern comic books in the 1950s) and anime *Hyouge Mono*, first released in 2005. Yasuke is heroically portrayed, saving people from the burning Honnō-ji and then hunting for Nobunaga's killer, who he finds out is not actually Akechi Mitsuhide, but—in a conspiratorial twist—Hideyoshi. Yasuke's role in the story ends when Hideyoshi grants him his freedom in exchange for his silence.

In 2008, again totally fictionalized, Yasuke became one of the main characters in the Japanese novel *Momoyama Beat Tribe* by the prolific author of light historical fiction, Amano Sumiki. In the story, after surviving the Honnō-ji battle, Yasuke is trying to earn money as a dockhand to return to Africa, but finds he is being exploited. He escapes and meets up with the other three protagonists, also down on their luck and escaping from precarious situations, to form...a street dance group! (Yasuke plays the part of a drummer.) The novel relays an optimistic message about self-renewal and overcoming status barriers. In 2017, it became a theatre production and was featured on morning television.

In 2013, Yasuke entered the cyber age and made his debut in the world of computer gaming as a character in the long-running series of turn-based strategy role-playing video games, *Nobunaga's Ambition*. This award-winning series was first published in Japanese by the video game publisher, developer and distributor Koei, now Koei Tecmo, in 1983. Several of the games have since been released outside Japan. Yasuke appears as a heroic warrior and playable character in *Spheres of Influence*, the fourteenth title in the series.

More recently, in 2017, Yasuke appeared as a character in the action role-playing video game *Nioh* which, only two weeks after its release, had sold over one million copies worldwide. Published by Koei Tecmo and Sony Interactive Entertainment, *Nioh* is based on an unfinished script by the legendary Japanese filmmaker Kurosawa Akira. The main character is the English samurai William Adams who is enlisted to fight by ninja in the service of Tokugawa Ieyasu. Players have to take Adams through levels, fighting supernatural and human enemies, one of whom is Yasuke, who can be challenged to a duel as a bodyguard of the game's villain, Edward Kelley. In the game, Yasuke is fighting to raise Nobunaga—to whom he is eternally indebted for manumitting him—from the dead.

In 2017, also, the best-known African American jazz musician in Japan, Marty Bracey, and fellow musicians, gave a special poetry-jazz performance based on Yasuke's life.

But Yasuke does not only appear strictly as himself.

Since 1999, Yasuke-inspired characters have also found their way increasingly into mainstream works of popular culture, many of which have crossed the seas, further widening awareness of his story around the world.

The best known of these Yasukes is the worldwide best-selling manga series *Afro Samurai*, first printed in 1999. *Afro Samurai* is set in a science fiction future and the hero, Afro, a black "samurai" is on a mission to revenge his father's killer. The anime version (anime is essentially the animated film form of manga comics) was first released in 2007 and the main character is voice-acted by Samuel L. Jackson.

Another example of a character based on Yasuke in another sci-fi setting is Young, who appears in the 2009 manga *Nobunaga Concerto*. The Japanese language manga has since been turned into an anime series in 2014, a television drama series, also in 2014, and a movie in 2016. The plot revolves around a Japanese high school student who travels back in time unintentionally and becomes Nobunaga. Episode 10 of the anime introduces Young, an African American teenaged baseball player who's also a trapped time traveler and who goes on to serve the counterfeit Nobunaga as a counterfeit Yasuke.

In 2007, Yasuke (or at least the idea of Yasuke) hit the big screen for the first time in the form of *Taitei no Ken* (The Emperor's Sword). It was a Japanese sci-fi samurai movie, based on a series of novels by best-selling author Baku Yumemakura. The lead character, Yorozu Genkuro (played by Abe Hiroshi), is Yasuke's grandson.

While writing this book, no fewer than four major Yasuke film projects set in Old Japan were launched. Sometime in the near future, we'll be seeing another representation of Yasuke on the big screen.

The town library in Lansdowne, Pennsylvania—a suburb of Philadelphia—is a small one-story building of faded sandstone quietly tucked in a residential street. In the spring and summer, it almost vanishes behind numerous thick trees so that one sees only its steps leading through the shade to the building hidden within. With a little stretch of the imagination, in appearance, it could be one of the residences leading up to Azuchi Castle. Inside, you'll find a collection of more than sixty-five thousand books, DVDs, magazines and CD-ROMs.

One of the many groups who gather within has been the Teen Reading Lounge, a nontraditional book club funded through grants provided by the Pennsylvania Humanities Council. Teens aged twelve to eighteen are treated to free books, field trips and guest speakers, and it provides the community with the opportunity to construct a course full of reading, discussion and activities that stretch far beyond a typical book club.

Keville A. Bowen, a facilitator for TRL, and his coordinator Ken Norquist, used the Lounge to explore the topic of positive black history *outside* of America. The story of Yasuke proved the perfect introduction. Keville is a professional American manga artist, who specializes in drawing and teaching about manga with black characters. He'd only recently found the story of Yasuke. Years after graduating with a BS in Media Arts and Animation, he'd started working per-diem as an art teacher for the local libraries as well as selling self-published comics at art conventions. It was at one of these conventions that he learned about Yasuke for the first time. Keville recalls, "A barber, who remembered me from a previous comics convention, was moving his busi-

This Teen Reading Lounge flyer includes some of the best known black mangu characters, including Yasuke.

ness to Japan. His excitement overflowed as he explained his love for Japanese black culture and a historical black samurai."

A tri-citizen himself—Trinidad, Canada and the United States—Keville understood the man's excitement all too well. He became fascinated with the story behind "black people in Japanese history." At the time, Keville explains, "Professor Lockley's paper ('The story of Yasuke: Nobunaga's African retainer,' published in 2016) was the most comprehensive account of the life of Yasuke and helped shaped the landscape of my course." For Keville, Yasuke's past was a reminder that "there was some light in a dark time of Black/African culture." Yasuke was one positive in a wider ethnic history "generally portrayed as containing suffering and victimhood."

After some research, Keville shared this story and pitched his idea for a Black History Month–themed anime club to the Lansdowne Public Library around the time they were considering applying to the TRL program. He came to Yasuke through

BY KIND COURTESY OF THE ARTIST, KEVILLE A. BOWEN

The cover of a Yasuke-themed manga book.

other Japanese-language "black manga" (manga with black people as the protagonists) and was inspired to base an art course for teenagers on the character of *Kurosuke* and Yasuke's life in general. He felt Yasuke's story would inspire and empower his students, who are mostly African American.

"Are there any black anime characters?" The simple question posed to a group of teen anime fans led to these same teens exploring examples of manga with black protagonists such as *Knights* by Murao Minoru. They also learned about successful black figures in the fifteenth and sixteenth centuries and then created their own comic about the black samurai, Yasuke.

To help with their ongoing research, in April 2017, around twenty students were present for a Skype interview with me from my home in Chiba, near Tokyo. The conversation started with questions about Japanese kids' familiarity with Yasuke's history, and ranged to black cultural influence in Japan, which then changed to the topic of the sense of being "other" and what it's like being a non-Japanese person living in Japan. The group also discussed anime/manga culture, its gravity within Japan and its unique portrayal of women compared to other animation traditions. The topic of exploring black history outside of the United States was such a success that the teens continued sharing more black characters in anime/manga throughout the year. "We enjoyed it so much that I continued with the Teen Reading Lounge, expanding our topics to include unknown black historical figures from America's Civil War and encouraging teens to continue to see the world from a different perspective," said Keville.

Yasuke's story is again seen as one of inspiration, inclusion and positive action. People do not need to feel trapped by the cards that life has dealt them or the circumstances of prejudice, expectation or seemingly limited horizons. There were many men and women like him, not just from Africa, but from all over the globe. His was a time when people could make a name and fortune for themselves pretty much anywhere in the world given the right circumstances. An age before deeply institutionalized racism, industrialized and hereditary slavery, before governments were powerful enough to control their citizens' international movements (no passports or ID cards), and before deep knowledge of the wider world removed from a narrow localized narrative. An age when to be exotic and to possess valuable information of the outside world, international contacts, foreign languages and different technological know-how was a saleable and desirable asset.

This was Yasuke's time.

He entered the halls of power, lived in grand castles and fought on the front line; he became a samurai, a member of one of the most famous warrior elites in history. He was a trendsetter and pioneer who allowed other foreigners to be employed in Japan in droves for decades after until Japan entered a new age of maritime restrictions in the 1630s.

The spirit of the most famous, indeed one of the only, Africans to play a serious part in Japanese history remains with us today, unforgotten centuries later. In the form of an ever-changing legend, an inspiration for people around the globe.

Yasuke vanished in 1594.

His story is only now beginning.

★ ★ ★ ★ ★

AFTERWORD

The Witnesses

Luís Fróis: Fróis, the most prolific writer on Yasuke, was born in Lisbon in 1532 and joined the Jesuits in 1548. He arrived in Japan in 1563. Fróis wrote more than one hundred letters and reports about Japan, many running to thousands of words, and several books during the 1580s and '90s. His work, which is highly observant, descriptive and even entertaining, formed much of the basis of European knowledge on Japan until the modern age and he was one of the greats of the Jesuit mission in Japan. He was remembered by a colleague as: "meritorious more than any others of the Japanese Christianism and for the deeds which during thirty-four years consumed him there, and for the memories of the successes of that Church of which he yearly gave news to Europe." He lived the last seven years of his life in Nagasaki where he died in 1597, aged sixty-five. His last published work was an eye-witness report of the Nagasaki martyrs' executions of 1597.

Ōta Gyūichi: The author of *The Chronicle of Oda Nobunaga*, and numerous other books recording the times in which he lived, was born in 1527 and died in 1613. As a samurai he was an expert archer who took part in many of Nobunaga's early battles. He was also a member of Nobunaga's falconry team. By the 1570s, he had been promoted to Nobunaga's administrative staff in both Kyoto and Azuchi. He proceeded to serve Hideyoshi and then his son in administrative capacities. Most of his writing was only published posthumously from manuscripts and diaries he wrote throughout his life. He recorded Yasuke twice, specifically covering his first audience and the fact of his promotion.

Matsudaira Ietada: Matsudaira Ietada, 1555–1600, who recorded Yasuke once in his diary after the Takeda campaign, went on to be promoted under his kinsman Tokugawa Ieyasu. In 1599, he was given command of Fushimi Castle near Kyoto. It proved a fatal appointment. In 1600, just prior to the decisive Battle of Sekigahara at which Tokugawa seized the reins of power for good, Matsudaira died defending the walls. His diary is one of the best historical sources of the times he lived in.

Lourenço Mexia: Mexia was a close aide to Valignano and *also* an informant on his activities to Rome, appointed by the Jesuit Superior General himself. He was a strong supporter of Valignano's policy of adaptation to Japanese norms, especially regarding diet. He observed that the Japanese judged people on what they ate and how they ate it, and the European's table manners and dietary choices, especially meat, were damaging their likelihood of converting the Japanese. He must have known Yasuke well, but chose to record his presence only once due to his surprisingly swift promotion in Nobunaga's service (this is one more mention than Valignano saw fit to include). It is from Mexia that we know about Yasuke and Nobunaga's close conversations and potential elevation to lord status. He died in 1599.

The Jesuits

Alessandro Valignano: Alessandro Valignano left Japan in 1582 bound for Europe with the first Japanese embassy to Rome, but he never got there. In India, new orders awaited him, appointing him to the post of *Provincial* of the Jesuits in Asia. He returned to Japan twice, in 1590–1592 and 1598–1603, but was based mainly in Macao where he carried on his educational tradition by, among other things, founding St. Paul's College of Macao (which claims to be the first European university in Asia) to train Jesuits in Chinese language and traditions to facilitate their mission there.

Valignano was said to be a man of "tremendous energy and boundless religious ardor, a born leader of men, who by the charm of his personality and the irresistible power of his example inspired the missionaries with ever-fresher and ever-greater enthusiasm for their work." This energy allowed him to write perhaps dozens of books and thousands of letters about the Far East, the best methods of missionary work and instruction manuals on how to logically refute "paganism" and numerous educational tracts. It is said he often worked into the early hours of the morning.

The Japanese embassy which he sent to Rome in the 1580s, normally referred to as the Tensho Embassy, met with two Popes and wowed both European rulers and their citizens. Their visit was quite the event of the decade, with crowds of thousands turning out to greet them. Even Queen Elizabeth of England, persona non grata in Catholic lands after her excommunication in 1570, demanded two reports a week from her European spies on their progress. The leader, Ito Mancio, was appointed *Cavaliere di Speron d'Oro* (Knight of the Order of the Golden Spur), and is possibly the only samurai to have concurrently been a European knight also.

After Valignano's death, an anonymous colleague wrote, "In him we lament not only our former Visitor and Father, but, as many will have it, the Apostle of Japan. For, filled with a special love for that Mission and burning with zeal for the conversion of that realm, he set no limit to his efforts on behalf of it."

Valignano died in Macao in 1606, at the age of sixty-seven.

There is no record of him having met Yasuke again.

Today, less than 1 percent of the Japanese population is Christian.

Father Gnecchi Soldo Organtino: After the destruction of Azuchi, including the seminary, Organtino was resident in the safety of Lord Takayama Ukon's Takatsuki fief until he managed to found a new mission in Osaka. Kyoto was still too volatile and uncertain. He tried to mitigate the fallout from Hideyoshi's banning of the Jesuits in 1587, but ultimately failed. However, he managed to stay in hiding within the proximity of Kyoto to support the Christian community there, and even baptized two of Nobunaga's grandsons. He died in 1609, in Nagasaki, at the age of seventy-six.

Father Gaspar Coelho: The mission superior who caused so much trouble by his scheming and overly aggressive attitude did not live long enough to be chastised by Valignano on his second visit. He died in disgrace in 1590 at Kazusa, near Kuchinotsu, where Yasuke had briefly lived ten years before and Valignano had set up his first seminary.

The Warlords

Hideyoshi: Hideyoshi came out on top in the brief battle to succeed Nobunaga and continued his work of unifying Japan.

He initially supported the Catholic missionaries in their work, but like Nobunaga had no serious religious conviction, unless it suited his political ends. The Jesuits eventually got on his nerves, and seeing them as a potential threat, like every other potential threat, he dealt with them. In 1587, he banished all Jesuits from his realm. He never seriously enforced the prescript, but it was the first warning sign that Catholicism would not have an easy future in Japan. By the early 1590s, there was nothing left to do in the unification of Japan, and Hideyoshi decided to invade Korea with the ultimate goal being to sit on the imperial throne of China and possibly conquer India as well. Although the war was clearly a failure, Hideyoshi refused to concede defeat and the worn-out samurai held on in isolated castles on the Korean coast until he died. This legacy strains East Asian relations to this day.

Tokugawa Ieyasu: Nobunaga's key ally. Following Hideyoshi's death, Tokugawa Ieyasu usurped power through acting as chief regent for Hideyoshi's infant son Hideyori (a similar ploy Hideyoshi had used when assuming Nobunaga's power). Tokugawa's ascent, however, was not without dissent, and resulted in a series of battles. The final conflict was one of the largest battles, globally, of the whole seventeenth century. The Battle of Sekigahara, in which approximately one hundred seventy thousand warriors took part (on the day itself; tens of thousands of others were delayed or fighting on related battlefields), was a decisive victory and decided Japanese politics for nearly three hundred years. Tokugawa's descendants would rule in peace, with virtually no challenge, until the 1860s. Ieyasu himself, described as one of the richest men in the world by an English merchant, founded Tokyo and left a legacy of laws and guidance that shapes Japanese society to this day. He is considered by some to be one of the greatest statesmen who ever lived.

Takayama Ukon: Takayama's support was crucial for Hideyoshi's usurpation of the national leadership of the Oda clan, but that did not mean he was invulnerable. In 1587, Hideyoshi ordered all Christian lords to renounce their faith. Takayama declined and was banished. He received a measure of forgiveness through being permitted to enter the service of the powerful Maeda clan but still refused to renounce Catholicism. After the definitive and final Jesuit expulsion edict in 1614, which included prominent Japanese Christians, he went into foreign exile, along with three hundred of his followers, in Manila. The colonial government of the Spanish Philippines saw an opportunity and offered to invade Japan to protect the Japanese Christians. Takayama refused to give his support and shortly afterward, in February 1615, breathed his last. The Spanish honored him with the funeral of a great lord and he is commemorated with a statue in the center of the old Japanese quarter of Manila, Plaza Dilao. With the support of Pope Francis in Rome, Takayama was beatified in his home town of Osaka in 2017, and became the Blessed Justo Takayama Ukon, only one step from sainthood.

Arima Harunobu: Arima Harunobu, who'd first welcomed Yasuke to Japan, regained some measure of autonomy after the Battle of Okitanawate, but remained in the shadow of his Satsuma clan allies. His people knew peace for the first time in decades. When the Shimazu were humbled by Hideyoshi, Arima bent the knee to him and subsequently was dispatched with two thousand troops to Korea in 1592. His run of picking the victors continued when he supported Tokugawa Ieyasu after Hideyoshi's death but his luck ran out when he failed in a mission to invade Taiwan and some sailors on one of his ships ran amok in Macao and were executed by the Portuguese authorities in 1608. The following year, he seized the Portuguese trading ship in Nagasaki in revenge and after a long battle, Captain Major André Pessoa blew up the whole ship rather than surrender. Although Ieyasu rewarded Arima for this, the "reward" was a marriage between Arima's son and Ieyasu's adoptive daughter, a problem because Arima's son was already married. The son capitulated and divorced his Catholic wife, apostatized and poisoned Ieyasu's mind against his own father, and in 1612, the senior Arima, already exiled, was ordered to perform seppuku. As a Christian, Arima could not commit suicide, so instead he accepted the death of a common criminal, beheading.

Katō Kiyomasa: The warlord who employed "Kurobo," and wrought havoc in Korea. After Hideyoshi's death, Katō chose to support Tokugawa Ieyasu, and was rewarded richly by becoming one of the most powerful lords in the land. He also remained loyal, however, to Hideyoshi's son Hideyori, whom Ieyasu had usurped, and attempted to act as a mediator between them. He died in 1611.

Ōtomo Sōrin: Ōtomo Sōrin became a vassal of the great conqueror, Hideyoshi, in 1587. He died of old age the same year. We know nothing more of his estranged wife "Jezebel," except that she also died in 1587.

Ōmura Sumitada: The lord who gifted the Jesuits Nagasaki. Despite Ōmura's questionable adherence to Catholicism at first (to obtain arms and outside support), he made strenuous efforts to understand the creed and remained a Christian until his death from tuberculosis, on June 23, 1587. There is no record of what happened to his daughter, who'd refused to marry Arima. His son Yoshiaki, however, made the politically astute move (in the climate of the early seventeenth century) to ban the Jesuits and Christianity from the Ōmura domain.

Hattori Hanzō: Nobunaga destroyed the autonomy of the Iga ninja once and for all. However, their leader, Hattori Hanzō, took them into the service of Tokugawa Ieyasu, and a large corps, around three hundred, formed a part of the guard at Ieyasu's new Edo Castle. Hattori has gone down in history and legend as the best known of the ninja, and as such has enjoyed a huge showing in popular culture, video games, movies, television, manga and books in Japan and overseas. Most famously the *Kage no Gundan* (Shadow Warriors) movie and TV series which depict him and his (semifictional) descendants. Hattori and his descendants of each generation, also named Hanzō, were played by Sonny Chiba in the series, and when Quentin Tarantino needed a Hattori Hanzō for the movie *Kill Bill*, he commissioned Chiba to play a fictional Hattori Hanzō XIV. The gate that Hattori guarded in Edo Castle, now the Imperial Palace, was named after him, and in turn the Hanzōmon metro line, is named after the castle gate.

The Places

Japan: Japan was reunified as a political unit by Hideyoshi in 1590 bringing an uneasy end to The Age of the Country at War. It turned out to be only a pause in the fighting. When Hideyoshi died in 1598, the struggle to succeed him led to the massive conflagration of the Battle of Sekigahara where the forces loyal to Hideyoshi's seven-year-old son, Hideyori, squared off against Tokugawa Ieyasu. On the battlefield, one hundred seventy thousand samurai fought. In the numerous sideshows, many tens of thousands more were involved. Ieyasu won a crushing victory, and shortly afterward founded a new shogunal dynasty which would rule until the modern age. The final conflict to secure Tokugawa's rule came in 1614–1615 when the last remaining supporters of Hideyori defiantly gathered at Osaka Castle. In a series of battles, Tokugawa Ieyasu again came out on top and this time it was definitive. There would be no real challenge to Tokugawa rule until the 1860s.

Unfortunately for the Jesuits, they'd backed the wrong side.

Crosses and Catholic banners had been held high in battle at Osaka. The punishment was permanent expulsion. Over the next two decades, Catholics—or, according to Tokugawa law, criminals and purveyors of pernicious teachings—were given the chance to recant their faith or face death. The majority apostatized sometimes under heavy torture. At the same time, the large Japanese diaspora around Asia, estimated to have perhaps been as high as one hundred thousand, were not always on their best behavior. Piracy and mercenary activity were rife, a great embarrassment for the shogunate which wished to look respectable in the eyes of the world. By the 1630s, the government had had enough, and promulgated a series of laws prohibiting Japanese citizens from travel abroad (and denying repatriation to those who did not come home quickly). All Catholic nations were forbidden to enter Japanese waters and the Dutch were the only non-Asian foreigners allowed to trade at all. They were restricted to a small man-made island called Dejima in Nagasaki bay. Chinese (and other Asian trade which was often conducted on Chinese ships), Korean, Ryukuan (modern-day Okinawa Prefecture, but then an independent nation) and Ainu (the indigenous inhabitants of Hokkaido and the islands further north) trade continued and formal diplomatic relations were maintained with Korea, Ryukyu and the Dutch East India Company. Yasuke could not have flourished in this world; he would not have even been able to travel farther than the Nagasaki dockside.

In the 1670s and '80s, economic issues, mainly to do with declining output from the silver mines that had funded Japanese imports for so long, forced a rethink and it was decided to further limit the amount of foreign trade each year to retain bullion as far as possible for domestic use. This, coupled with a drive to improve domestic industry to make up for reduced imports, had the effect of boosting the national economy. Production of products like silk, sugar and tea boomed and the quality rivaled that of the old Chinese imports for the first time as techniques were perfected, often with the help of Chinese experts.

Following the disaster of Hideyoshi's Korean war, the Tokugawa declined any serious foreign military activity and imposed a national peace. The energy that the samurai had once expended in war was spent on the arts and scholarship. Drama, printing, pornography, writing, painting, pottery, in fact just about any form of art you can think of, flourished. Philosophy and ethical studies took over from military strategy as the learning of preference, although the martial arts were never forgotten and were assiduously practiced and perfected. Not everything was rosy, however. Natural disasters and famine were never far away from the growing population. The strict caste system—samurai, peasants, artisans and merchants, in that order—implemented by Hideyoshi and continued by his successors meant that social mobility was difficult.

By the nineteenth century, social pressures were mounting, pressures that would lead to a very different future for Japan. But that is another story.

Macao: Macao continued to thrive on trade with Japan until the Japanese government definitively expelled all Portuguese residents and their Japanese families in 1638 due to perceived Portuguese support for a rebellion in the Shimabara peninsula, Arima's old lands. The Macanese were devastated; the Nagasaki trade was the cornerstone of their economy. And so they sent four of their leading citizens to beg for the restoration of trading rights in 1640. The shogunate was not amused with the "worm-like barbarians of Macao," and executed sixty-one of the ship's complement. A skeleton crew of thirteen was left alive to sail back to Macao with the message that if they ever darkened Japanese waters again, no one would be spared. Macao made do with other inter-Asian trade, especially with Manila, but was never as important an outpost of Portugal as it had been during the early days. By 1999, when it was reunited with mainland China, it had become the last European colony in Asia.

Nagasaki: Nagasaki grew and grew. It blossomed on trade and Christian faith until the Tokugawa government banned that religion. Most of the good burghers of Nagasaki quickly recanted Catholicism; the rest died by execution or torture. In the 1630s, it became the sole port authorized to trade with Europeans, and after 1641, the Dutch became the sole Europeans to be permitted trade there. The trade with China and Southeast Asia continued at a regular pace, around one hundred to two hundred Asian ships per year and one, sometimes two, Dutch ships. It remained the most multicultural city in Japan until the modern age. For the next two hundred years, around 10 percent of the population were Chinese.

August 9, 1945—the city of Kokura was half-covered in smoke from fires started by a firebombing raid of more than two hundred United States B-29s on nearby Yahata the previous day. With such low visibility over Kokura, another B-29, *Bockscar*, decided on its backup target and dropped the second atomic bomb on Nagasaki, ultimately killing as many as two hundred thousand people.

Tottori Castle: Tottori Castle remained a formidable fortress and center of local government until the modern age. In 1943, it was badly damaged in a massive earthquake, and the old noble family which had been in residence since 1600, the Ikedas, donated what was left to the people of the city. Today, the walls have been restored and you can attempt to climb the sheer slopes to the mountain summit. Beware of bears.

Azuchi: Azuchi was Nobunaga's city. There was little there before him and little left after him. Today it is a sleepy town of around ten thousand. The castle ruins are well preserved and can be visited; there is even a reconstruction of the top two floors of Nobunaga's glorious seven-floor donjon. Sadly nothing of the original building remains.

Sakai: Sakai lost its international verve and vibe when Japan restricted foreign trade to designated ports, of which it was not one. It continued, however, to be a major center of national shipping and trade, particularly known for its weapon and knife manufacture. Today, it has been all but swallowed up by its larger and louder neighbor, Osaka.

Kyoto: Kyoto remains the spiritual capital of Japan, even if it gave up the title of Imperial Capital in 1868 when the emperor moved to his current home in Tokyo. It is one of the world's great cities, bursting with energy both ancient and modern and tourists from around the world flock there. Temples and shrines neighbor markets and department stores. It boasts the second most Michelin stars of any city in the world. Number one is Tokyo.

The Satsuma and Mori clans: Although these two hugely powerful clans bent the knee to Hideyoshi, they were never destroyed in the way that Nobunaga had destroyed the Takeda. Both fought against Tokugawa Ieyasu at the Battle of Sekigahara, but again were allowed to survive as coherent entities. The Satsuma clan were even permitted to carry out an invasion of the Ryukyu Kingdom (modern Okinawa) in 1609 and rule those islands as a colony for the next two centuries. This gave them direct access to the hugely lucrative trade with China, something that no other clan other than the ruling Tokugawa enjoyed.

Both clans were mortal enemies of the other, but when they combined forces in 1866, they were powerful enough to remove the Tokugawa from power and usher in a new era for Japan known as the Meiji Era; they founded modern Japan.

The Mori were cannier than the Satsuma, and slowly edged them out of power in the late nineteenth century. To this day, a large number of prime ministers, including the current incumbent (in 2018) Abe Shinzo, are from what was Mori clan territory, the modern Yamaguchi Prefecture.

AUTHOR NOTE

In 2009, quite by chance, I first happened upon the extraordinary, and little-known (especially then), historical character of Yasuke.

I'd moved to Japan a decade before from Britain for a teaching opportunity. Like Yasuke, I was a stranger in a strange land but learning every day: the Japanese language, patience and staying quiet, Japanese cookery, the beauty of *onsen* (hot springs), what snow really is, the true value of central heating and how to teach and to deal with personal relationships cross-culturally. These first years in Japan, coincidentally in the tiny town of Shikano in Tottori Prefecture (where Yasuke may have visited), changed my life. I grew up, learned a fascinating language (my favorite), determined the shape of my future.

When I stumbled upon the Yasuke story online, I instantly became fascinated by this man who'd traveled so far from his homeland to appear directly beside the dominant warlord in Japan and be granted another culture's highest opportunity and honor. Although I'd initially assumed, blithely, that men and women like Yasuke were all slaves, in grave conditions, I came to see there was actually a far more complex and inspiring story to tell. Here was a slave soldier from Africa who'd most likely worked for royalty in India, then for one of the most prominent Jesuits in Christendom and ultimately for a mighty Japanese warlord. It was remarkable, epic even. A true-life tale of great adventure.

Migrants like Yasuke, however, have generally managed to slip through the cracks of historical research and I soon decided to find out more and begin the study necessary to write a book based on his story.

Over the next six years, I investigated primary sources (diaries, letters, histories written more than three hundred years ago) for any mention of Yasuke or men and women like him. The internet provided new means to access highly obscure European accounts of sixteenth-century Japan, old Japanese chronicles, and dusty volumes about ancient African kingdoms which could not have been easily obtained by one person without a great deal of travel only a decade ago. I also found relevant material in university libraries.

Here, he's mentioned escaping death at the hands of a curious crowd who perhaps craved a piece of clothing as some form of celebrity trophy. Here, a diary entry where Yasuke was witnessed performing feats of strength and chatting convivially alongside the sons of Japan's most powerful warlord. It was not long before I could imagine Yasuke walking the wide boulevards of Kyoto, dressed in exotic garb from China, India and Europe, an intimidating spear in one hand, a gently curved Japanese sword thrust through a sash at his waist. I also began seeing links between Yasuke's story and others, both in Japan and around the world. A new remarkable world of international exploration, soldiery and trade opened up to me, beyond the notion of the only pioneers being western Europeans searching for glory and "new worlds," but, rather, a far more nuanced story of all those moving around the contracting globe in the sixteenth and seventeenth centuries. Those whose talents, the vicissitudes of fate, and perhaps a guardian angel or two, determined where and how far they could, and would, go.

All this detective work merged to paint a picture of Yasuke's life. But the picture had yet to become a complete narrative where all the gaps were filled. Fortunately, I was able to directly contact many researchers and historians around the world who generously answered my requests appealing for otherwise unobtainable research leads, informed opinions and material. Even the first secretary of the Embassy of the Republic of Mozambique in Tokyo granted me an interview to assist in crossing the *Ts* and dotting the *Is* on some outstanding questions on a country about which it is still difficult to find much historical information.

This work gave my career a very new and specific focus: I began to teach courses concentrating expressly on the Japanese discovery of the world and the world's discovery of Japan. Yasuke shaped me. And, by 2015, flush with an abundance of genuine historical material and evidence of this amazing man, I set out on the task of writing his life story, thinking it would cover a few thousand words. At fifteen thousand words—and five thousand over my intended academic publication's word limit—I realized Yasuke had a much larger story to tell than I'd first dreamed.

And so this book came to be written, almost by mistake. But the more stones I rolled over, the more fascinating life stories, yet to be told, emerged: stories of mercenaries, sailors, explorers, travelers, sultans, viziers, concubines, pirates, missionaries, cooks, warlords and adventurers, as well as heartrending tales of hard, anonymous, unrelenting slave labor. It became the story of not just Yasuke, but that of people like him, whose deeds do not normally enter the history books, either because they cannot themselves write their stories, or because the dominant sections of global society tend to concentrate on the great exploits of their own classes and castes, and not the "little people"—or in Yasuke's case, the giants—who hold them up. I realized, or hoped, that millions of people could speak through Yasuke. He could, when brought fully back to life, perhaps give a voice to those whom history has often forgotten.

But then I made an even more remarkable discovery. Yasuke lives on today. The African samurai actually seems very much to be a character of the internet age as much as the sixteenth century. Hundreds have been inspired to produce documentaries, make computer games, write novels, draw manga and use Yasuke's legend as a base for educational and cultural programs. As the final chapter shows, Yasuke has taken the step bestowed to only a few people in history, from mere mortal to an adaptable and still-growing legend. This remarkable man's story seems to attract people for a variety of reasons as I discovered in 2016 from the feedback to my first academic paper about him: "The story of Yasuke: Nobunaga's African retainer." The reactions that came in via email and other platforms from all over the world were in some ways shocking in what they revealed about modern humans and our relationship to history in general *and* to the historical character of Yasuke in particular.

For Sarah, a television producer, Yasuke was representative of an alternative view of history which does not place white European males at its center, but tells a soaring success story of a non-European, without placing them as a victim. As an American Caucasian female working in a largely male profession in Japan, she'd personally experienced many instances of sexism and racism directed toward her, including jibes at her Japanese husband by a white American coworker. For her, Nobunaga's regard for Yasuke legitimized his worth in her eyes far more than the Jesuit disregard for their African servants casts them as victims of their age.

A British author and fellow Japanese history enthusiast wrote to suggest Yasuke proves that, despite what some people across all nations now claim, the world "has *always* been a lot smaller than it appears." Looking through history, we find tens of thousands of Yasukes—immigrants in places you'd never think a person of that race, color, nationality or creed would be. "In times when the world faces a refugee crisis, Yasuke proves there have always been exceptional people who've adapted and become part of a culture completely alien to their own."

For many correspondents, Yasuke represented the outsider who achieves success, a lesson for everybody on how to deal with modern-day issues of multiculturalism and alienation in a global-ized world—a world where homogenous societies no longer exist and political states rarely follow ethnic or tribal borders. There is no reason why a man like Yasuke was any less likely to rise to prominence in a medieval Japanese setting than Handel was in a British setting or Son Masayoshi, son of postwar Korean immi-grants and one of the world's richest self-made men, in modern-day Japan. Yasuke's story is one to provoke inspiration, inclusion and positive action.

The fact that published history has traditionally been written from an ethnocentric, and predominantly Eurocentric, perspec-tive is probably the most likely reason why Yasuke's story has re-ceived so little serious attention up to now. It does not fit into any national box, nor does it identify with major national diasporas or cross-national community relationships, as no one knows for certain from where he originates. The academic research on the black presence in Japan and East Asia, in any language, is tiny, and research on non-Japanese and immigrant communities, with the exception of European traders in the sixteenth and seventeenth centuries and current-day immigration issues, in the Japanese con-text is also highly limited. Yasuke has essentially managed to slip through the cracks of historical research and therefore historical storytelling until now. Whether this attempt at telling the full and comprehensive story of the African samurai will change it, or in fact reduce the fascination that comes with his mystery is hard to tell. I, for one, hope not. I hope Yasuke's story lives on for a long time and continues to provide a source of inspiration for whoever needs or wants it.

This book is about one young man of African origin whom the tides of history washed up in Japan. The central theme is his life, but to understand and analyze that life, it became necessary to illuminate the maritime and migratory lives of Africans and other peoples who had contact with them in the sixteenth century. As such, it covers a wide swathe of the globe illuminating his journey and likely life from Africa to Japan. After the more than eight years that this book eventually took to create, I apologize for any errors that may have slipped in, and the fault is entirely mine.

This book does not in general attempt a critical look at the African slave trade and its global consequences, nor does it attempt any particular cultural criticism of any who engaged in what we might now regard as dubious practices. It tries to look at facts and possibilities and present them as such. To the contemporary mind, many of the activities and beliefs of people of all ilk herein seem strange and perhaps even horrific. However, at the time they were not necessarily seen in the same light. In a world where Christians and Muslims, and indeed militant Buddhist monks, saw it as their prime and sacred duty to spread the word of "their" god, enslaving and even killing people was often justified as an act that would save the victim's eternal soul. It is easy to look back and judge, and likely people in the future will look back at our world with disbelief and horror. I have tried to resist the temptation to write as a modern judge and, instead, write as a dispassionate observer so as to give a better feeling and flavor of the times. I hope not to appear callous for doing this.

I would like to express profound gratitude to Manami Tamaoki, my agent, and her team and colleagues, particularly Ken Mori and Alex Korenori, at the Tuttle-Mori Agency in Tokyo, for having the faith, and dedicating the energy and time to help develop a very rough idea from a first-time author. None of this would ever have happened without them. They helped shape an earlier rendering of this book (published in Japan) and found the right agency in the United States to facilitate the version which you now find in your hands. Thank you to Peter McGuigan and the whole team at Foundry Literary + Media for your commitment to the project, shaping its future and helping to bring Yasuke to a wider audience.

Peter teamed me with Geoffrey Girard, an experienced, inspiring and innovative author and collaborator. Throughout the time we were writing together, I never failed to pick up new tips, ideas and techniques which continuously strengthened and took the book in new, exciting and often unforeseen directions. Geoffrey and I traveled several thousand miles together across Japan in the summer of 2017, investigating Kyoto, Azuchi, Lake Biwa, Mount Fuji and the former Takeda domains and the routes of Yasuke's principal travels. Not to mention more general background work in various regions—from research in the National Diet Library in Tokyo to an unforgettable dinner in Kyoto at one of the oldest restaurants in Japan, which amazingly existed during Yasuke's lifetime. It was a fast-paced and tiring expedition, and my leg muscles felt ten years younger at the end. Geoffrey, thank you for all your hard work, your great questions and your discerning eye.

Thanks to our publisher and editor, Peter Joseph, and his team at Hanover Square Press who embraced this story and its message, put their full support behind it from Day One, and then kept a close eye on proceedings and provided valuable guidance and collaboration throughout.

At Ohta Publishing in Tokyo, I am deeply grateful for the work of Junko Kawakami, the original editor, and Yoshiko Fuji, the translator of the original Japanese edition, for their hard work, input and modification suggestions which were crucial to the book. And also to Sakujin Kirino sensei, a venerable expert on the Honnō-ji Incident, whose kind reading of the first book and comments were highly constructive. Here is also an appropriate place to thank my friend, the Master Calligrapher Ponte Ryuurui, who crafted the characters at the beginning of each part of the book. He can be found at www.ryuurui.com.

For academic advice, support, ideas and friendship, a thousand thanks to Professor Akira Mabuchi of Nihon University College of Law, Professor Timon Screech of the School of Oriental and African Studies University of London, Professor Lúcio de Sousa of Tokyo University of Foreign Studies, Cliff Pereira, Fellow of the Royal Geographical Society, Dr. Onyeka Nubia, Writer in Residence at Narrative Eye, Dr. Ryan Hartley, and Philip Lockley (my brother).

Other notable contributions came from my old friend Akinori Osugi, the artist Keville Bowen and from Heidi Karino's lovely translations and language advice. And of course thanks to the old guys at the gym, who were happy to spend hours debating Yasuke's life, while I listened and noted their interesting takes in my head. In particular, Isao Hashizume, who generously hosted Geoffrey Girard and myself at his ancient family home in Hikone on the shores of Lake Biwa, and accompanied us in Azuchi. And thanks to all the other people who have helped in a hundred ways.

Finally I also would like to thank my wife, Junko, for putting up with endless Yasuke talk, probably not quite finished yet. Her ideas, advice and translation help with obscure texts were key to conceiving of Yasuke's life. Secondly, my children, Eleanor and Harry, who wanted to hear the stories and play sword fighting (I always had to be Akechi) in the park. Masae and Yusaburo Kinoshita and Andrew (my dad) and Caryl Lockley for looking after us and the kids so well to give me time to work on the book.

Yasuke's story continues—it sometimes seems that it's only now getting started—and continues to provide a source of inspiration for all who meet him.

弥助殿、幸あれ。バンザイ！

Thomas Lockley, Tokyo, 2018

NOTES

The selected bibliography at the end of each chapter section is a collection of the best resources available for readers to find out more about Yasuke's world. For more, please visit AfricanSamurai.com.

PART 1

Chapter 1

Yasuke's name: The word *Yasuke* (pronounced *Yas-kay*) is almost definitely a Japanese rendering of a foreign name, although "Yasuke" is not a wholly unknown name in Japan. Slaves and freedmen in the Portuese world were generally known by the Portuese names their masters gave them, and Yasuke would have been introduced by Father Organtino, the priest who accompanied him to his audience with Nobunaga, by this name. Similar-sounding names can be found in many of the variants of the biblical name "Isaac" from around the Indian Ocean. It is Yisake in Amharic (Ethiopian), Isaque in Portuese (pronounced something like "Yi-saa-ki") and Ishaq in Arabic (pronounced "Yi-shak"). Any of these three variants would quite likely be rendered into Japanese as "Yasuke," as the sounds of Japanese do not exactly match those of an Ethiopian language, Arabic or Portuguese.

Guns: Warfare in Japan changed forever in the years following 1543, when a Chinese pirate ship with several Portuguese merchants onboard was accidentally blown to the tiny island of Tanegashima just south of Kyushu. The local *daimyō* was predictably fascinated by the harquebus muskets which the merchants were only too happy to demonstrate the use of, and it didn't take long for his craftsmen to copy the innovative and effective killing machine. From there on, gun usage and manufacture spread like wildfire, and one of the earliest proponents was a youthful Oda Nobunaga. The gun transformed warfare, and society, as even peasants could be trained to use them cheaply, easily and quickly. No longer did samurai have to train from birth with sword and bow before they stood a chance on the battlefield. Armies could and did expand quickly and cheaply. While the older weapons still had their role, guns quickly became the center of strategy and battle plans. The Japanese soon copied and started to mass-manufacture them on a massive scale. These guns were of the harquebus type, early matchlock muskets, fired by pulling a trigger which touched a lighted match to the ignition. A charge of gunpowder was inserted into the muzzle, followed by the lead shot. The load was placed securely at the correct end of the barrel with a ramrod. These were light guns, easily carried by fast-moving armies and required little training to use. Shot was made on the battlefield by molding molten lead with a bullet mold carried at the warrior's belt. It is said that by the turn of the seventeenth century, there were more guns in Japan than in the whole of the rest of the world combined.

Dates: The dates used throughout this book are derived from historical documents. The calendar used in Japan at the time was the Chinese lunar calendar, hence New Year occurring in February of the European calendrical system. The calendar used by the Jesuits during this time was the Julian calendar. The Gregorian calendar, used in much of the world today, was adopted by the Catholic Church in 1582, but news of this would not have reached Japan for some time. Some dates in the historical documents are unclear, others contradictory. Every effort has been made to be as accurate as possible.

Burial at sea: Death at sea on Portuguese ships, when it came, was a simple matter. The master blew his silver whistle, the survivors bowed their heads and prayed to their god or gods that they were not next, and then their comrade, still warm, was tossed into the sea wrapped—*if* he was rich enough to have one—in his own sleeping mat.

Missing travelers: Many of the other Europeans employed on the initial India run—a mishmash of Portuguese, Germans, Flemings, Italians and Spaniards; mostly fugitives or desperate treasure hunters—had either perished on the earlier voyages or been content to stay in Goa or other parts of India after the arduous trip east, hence the high proportion of African and Indian sailors on Portuguese ships in Japanese art of the period. Any remaining European officers onboard had specific orders from the Portuguese Crown, who made all appointments, to carry on eastward. The Jesuits were sent by their order, and the mission superior (or Visitor if one were present) could decide where to send them upon their arrival in India. The common sailors had no such orders. They were short-term hires, engaged only for the one-way voyage, and then laid off or rehired once they landed at a new dock.

Trade in Japan: Systems and industries often thrive in strife-torn times, and sixteenth-century Japan was no exception. People from across the world, from Rome to China and Mexico, recognized those opportunities in Japan and went to answer that call. Japan had been focused for hundreds of years on a China-based trade system which had transferred extraordinary wealth from the center of Asia (China) to the periphery (Japan, Korea, Thailand and the smaller kingdoms and sultanates which now make up Vietnam, Malaysia, Indonesia and the Philippines), yet demanded subordination to the power, munificence and spiritual centrality of the Chinese emperor: The Son of Heaven. Trade was only allowed in controlled amounts, permitted to specially appointed trade partners (e.g., the governments that the Chinese throne recognized as legitimate) in the subordinate countries. Officially, it was designated as "tribute" but the Chinese government paid handsome sums for this tribute, and it was, in effect, a complicated closed-trading system which also conferred recognition on native rulers who engaged in the tribute trade. As civil war shook the Japanese islands, just who "the government" was became increasingly unclear. In the early sixteenth century, two noble Japanese families arrived in China claiming that status, and the Chinese agreed to trade only with one of them: the Hosokawa (later to refuse Akechi's advances and take Hideyoshi's side). Those left out, the Ouchi, were enraged and started a violent rampage in the Chinese port of Ningbo, killing locals and plundering property.

The Chinese were unsurprisingly not pleased with a foreign conflict spilling over onto their own soil, and demanded the perpetrators be delivered to them for justice. With no government in Japan able to enforce the Chinese request, nothing happened and eventually the Chinese cut both trade and diplomatic ties with all of Japan. Japanese people were henceforth forbidden to enter Chinese territory on pain of death, and Chinese merchants would suffer the same penalty if they traded with Japan. China's allies in Korea also tightened trading relations and restricted Japanese traders to one port, Busan, in the southeast. Products that had previously come from China, either from China itself or traded through Chinese ports from destinations as far away as Africa, became hard to come by in Japan. Korean trade also dropped dramatically. The products which became scarce ranged from luxuries such as silk, tiger skins, art, books, ivory and sugar to necessities such as coins, medicines and tea. There were two solutions to this problem: piracy and finding other sources of trade. Asian states that had previously traded only via China now opened up direct relations with Japanese lords and merchants. (As there was no effective central power in Japan, the regional lords and private entities were the ones to negotiate with.) Japanese ships started to range far afield for the first time, establishing Japanese communities and trading hubs in places like Siam (Thailand) and Manila. These "Japan Towns" facilitated a dramatic increase in trade, foreign relations and knowledge of the wider world. There were fortunes to be made in trading both local goods and products such as leather and sugar, and also China-derived goods such as silk; war machines needed to be oiled, and trade was the ideal way to make money. Furthermore, contacts were made now with strange pink and black peoples, the likes of whom many Japanese had never seen or conceived of before. Europeans, having met Chinese and Japanese merchants in Indian and other Asian ports, pressed east to find the source of the silver and silk these men carried with them. Setting foot in East Asia for the first time, they jumped into the intermediary trade too; it was much easier to trade spoilable products over short distances and then simply export the silver westward to trade again for products like spices which fetched a king's ransom in Europe. The Portuguese were allowed to trade in the Chinese port of Guangzhou (Canton) and eventually founded a base at Macao nearby. They could therefore get those products from the Chinese that the Japanese so desired, and make a huge profit by trading them. The riches to be made became legendary and the Portuguese Crown reaped the rewards. Thousands of Chinese entrepreneurs, or pirates, also avoiding the shackles of central trade quotas, disappeared mysteriously from southern Chinese ports each year, only to return later with ships filled with Japanese silver, gold, sulfur, art works and weapons. By Yasuke's time, Japan had become a place central to trade networks in its own right, not only a peripheral state in the Chinese hegemonic sphere. Local extractive and manufacturing industries expanded massively to pay for these imported products: silver, sulfur, copper and manufactures—mainly traditional weapons, but increasingly guns too. As these were new industries, expert help from abroad was sometimes engaged to share knowledge of exploitation techniques. These engineers came mainly from China clandestinely, but Portuguese and others, like Yasuke, also found their skills and knowledge in high demand. The increased income from silver and foreign trade enabled Japanese people to pay for ever more foreign products, thus increasing sophistication and cosmopolitanism in the islands. It also attracted ever more foreigners and adventurers, with skills to sell, to take part in the trade. As the foreign population and their power and influence increased, perceptions of the usefulness of, and of course the threat they posed, changed accordingly. The Japanese became far more open to ideas beyond their traditional mixture of native and Chinese roots and interest in other foreign ideas, products, culture and concepts—such as Christianity—became de rigueur among the ruling classes ("interest in" did not automatically mean acceptance, of course). Yasuke was, in Nobunaga's mind, a representative of this new feeling in Japan that the world was smaller, more relevant. Yasuke symbolized this in himself, simply by being at Nobunaga's side, but also proved it to others. It raised Nobunaga up in his supplicants' eyes and gave him the legitimacy of foreign as well as domestic recognition, connecting him in everyone's eyes with a world that stretched far beyond a horizon that any Japanese ruler had before conceived of.

Japão: The Jesuits' lingua franca was Portuguese and they called Japan "Japão," but they also sometimes used *Iapam* which is a name from old Portuguese, a word nearer to modern Galician (a Spanish regional language) than modern Portuguese. The revolutionary 1603–1604 dictionary *Vocabvlario da Lingoa de Iapam* (written by the Jesuits and comprising 32,293 entries) contains two more entries for Japan: *nifon* and *iippon*. The title of the book, however, clearly gives the name: *Iapam*. (The letter *J* was not used regularly until the seventeenth century.) The early Mandarin Chinese name for Japan was *Cipan* (sun origin), but the Portuguese likely first heard the word *Jipang* being used in the islands of Southeast Asia. Over time, *Cipan* and *Jipang* and *Iapam* morphed into the English word "Japan."

The Black Ships: When the Portuguese conquered Goa, they conveniently found a thriving shipbuilding industry combined with an abundance of timber resources within easy distance. They co-opted, probably enslaved, the shipwrights and established secure supplies of wood, and were soon building all shapes and sizes of vessels, from longboats to galleys to huge ocean-going *naos*, which eventually exceeded one thousand tons. This meant that they did not have to rely on ships arriving from Europe and could swiftly increase their local maritime power through exploiting colonial resources and manpower. It is said these ships were constructed of a black-colored wood, hence the name by which they became known in Japan—*"kurofune,"* or "black ships."

Selected Bibliography

Boxer, C. R. *Fidalgos in the Far East, 1550–1770.* The Hague: Martinus Nijhoff, 1948.

Cooper, Michael. *They Came to Japan: An Anthology of European Reports on Japan, 1543–1640.* Berkeley, CA: University of California Press, 1965.

Correia, Reis and Lage, Pedro. "Francisco Cabral and Lourenco Mexia in Macao (1582–1584): Two Different Perspectives of Evangelization in Japan." *Bulletin of Portuguese-Japanese Studies* 15 (2007): 47–77.

Fujita, Niel. *Japan's Encounter with Christianity: The Catholic Mission in Pre-Modern Japan.* New York: Paulist Press, 1991.

Gill, Robin. *Topsy-Turvey 1585.* Key Biscayne, FL: Paraverse Press, 2005.

Moran, J. F. *The Japanese and the Jesuits: Alessandro Valignano in Sixteenth Century Japan.* Abingdon: Routledge, 1993.

Russell-Wood, A. J. R. *The Portuguese Empire, 1415–1808. A World on the Move.* Baltimore and London: The Johns Hopkins University Press, 1992.

Souza, George. *The Survival of Empire: Portuguese Trade and Society in China and the South China Sea 1630–1754.* Cambridge: Cambridge University Press, 1986.

Chapter 2

"Italian": Valignano would have called himself Neapolitan. The ruling family of the Spanish Empire, the Hapsburgs, had territories dotted throughout Europe and the world that they'd acquired by marriage or conquest, and Naples happened to be one of them at this time. Subjects from different parts of the Hapsburg empire would not have referred to themselves by their emperor's name; they used their local appellation. The people we now know as "Spanish" would have said Castilian or Catalan, etc., depending on their birthplace.

Jesuits in Japan: The Jesuits, during Yasuke's time, were the only missionary order Rome allowed to preach in Japan. To keep this monopoly, they argued that other orders—the Franciscans, Dominicans, and Augustinians—would only misunderstand the knotty complications of the Japanese mission; that their meddling would simply undo all the good work already done by Ignatius's followers. While other orders, particularly the Franciscans, were lobbying to be let in, Rome steadfastly maintained the Jesuit monopoly. When other orders did eventually gain access, the Jesuits were proved right, among the first martyrs of Japan were a group of Franciscans who had overstepped the mark in what was considered decent and respectful behavior. They were crucified for their actions.

Jesuit missionaries had been active in Asia—what Europeans knew as "the Indies"—since only 1542. Attracted by travelers' and merchants' tales, Francis Xavier first landed in the port of Kagoshima, in the southernmost of the main Japanese islands, Kyushu, in 1549. While he may have held the Japanese in high regard and referred to them as "the best race yet discovered," as a good Jesuit, he still made great efforts to debate and supplant their religion and philosophy. The Buddhist establishment there was not initially hostile to these strange foreigners with their outlandish ideas and dogma, merely curious to know more, wondering perhaps if it was a new version of their own beliefs. Although Xavier made only a few hundred converts prior to his death in 1552, he'd efficaciously laid the foundations of the Jesuit mission in Japan and sent back the first reliable information on the "mysterious country" to Europe. He also left behind several priests, brothers and converts to carry on his work, which they duly did. By the time Yasuke arrived in 1579, there'd been one hundred thousand conversions. It was a fine start to the business of soul saving. The Jesuits worked hard at their mission in Japan, and from the first, they made genuine efforts to spread the Word of God through helping the poor and people in need of aid. While this endeared them to those people, much of the wider population remained puzzled by this and tended to look down on such charitable actions: *Why would people who claimed such high status demean themselves by consorting with lepers and outcasts?* Well, the poor themselves clearly didn't all see it that way and the Jesuits founded hospitals, leper colonies and orphanages to which many flocked.

Jesuit schools: By the 1570s, in the Japanese domains that had been touched by the missions, there were Jesuit catechism classes, informal schools which taught their students to be good Catholics, and also literacy in Japanese and basic Latin prayers. It was a rare opportunity to read and write in an age of war and chaos when only those upon high normally could. Although they would receive occasional attention from visiting priests and brothers, it was the Japanese lay helpers who carried on the teaching day to day. The Jesuits believed in education, and extended it wherever they could. They'd opened their first school in Europe in 1548 and founded more than thirty within the next decade. These schools were intended to counter the Reformation, act as missionary beachheads and promote the veracity of Roman Catholic thought. They also sponsored and facilitated scholarship, knowledge creation and global publishing on a massive scale. Jesuit dictionaries and lexicons of native languages in seventeenth-century Asia and the Americas were the first resources Europeans used to understand these ancient tongues, and still provide modern scholars with many of the earliest reliable phonetic transcriptions. In the next two hundred years, they'd found more than eight hundred formal educational institutions worldwide, in addition to countless unrecorded community classes, becoming one of the largest nongovernmental educational organizations in human history. Today, they still run more than five hundred institutions. (Jesuit schools have educated, among others, Descartes, Voltaire, Molière, James Joyce, Peter Paul Rubens, Arthur Conan Doyle, Fidel Castro, Alfred Hitchcock and Bill Clinton.)

Hinoe Castle: The fact of Arima's castle of Hinoe containing building material from plundered Buddhist temples is taken from archeological records. It is estimated that some forty temples and shrines were destroyed while Valignano stayed in Kuchinotsu. What had been a rough fortress in 1579 had, by 1590, become a palace, with both Chinese and European influence worked into its Japanese splendor. The sliding doors were painted with gold leaf, and summer scenes of the flora and fauna of the mountains adorned them. After opening these doors, the beautiful scenery of the Sea of Ariake dotted with islands could be seen. Fróis's assessment in 1590 was: "All rooms, big and small, were decorated with golden objects and resplendent and gorgeous paintings. This mansion is located within a brilliantly completed castle that was recently built by Arima Harunobu."

Selected Bibliography

Braga, J. M. "The Panegyric of Alexander Valignano S. J." *Monumenta Nipponica* 5, no. 2 (1942): 523–535.

Cooper, Michael. *Rodrigues the Interpreter: An Early Jesuit in Japan and China.* New York and Tokyo: Weatherhill, 1974.

Elison, George. *Deus Destroyed: The Image of Christianity in Early Modern Japan.* Cambridge, MA: Harvard University Press, 1973.

Farris, William Wayne. *Japan to 1600: A Social and Economic History.* Honolulu: University of Hawaii, 2009.

Hesselink, Reinier. *The Dream of Christian Nagasaki: World Trade and the Clash of Cultures, 1560–1640.* Jefferson, NC: McFarland & Company, 2016.

Massarella, Derek (Ed.) and J. F. Moran (trans.). *Japanese Travelers in Sixteenth-Century Europe: A Dialogue Concerning the Mission of the Japanese Ambassadors to the Roman Curia.* London: The Hakluyt Society, 2012.

Chapter 3

Yasuke's origins: Pinpointing Yasuke's origins in Africa is difficult. One secondary source, Solier, who wrote about Yasuke in the 1620s, stated that Yasuke was from Mozambique. There is no other evidence for this, and no prior source mentions it. The tribe in the immediate vicinity of the Portuguese-occupied island of Mozambique were called the Makua, a relatively peaceful agricultural people who had only migrated to the region in the 1570s. Until circa 1585, long after Yasuke had left, they managed a relatively conflict-free coexistence with the Europeans. While there were probably a few Makua slaves at Yasuke's time, the record is unclear, and the peaceful nature of relations makes it less likely that he was a Makua, as slaves would more likely have come from people who were unfriendly to the Portuguese. The possibility that he was sold because his family was in dire straits and he was an unneeded mouth to feed also exists. There were several famines during the decade, but a family would normally sell a young child not a strong young man. Another problem with this theory is that slavers preferred children because they were easier to control and manipulate; Yasuke would have been eighteen or nineteen when Valignano passed through Mozambique, late in the day to be enslaved. Finally the Makua had a very distinctive culture of filing teeth into points, this would surely have been a remarkable fact to the Japanese of the time, and would probably have been mentioned. It is not. One final problem with the Mozambique origin theory is that the Portuguese slave trade from Mozambique was relatively small at this time; only around two hundred to five hundred people a year were forcibly transported to India. Between 1500 and 1850 the total number of enslaved people transported to Indian Ocean destinations from Mozambique is estimated to have been between forty thousand to eighty thousand. The Arab, Jewish, Guajarati and Turkish slave trade from Northeast Africa by contrast was far larger; over the course of history, an estimated eleven to fourteen million Northeast African people were sold.

And then there is his height and extremely dark skin. Neither are characteristic of the peoples of the Mozambique region, who are generally smaller and have lighter-colored skin. But the north of Africa, integrated into the Indian Ocean slave trade, provides a people who sound far more like our description of Yasuke. The Dinka, for example, from what is now (in 2018) the world's youngest state, South Sudan, are famously, on average, the tallest people in the world. They are also strong warriors who hold themselves well and are much darker skinned (all things said of Yasuke) than their neighbors, in modern-day Ethiopia, Eritrea and Somalia. The Dinka are cattle herders and fierce warriors who in those days lived slightly farther north of their current lands on the banks of the Nile. They partake of distinctive facial scarring upon reaching adulthood, but Yasuke would probably have been taken before his coming of age rituals and therefore lacked these features. Slave raiders from what is now northern Sudan also raided the Dinka people at this time. The Dinka people only got their modern name in the nineteenth century, probably after being randomly assigned it by a British explorer or administrator. They call themselves the

Jaang. Through process of elimination, I have concluded that Yasuke was a member of the Jaang people.

The Age of the Country at War: The name *The Age of the Country at War* (*Sengoku jidai* in Japanese) harks back to an era of intensive warfare in ancient China which concluded with the victory of the state of Qin in 221 BCE and the submission of the other six independent states which formed the Chinese world of the time. This was the first unified Chinese Empire, and the Japanese historians who named *their* "Age of the Country at War" after it were seeking to legitimize their own state's unification and nationhood by alluding to a classical example of a state forged in bloody conflict.

The Jesuit printing press: The Jesuit press in Japan, exported from Lisbon on Valignano's orders, became their most globally prolific in the final years of the sixteenth century, producing copies of European *and* Japanese texts in the thousands when most print runs in Europe only ran into the hundreds. It was removed to Macao when the Jesuits were expelled.

Jesuit plotting: Despite their worldwide reach and willingness to do virtually anything to meet their goals in Christ, the Jesuits truly didn't have any plots more sinister than the saving of souls at this time. Any dreams of global European imperial domination are retrospective. That said, the leading Jesuits were often members of the most exalted families of Europe, and political intrigue came quite naturally. They could rarely avoid using worldly means to achieve their otherworldly ends. Some of the more blusterous Portuguese *Fidalgos* and Spanish conquistador types, however, did have more nefarious plans, perhaps, seeing the Jesuits as their tools to make a beachhead in East Asia. But such plans were pie in the sky and the Spanish throne absolutely forbade them; provoking the Chinese or Japanese to war would result in the loss of the Philippines and other imperial territories and valuable trade, not to mention the potential massacre of tens of thousands of Christians. The King of Spain specifically forbade his subjects to fight with Japanese samurai.

Selected Bibliography

Allen, Richard B. *European Slave Trading in the Indian Ocean, 1500–1850.* Athens, Ohio: Ohio University Press, 2014.

Chatterjee, Indrani, and Eaton, Richard. *Slavery and South Asian History*. Bloomington: Indiana University Press, 2006.

De Sousa, Lúcio. *Daikokaijidai no nihonjin dorei (Japanese slaves of the Maritime Age)*. Tokyo: Chuko Sosho, 2017.

Farris, William Wayne. *Japan to 1600: A Social and Economic History*. Honolulu: University of Hawaii, 2009.

Fogel, Joshua A. *Articulating the Sinosphere. Sino-Japanese Relations in Space and Time*. Cambridge, MA: Harvard University Press, 2009.

Gordon, Murray. *Slavery in the Arab World*. New York: New Amsterdam Books, 1989.

Jansen, Marius. *The Making of Modern Japan*. Cambridge, MA: Belknap Press, 2000.

Lorimer, Michael. *Sengokujidai: Autonomy, Division and Unity in Later Medieval Japan*. London: Olympia Publishers, 2008.

Madut-Kuendit, Lewis Anei. *The Dinka History: the Ancients of Sudan*. Perth, Australia: Africa World Books, 2015.

Pacheco, Diego. "The Founding of the Port of Nagasaki and its Cession to the Society of Jesus." *Monumenta Nipponica* 25, no. 3/4 (1970): 303–323.

Saunders, A.C. de C. M. *A Social History of Black Slaves and Freedmen in Portugal, 1441–1555*. Cambridge: Cambridge University Press, 1982.

Üçerler, Antoni. J. "Alessandro Valignano: man, missionary, and writer." *Renaissance Studies* 17, no. 3 (2003): 337–366.

Chapters 4 & 5

The Jesuits and drama: All around the world, the Jesuits were also strong believers in the power of theatre and drama to bring bible stories and the message of Jesus to life. Japan was no exception. The plays were performed with elaborate music and dance, often in the dark, and lit by lanterns and torches that were designed to keep out the literal and spiritual dark and show the central message of The Light. These became key events, cementing the central role of missions in lives and communities, even "uncomprehending converts," perhaps simply following their patriarch's, headman's or lord's order to become Christian, got a living, breathing expression of the concept they had professed during baptism. Divinity lived and could be seen. School children and local people took part and viewed them with gusto. One can only imagine the wonder with which these plays were greeted in small and extremely remote Japanese fishing villages and mountain communities, far from the domainal capitals and castle towns where any form of theatrical entertainment normally took place. In a way that would not be possible through words or pictures, until the printing press arrived from Europe in 1590 and started to mass-produce Christian artwork and texts, the message of the Bible came to life and worked its way into the people's hearts through performance. The most popular were the Christmas plays, and the first one took place in Ōtomo Sōrin's territory, Bungo, in 1560. Thousands are said to have traveled from miles around to see the story of Adam and Eve enacted by local Japanese Catholics. A tree sporting golden apples was placed in the middle of the stage, and so "real" was the performance that when Lucifer tempted Eve beneath the apple tree, the audience burst into tears. Things only got worse when an angel appeared and led Adam and Eve out of the Garden of Eden. As a finale, the angel reappeared and consoled the weeping playgoers with news of a distant day of salvation. Yasuke must also have enjoyed these dramas, a release from the often-monotonous work of his everyday life; perhaps he even joined in the acting. Balthazar, one of the three kings who attended Jesus' birth was, after all, traditionally depicted as a black-skinned Ethiopian.

Nagasaki prostitution: The Nagasaki region was poverty stricken and difficult to farm, and has borne the reputation for centuries of selling its daughters, and sometimes sons, into prostitution, permanently or as a temporary measure to raise a dowry for a good marriage, a reputation that continued well into the twentieth century. The fact that it was a port city with a widely fluctuating and wealthy population from all over the world simply added to a historic issue of poverty, capitalistic craving and human trafficking. Both poor women themselves and their families often jumped at the chance to escape poverty and to share in the riches of trade seemingly everywhere around them.

Images of Africans: Later, pictures of non-Asians (both black and white) would show them as devil-like beings, but the pictures during this early period of contact before Christianity came to be seen as a threat, seem to be remarkably nondiscriminatory. They show the "exotic" habits, clothes, behavior and racial characteristics in a generally unprejudiced, but clearly fascinated, way.

Blackened teeth: *Ohaguro* was a fashion for painting teeth black for cosmetic purposes, which persisted in Japan from ancient times until the nineteenth century. Women normally did it after they had married, but high-class men such as imperial court aristocrats and senior samurai also partook. Teeth were varnished with a lacquer made of iron filings, which needed to be reapplied several times a week, somewhat similar to modern-day nail varnish. The people in the fishing villages where Yasuke initially lived would probably not have had the time or resources to paint their teeth in this manner, but Lord Arima's court would have, so Yasuke would have encountered it soon after arriving in Japan. He may also have been aware of the practice from living in Macao.

Selected Bibliography

De Sousa, Lúcio. *Daikokaijidai no nihonjin dorei (Japanese slaves of the Maritime Age)*. Tokyo: Chuko Sosho, 2017.

De Sousa, Lúcio. *The Jewish Diaspora and the Perez Family Case in China, Japan, the Philippines, and the Americas (16th Century)*. Macao: Macao Foundation, 2015.

Fogel, Joshua A. *Articulating the Sinosphere. Sino-Japanese Relations in Space and Time*. Cambridge, MA: Harvard University Press, 2009.

Josephson, Jason Ananda. *The Invention of Religion in Japan*. Chicago: University of Chicago Press, 2012.

Kang, David C. *East Asia Before the West: Five Centuries of Trade and Tribute*. New York: Columbia University Press, 2012.

Keevak, Michael. *Becoming Yellow: A Short History of Racial Thinking.* Princeton: Princeton University Press, 2011.

Nelson, Thomas. "Slavery in Medieval Japan." *Monumenta Nipponica* 59, no. 4 (2004): 463–492.

Turnbull, Stephen. *Ninja: AD 1460–1650.* Oxford: Osprey, 2003.

Turnbull, Stephen. *Katana: The Samurai Sword.* Oxford: Osprey Publishing, 2011.

Turnbull, Stephen. *Ninja: Unmasking the Myth.* Barnsley: Frontline Books, 2017.

Chapter 6

Marriage and divorce: Yasuke arrived in 1579, an interesting time for sexual relations in Japan. Despite long exposure to Chinese ideas, society was still in the process of absorbing Confucian ethics of human relations, where women take a decidedly inferior role to males. However, it had probably not taken serious root among the less educated lower and rural classes. Buddhism had also become popular among the lower classes over the previous few centuries and the emphasis it puts on perceived female pollution may also have been having an effect on how society saw women and women saw themselves. In 1579, Catholicism, with its very foreign concepts of marriage and sexual relations had only a small hold in Kyushu and central Japan, but there seems to be little indication that many male converts chose to follow the missionaries' teachings on having only one sexual partner. Indeed, why should they when they saw the foreign Catholics, and later Protestants, who lived in Japan ignoring them with abandon? Following local custom, the foreign visitors commonly took a temporary wife and then paid her off at the end of their stay. That money became her dowry, so that she could take a more stable local husband and have a family. At this point in history it does not seem to have damaged a woman's reputation to have had sexual relations with a foreigner. Japanese and foreign men who could afford it often had one principle wife (polygamy in Japan was rare), but several concubines; the temporary visitors to Japan simply took temporary "wives," and "divorced" them at the end of their stay. Japanese men who had reason to live in different places at different times of the year, for example, merchants, also followed this practice and men who were exiled abroad, for example to the Amami Islands (which only officially became part of Japan much later) often took a local wife too.

But what were the older ways that still exercised such a powerful influence on the Japan that Yasuke knew? Firstly, this era was seeing the lowest point in the status of women in a region that had since ancient times held female status in high regard with clear property and inheritance rights, wide participation in economic and military activity, relative sexual freedom for both sexes, high rates of divorce by either party and remarriage among other things. We now see these things as modern, but various parts of the world knew them of old. From 1300, this status began to fall and by Yasuke's time, wives were going to live with their husband's family instead of staying with their own, inheritance was largely the privilege of an elder son, divorce by the male was more common than the female, property rights were reduced and dowries became commonplace. Still the Jesuits were shocked at the degree of freedom that women enjoyed, freedom of movement without a husband's permission, high levels of female literacy (the fact that literate females were respected), the commonplaceness of makeup and beautification (among both sexes), and the degree to which women were able to refuse an arranged marriage and enjoyed certain sexual freedoms. They also noted that the higher up the social scale, the less equal intersex relations were; i.e., a peasant couple were basically equal but aristocratic ladies far from it.

Mori clan drives out Catholics: The Mori clan of western Honshu were originally a minor family in the shadow of the far more powerful Ouchi clan, but by the 1550s, the Ouchi's day was done and in 1557 they were destroyed by the Mori, who took their capital city of Yamaguchi. The Jesuits had set up one of their first missions there in 1550 and around five hundred local people had been quickly converted. It is said that the first ever Japanese Christmas mass was in fact celebrated there in 1552. The Mori were not impressed with the headway made by the Jesuits and expelled them forthwith.

Sea Lords: They were often simply called pirates in Japanese, *kaizoku*, and *wokou*, Japanese bandits (or dwarf bandits) in Chinese. Sea Lords is the name modern scholarship assigns them. They were generally peripheral peoples who came under no central land-based control until just around Yasuke's time when they were being co-opted into "legitimate" state structures and land-based norms of hierarchy.

Selected Bibliography

Cooper, Michael. *Rodrigues the Interpreter: An Early Jesuit in Japan and China.* New York and Tokyo: Weatherhill, 1974.

Fujita, Niel. *Japan's Encounter with Christianity: The Catholic Mission in Pre-Modern Japan.* New York: Paulist Press, 1991.

Nawata Ward, Haruko. *Women Religious Leaders in Japan's Christian Century, 1549–1650.* Surrey and Burlington, VT: Ashgate, 2009.

Ōta, Gyūichi (J. S. A. Elisonas & J. P. Lamers, Trs. and Eds.). *The Chronicle of Lord Nobunaga.* Leiden, NL: Brill, 2011.

Shapinsky, Peter. *Lords of the Sea: Pirates, Violence, and Commerce in Late Medieval Japan.* Ann Arbor, MI: University of Michigan Press, 2014.

Turnbull, Stephen. *The Samurai: A Military History.* London: Routledge, 1977.

Turnbull, Stephen. *Pirate of the Far East 811–1639.* Oxford: Osprey, 2007.

Chapter 7

Pirate attacks: Japanese pirates were known around the world to be tenacious fighters. One particular example was that of an English ship overrun off Singapore in 1604, which managed to contain a raiding pirate band in its own main cabin after a long fight. After four further hours of siege, the English realized the pirates would fight until the last man and ultimately used their cannon to destroy a portion of their own ship; all the pirates died.

Rocket launchers: After the end of the Korean invasions in the 1590s, the Japanese government sent a mission to find out more about these rockets, as they'd been so deadly. They also appear to have been on Japanese pirates' ships in Yasuke's time, a decade or more earlier.

Ama: The "sea women" divers have a history stretching back several millennia. Until the 1960s, the ama dived wearing only a loincloth or nothing at all, and still today dive without the aid of scuba equipment. They are best known for pearl diving, but traditionally dived for anything of saleable value, including shellfish, coral and kelp. Today, a few women still ply this trade, but the harshness of the environment and availability of other work means this ancient profession dwindles by the generation.

Selected Bibliography

Cooper, Michael. *They Came to Japan: An Anthology of European Reports on Japan*, 1543–1640. Berkeley, CA: University of California Press, 1965.

Kataoka, Yakichi. "Takayama Ukon." *Monumenta Nipponica* 1, no. 2 (1938): 451–464.

Petrucci, Maria Grazia. *In the Name of the Father, the Son and the Islands of the Gods: A Reappraisal of Konishi Ryusa, a Merchant, and Konishi Yukinaga, a Christian Samurai in Sixteenth-century Japan*. Unpublished Masters thesis, The University of British Columbia, 2002.

Shapinsky, Peter. *Lords of the Sea: Pirates, Violence, and Commerce in Late Medieval Japan*. Ann Arbor, MI: University of Michigan Press, 2014.

Society of Jesus. *Cartas que os padres e irmãos da Companhia de Jesus escreverão dos reynos de Japão e China II (Letters written by the fathers and brothers of the Society of Jesus from the kingdoms of Japan and China—Volume I)*. Evora, Portugal: Manoel de Lyra, 1598.

Turnbull, Stephen. *Pirate of the Far East 811–1639*. Oxford: Osprey, 2007.

Watsky, Andrew M. "Politics, and Tea. The Career of Imai Sokyu." *Monumenta Nipponica* 50, no. 1 (1995): 47–65.

Chapter 8

Kyoto: Kyoto's location and layout were originally chosen primarily for their auspicious properties as defined by geomancy (*hōgaku* or *hōi* in Japanese) and magical divination. These spiritual antecedents had practical advantages too—the rivers and southern valleys allowed positive energy to flow from favorable directions, but *also* trade goods. The mountains blocked malicious spirits and evil from disrupting human affairs but *also* acted as defensive barriers and sources of natural resources. Water to the north, Lake Biwa, channeled prosperous currents, and *also* acted as a useful trade route connecting Kyoto with the north coast and abundant marine produce, particularly mackerel. Kyoto was founded as Heian-kyo, meaning the "capital of tranquility and peace" when Emperor Kammu moved the Japanese capital there from Nagaokakyō in 794, and was conceived on an enormous scale. Chinese and Japanese architects, engineers, soothsayers and diviners laid out the city and its key buildings on Chinese lines, an enormous grid system of twelve hundred blocks of uniform size. The main entrance to the south was the great Rajōmon gate, which opened on the imposing Suzaku Avenue that bisected the city. The wide boulevard's northern terminus was the Imperial Palace, whose compound housed both ceremonial and residential buildings and additional structures, such as the Court of Abundant Pleasures, a pavilion designed for banqueting and entertainments. Civil space was reserved for two large public markets, as well as for merchant and artisan quarters in the lower city. High nobility and other aristocratic families were allotted land for residences according to rank in the upper city. Over the next ten centuries, the city had been called Kyo, Miyako, or Kyo no Miyako; and in the eleventh century, the city was renamed Kyoto (capital city), and all of the appellations essentially used Japanese renderings of the Chinese character for capital. Even after the seat of imperial power was moved to Tokyo in 1868, there remained a view—persisting to modern times—that Kyoto was still the spiritual and cultural capital of Japan. In 1945, Kyoto was the initial target for the "Fat Man" atomic bomb. However, several senior American generals knew Japan well enough to argue that destroying cherished Kyoto would make it impossible for the Japanese to ever forgive or work beside the Americans. The room agreed and another target on the list was ultimately chosen instead: Kokura, to the northeast of Nagasaki.

African or African American women in Japan: When Yokohama, near Tokyo, opened for foreign trade in 1859, large numbers of foreign people came to live there. Among these were the first black, believed to be African American, women who were described by the famous artist Utagawa

Sadahide as very hard workers. He portrayed them in a famous artwork in which he anthropologically recorded conditions among the foreigners.

The Kanō School: The Kanō School was the dominant Japanese school of artistry from the late fifteenth to the mid-nineteenth centuries. Artists often created work in teams, under a leading artisan, often a member of the original Kanō family. Typical subjects were scenes from nature and from Chinese classics. However, around 1590, the great works that we can see in this chapter, depicting multicultural life in Nagasaki, began to be created. They are called *nanban byobu*, or "southern barbarian folding screens."

Southern barbarian was the term by which the Japanese referred to southern Europeans, Africans and Indians because they approached Japan from the south on Portuguese ships. The term *southern barbarian* itself is Chinese and originally referred to the people of the South China Seas. The Japanese were often called "eastern barbarians" by the Chinese.

Selected Bibliography

Berry, Mary Elizabeth. *The Culture of Civil War in Kyoto*. Berkeley: University of California Press, 1994.

Curvelo, Alexandra. *Nanban Folding Screen Masterpieces: Japan-Portugal XVIIth Century*. Paris: Editions Chandeigne, 2015.

Elison, George and Bardwell Smith (Eds.). *Warlords, Artists and Commoners: Japan in the Sixteenth Century*. Honolulu: University of Hawaii Press, 1981.

Farris, William Wayne. *Japan to 1600: A Social and Economic History*. Honolulu: University of Hawaii, 2009.

Screech, Timon. "The Black in Japanese Art: From the beginnings to 1850." In *The Image of the Black in African and Asian Art*, edited by David Bindman and Suzanne Preston Blier, 325–340. Cambridge, MA: Harvard University Press, 2017.

Society of Jesus. *Cartas que os padres e irmãos da Companhia de Jesus escreverão dos reynos de Japão e China II (Letters written by the fathers and brothers of the Society of Jesus from the kingdoms of Japan and China—Volume II)*. Evora, Portugal: Manoel de Lyra, 1598.

Chapter 9

"The Black Monk from Christian": By the time of Yasuke's arrival in Japan, the Japanese concept of the world had developed quite considerably, or rather that of the ruling classes had. The Jesuits brought the first globes to Japan sometime in the 1570s, and it is known that Nobunaga treasured his and spoke at length with foreigners like Yasuke whom he met. Nobunaga (and the other high-ranking Japanese lords) would have been as conversant in world geography as most European rulers of the time.

However, outside ruling circles, old parlances would have proliferated. The traditional way of describing the world was *gosankoku* (The Three Countries), meaning China, Japan and India, and by extension, in traditional thought, the known "civilized" world. The traditional Asian worldview was centered on China, the *Sinosphere*. The very characters that make up the name for China in the Japanese language (and in many other East Asian languages) mean "the central kingdom" or sometimes "central effervescence." Culturally, this remained the same in Yasuke's day, but the Chinese had definitively rejected Japan diplomatically and commercially in the first half of the sixteenth century and this forced Japan to look farther afield for resources and markets. Ironically, this set off a great century of Japanese seafaring and commercial expansion in Asia and, to a certain extent, as far afield as Europe, with the help of the Jesuits. Introducing Yasuke as the "Black monk from Christian," was a sign that the speaker, and Nobunaga's court, were aware of a world outside the Sinosphere, and that Yasuke was from Christendom.

Seppuku: The act of ritual suicide by cutting open one's own belly (and then being swiftly beheaded by a friend to shorten the pain) was first recorded in 1180. It was used by warriors, male and female, to avoid falling into enemy hands, to mitigate shame or as the basis of a peace agreement. It was usually performed before spectators. If the initial cut to the gut is performed deeply enough, it can sever the descending aorta and bring death even before the beheading. An expert beheading (done by a second) includes *dakikubi* (the embraced head), in which a strip of flesh is left attaching the head to the body. Seppuku was forbidden as judicial punishment in 1873. Voluntary seppuku has continued, although it is extremely rare. The most famous example in the twentieth century was that of Nogi Maresuke, who performed seppuku with his wife on the day of the Emperor Meiji's funeral in 1912, to follow his liege lord in death. Due to his victories in the Russo-Japanese War, Nogi was a well known military figure around the world and his loyalty was widely praised on newspaper front pages globally including *The New York Times*. In 1970, famous author Mishima Yukio performed seppuku in protest during a failed coup to bring the emperor back to power.

Honnō-ji Temple: Honnō-ji Temple was originally founded in 1415 but moved several times due to destruction by fire. In 1582, it was situated to the south of what is now Nijō Castle and was again destroyed by fire during the battle that claimed Nobunaga's life and ensured Yasuke's place in history. It was rebuilt in 1592 in its current location near Kyoto City Hall and has managed to survive there to this day. It is a significant pilgrimage site for Nobunaga fans and houses a small museum with artifacts related to Nobunaga including Mori Ranmaru's sword. The temple has been the subject of numerous movies, manga and novels due to the cult of Nobunaga that lives to this day in Japan and around the world.

Selected Bibliography

Caraman, Philip. *The Lost Empire: The Story of the Jesuits in Ethiopia*. Indiana: University of Notre Dame Press, 1985.

Christensen, J. A. *Nichiren: Leader of Buddhist Reformation in Japan*. Fremont, CA: Jain Publishing Company, 1981.

Cieslik, Hubert. *Soldo Organtino: The Architect of the Japanese Mission*. Tokyo: Sophia University, 2005.

Fujii, Manabu. *Hokeshudaihonzanhonnouji (The Honnō Temple of the Lotus Sect)*. Kyoto: Honnō Temple, 2002.

Lorimer, Michael. *Sengokujidai: Autonomy, Division and Unity in Later Medieval Japan*. London: Olympia Publishers, 2008.

Ōta, Gyūichi (J. S. A. Elisonas & J. P. Lamers, Trs. and Eds.). *The Chronicle of Lord Nobunaga*. Leiden, NL: Brill, 2011.

Salvadore, Matteo. "The Jesuit Mission to Ethiopia (1555–1634) and the

Death of Prester John." In *World-Building and the Early Modern Imagination* edited by Allison B. Kavey, 141–172, New York: Palgrave Macmillan, 2010.

Society of Jesus. *Cartas que os padres e irmãos da Companhia de Jesus escreverão dos reynos de Japão e China II (Letters written by the fathers and brothers of the Society of Jesus from the kingdoms of Japan and China—Volume II).* Evora, Portugal: Manoel de Lyra, 1598.

Turnbull, Stephen. *The Samurai: A Military History.* London: Routledge, 1977.

Chapter 10

The height of Japanese doors: The standard height of a Japanese door, and length of tatami mat, even today, is six *shaku*, about six feet. Yasuke was described as *6 shaku 2 sun*, and may have been taller. One *sun* is approximately equivalent to an inch.

Bowing in Japan: Bowing probably arrived in Japan with prehistorical immigrants from China and Korea, and has remained a crucial part of culture and etiquette ever since. Today, it is second nature to Japanese people, and done virtually unconsciously in any number of situations, both formal and informal. Bowing, especially formal bowing, is governed by a myriad of rules. The bows that Yasuke and other supplicants to Nobunaga, including Valignano and Nobunaga's other senior vassals perform in this book were predominantly from a kneeling position on the floor. The bower would have knelt, placed both hands in front of themselves and lowered their head to the floor. Depending on the situation and the bowing person's rank, they might remain in that position, or return to an upright kneeling posture.

Massed musket volley fire: This deadly modern tactic became one of the most potent tools of armies throughout the world over the next centuries until machine guns were invented in the late nineteenth century. The contemporary popular imagination probably associates the technique most with the massed squares of infantry musketeers in the Napoleonic Wars more than two hundred years after Yasuke and Nobunaga lived. It was probably invented independently by the Chinese, Nobunaga and the Dutch, who are first recorded as using the tactic in Europe in the early seventeenth century.

Kiyomizu Temple: Entering the famous Kiyomizu Temple in Kyoto today, one of the first statues you see is a larger-than-life image of Daikokuten, the Japanese incarnation of the Indian god Shiva. He is the color of black ink, the same color as Yasuke is described as being.

Dinka cattle rearing culture: Dinka customs of dying hair with cow urine and stimulating the cows to produce more milk are well documented and can be seen on YouTube as well as read about in books.

Yasuke gifted: An episode from a decade later may also give a clue to Valignano's feelings toward the situation he found himself in when Nobunaga expressed an interest in Yasuke. Nobunaga's successor, Hideyoshi, had been mostly generous with the Jesuits, partly because he hoped for their help with obtaining two Portuguese warships. One day, he paid them a visit, a singular honor, on the huge galley that the Jesuits had built to defend Nagasaki and to transport themselves in style around the country. Hideyoshi "expressed great interest" in the galley, the implication being that he wanted it for himself, but the Jesuits declined to simply give it to him, and foolishly tried to bargain some extra land for it. Hideyoshi did not react with any outward rage; he appeared to simply forget about the galley. However, mere days later, he banned the Jesuits from propagating and forbade lords to force their vassals to convert, naming Christianity "a great evil for Japan." This was the first step on the path to eventual Jesuit expulsion and wider Catholic persecution. Had Nobunaga expressed a desire to take Yasuke into his service at Valignano's audience, as Hideyoshi did the galley, the Jesuit would have found it hard to decline the request. In fact, he may have even *offered* Yasuke's services before they were requested as he could see how taken Nobunaga was with the warrior. This was the diplomatic thing to do and Valignano was a highly seasoned diplomat. Had later Jesuits followed his example, the future of Japanese Christianity may have turned out very differently.

Selected Bibliography

Cardim, Antonio Francisco. *Elogios, e ramalhete de flores borrifado com o sangue dos religiosos da Companhia de Iesu, a quem os tyrannos do imperio do Iappão tirarão as vidas por odio da fé catholica* (Praises, and Bouquets of Flowers Sprinkled with Jesuit Blood, to Those Whom the Japanese Tyrants Kill Through Hatred of the Catholic Faith). Lisbon: Manoel da Sylua, 1650.

Cieslik, Hubert. *Soldo Organtino: The Architect of the Japanese Mission.* Tokyo: Sophia University, 2005.

Cooper, Michael. *They Came to Japan: An Anthology of European Reports on Japan, 1543–1640.* Berkeley, CA: University of California Press, 1965.

Crasset, Jean. *Historie de l'eglise du Japon (History of the Japanese Church).* Paris: Francois Montalent, 1669.

Elison, George & Smith, Bardwell (Eds.). *Warlords, Artists and Commoners: Japan in the Sixteenth Century.* Honolulu: University of Hawaii Press, 1981.

Madut-Kuendit, Lewis Anei. *The Dinka History: the Ancients of Sudan.* Perth, Australia: Africa World Books, 2015.

Ōta, Gyūichi (J. S. A. Elisonas & J. P. Lamers, Trs. and Eds.). *The Chronicle of Lord Nobunaga.* Leiden, NL: Brill, 2011.

Otsuki, Fumihiko. *Daigenkai.* Tokyo: Fusanbou, 1935.

Society of Jesus. *Cartas que os padres e irmãos da Companhia de Jesus escreverão dos reynos de Japão e China II (Letters written by the fathers and brothers of the Society of Jesus from the kingdoms of Japan and China—Volume II).* Evora, Portugal: Manoel de Lyra, 1598.

Solier, François. *Histoire ecclésiastique des îles et royaume du Japon (The Ecclesiastical History of the Islands and Kingdoms of Japan).* Paris: Sebastian Cramoisy, 1627.

PART 2

Chapter 11

The Emperor Ōgimachi: Ōgimachi reigned from October 27, 1557, to his abdication on December 17, 1586, His personal name was Michihito. His reign saw a modest revival of imperial fortunes as Nobunaga and

427

Hideyoshi were happy to exchange funds for imperial recognition and political legitimacy.

The ancient nobility of Kyoto: The imperial nobility of Kyoto, known as *kuge*, were the ancient aristocracy of the imperial court dating back to the eighth century, and sometimes longer. When the emperor actually ruled, they held power, but with the coming of the first military government, shogunate, in the twelfth century, their power-wielding days were over, as were the emperor's. In Yasuke's time, they were mostly powerless figureheads, puppets in the hands of the warlords who fed them crumbs. Their only political function was to give legitimacy to the warlord governments by bestowing meaningless but prestigious "imperial" rank upon men like Nobunaga.

The former shoguns, the Ashikaga Dynasty: The Ashikaga shogunate governed Japan from 1338 to 1573 when Oda Nobunaga deposed the last member of the dynasty, Ashikaga Yoshiaki. Each shogun was a member of the Ashikaga clan who'd originally come from northern Japan, what is today Tochigi. This period is also known as the Muromachi period and gets its name from the Muromachi district of Kyoto, where the government was based.

Selected Bibliography

Berry, Mary Elizabeth. *The Culture of Civil War in Kyoto.* Berkeley: University of California Press, 1994.

Farris, William Wayne. *Japan to 1600: A Social and Economic History.* Honolulu: University of Hawaii, 2009.

Fujii, Manabu. *Hokeshudaihonzanhonnouji (The Honnō Temple of the Lotus Sect).* Kyoto: Honnō Temple, 2002.

Kure, Mitsuo. *Samurai Arms, Armor, Costume.* Edison, NJ: Chartwell Books, 2007.

Ōta, Gyūichi (J. S. A. Elisonas & J. P. Lamers, Trs. and Eds.). *The Chronicle of Lord Nobunaga.* Leiden, NL: Brill, 2011.

Society of Jesus. *Cartas que os padres e irmãos da Companhia de Jesus escreverão dos reynos de Japão e China II* (*Letters written by the fathers and brothers of the Society of Jesus from the kingdoms of Japan and China—Volume II*). Evora, Portugal: Manoel de Lyra, 1598.

Turnbull, Stephen. *The Samurai: A Military History*. London: Routledge, 1977.

Chapter 12

Akbar the Great: Akbar the Great was the third Mughal emperor of India, who reigned from 1556 to 1605. During his rule, the Mughal Empire tripled in size and wealth, due mainly to his unstoppable conquests, but also connected to a centralized system of administration he established. Although a Muslim himself, he was careful to respect the different faiths of his subjects and it's believed he had a Catholic wife. Although dyslexic and illiterate, he was fond of literature, and created a library of more than twenty-four thousand volumes written in multiple languages from around the world. His reign significantly influenced the course of Indian history.

Habshi: *Habshi* slave soldiers essentially became members of mercenary bands—the commanders were senior African generals, who'd risen through the ranks to become men of wealth and power. It was they who bought the slave boys in the market and schooled their new "recruits" in the deadly arts. The training endured by the newly bought *Habshi* slave soldiers was brutal, the young men, as with child soldiers today, forced to kill and maim to become inured to the deeds they had to soon carry out. This inculcated a fierce loyalty to their generals, their employers and to each other. As outsiders in a foreign land, they had few other ties, and perhaps for Yasuke, becoming *Habshi* would have given him a sense of brotherhood and belonging. After the long and probably terrifying period of his life post-capture, he would finally have felt secure again, with comrades, friends and a family of sorts. Military slaves were normally paid a salary, fed and clothed. They also enjoyed spoils of war and generous bonuses, which boosted their loyalty further. Slaves were freed on their master's death, following Islamic tradition, so many did not feel that slavery was a life sentence—they had a future beyond servitude. If they survived. Many *Habshi* mercenaries are recorded as having been in Portuguese service (for example, six hundred formed a defensive force in the constantly beleaguered Portuguese fort of Diu, one hundred eighty miles north of Mumbai), and many others worked as sailors on Portuguese ships.

The Goa-Africa slave trade: Goa was renowned as a beautiful city of Muslim, Hindu, European, Persian and Turkish influences. And it was based, as all Portuguese outposts were, on easily defendable islands. As well as *Habshis*, many of the African slaves in Portuguese India came on Portuguese ships from the regions with maximum Portuguese contact in Africa, the area of the southeastern seaboard that now comprises Mozambique, Tanzania and Kenya. Some slaves, however, were brought from the deep interior in caravans that also transported ivory and gold for external markets. Portuguese slaving from southeast Africa to India comprised around 200–250 people a year in the late sixteenth century; in later centuries it would increase exponentially when demand from other Europeans, particularly the French, increased. Between 1500 and 1850, the Portuguese transported around forty to eighty thousand of an estimated five to six hundred thousand African people enslaved by Europeans in areas around the Indian Ocean. Northeast Africans also ended up with the Portuguese, especially in Yasuke's time, when the Portuguese did not have the capacity to directly traffic as many slaves as the conditions and booming economies of their Indian territories required. For many Africans, the Indian subcontinent proved not to be the final destination; their odysseys—like Yasuke's—continued farther to China and even Japan. The Dutch seafarer Linschoten noted that "these Abexiins (Africans) such as are free do serve in all India for sailors and seafaring men with such merchants as sail from Goa to China, Japan, Bengala, Mallaca, Ormus and all the Oriental coast."

Africans in Japan: That Japanese people were fascinated by and hospitable to Africans is noted multiple times. The Spaniard Sebastian Vizcaino considered offering public concerts when he saw how much excitement an African drummer among his attendants caused. Toyotomi Hideyoshi specifically asked Africans to entertain him on at least two occasions in the 1590s. On the first occasion, a single man, a sailor on a Portuguese ship, attended court to dance and sing; on the second occasion, a group of African guards dressed in red and armed with golden spears danced "a wild dance of fife and drum." They were gifted white robes for the performance.

Selected Bibliography

Azzam, Abdul Rahman. *The Other Exile: The Story of Fernão Lopes, St Helena and a Paradise Lost.* London: Icon Books, 2017.

Berry, Mary Elizabeth. *The Culture of Civil War in Kyoto*. Berkeley: University of California Press, 1994.

Cardim, Antonio Francisco. *Elogios, e ramalhete de flores borrifado com o sangue dos religiosos da Companhia de Iesu, a quem os tyrannos do imperio do Iappão tirarað as vidas por odio da fé catholica* (Praises, and Bouquets of Flowers Sprinkled with Jesuit Blood, to Those Whom the Japanese Tyrants Kill Through Hatred of the Catholic Faith). Lisbon: Manoel da Sylua, 1650.

Cieslik, Hubert. *Soldo Organtino: The Architect of the Japanese Mission*. Tokyo: Sophia University, 2005.

Cooper, Michael. *They Came to Japan: An Anthology of European Reports on Japan, 1543–1640*. Berkeley, CA: University of California Press, 1965.

Crasset, Jean. *Historie de l'eglise du Japon (History of the Japanese Church)*. Paris: Francois Montalent, 1669.

Edalji, Dosabhai. *History of Gujarat from the Earliest Period to the Present Day*. Ahmadabad: The United Printing and General Agency Company's Press, 1894.

Edward James Rapson, Sir Wolseley Haig, Sir Richard Burn (Eds.), 1922. *The Cambridge History of India*. Cambridge: Cambridge University Press.

Elison, George & Smith, Bardwell (Eds.). *Warlords, Artists and Commoners: Japan in the Sixteenth Century*. Honolulu: University of Hawaii Press, 1981.

Gill, Robin. *Topsy-Turvey 1585*. Key Biscayne, FL: Paraverse Press, 2005.

Ōta, Gyūichi (J. S. A. Elisonas & J. P. Lamers, Trs. and Eds.). *The Chronicle of Lord Nobunaga*. Leiden, NL: Brill, 2011.

Otsuki, Fumihiko. *Daigenkai*. Tokyo: Fusanbou, 1935.

Society of Jesus. *Cartas que os padres e irmãos da Companhia de Jesus escreverão dos reynos de Japão e China II (Letters written by the fathers and brothers of the*

Society of Jesus from the kingdoms of Japan and China—Volume II). Evora, Portugal: Manoel de Lyra, 1598.

Solier, François. *Histoire ecclésiastique des îles et royaume du Japon (The Ecclesiastical History of the Islands and Kingdoms of Japan)*. Paris: Sebastian Cramoisy, 1627.

Chapter 13

The Samurai as caste: In Yasuke's time, the word *samurai* simply described a profession: warrior (albeit a very specialized one). Shortly afterward, it became a caste name. At the end of The Age of the Country at War, around the end of the sixteenth century, most of those who'd fought on the samurai side in the civil wars, even some of the peasants, pirates and ninja, were classified as "samurai" in a formalized caste structure with the samurai at the top—a hereditary warrior/administrator/ruling class. The caste ranking continued with peasants, artisans and merchants, who took the lowest status (because they lived off everybody else's hard work). Outside of the scope of the caste system were *eta*, impure people who dealt with death, and *hinin*, nonpersons such as ex-convicts and vagrants who worked as town guards, street cleaners or entertainers. Legally speaking, an *eta* was worth one-seventh of a human being. The Age of the Country at War had been probably the most socially fluid period since the eighth century. Able men and women, like Yasuke, were able to rise through the ranks due to the chaos. No more. From this time until their caste was abolished by law in 1873, the samurai were forbidden (in most of the country) to farm or engage in mercantile activity and had to live in castle towns rather than country villages. This was the time when the word *samurai* takes on its modern meaning of a warrior *caste* rather than actual warrior *role*. In the virtual absence of war or any challenge from below between the seventeenth and nineteenth centuries, the samurai caste had little warring to do and the martial arts we now associate with this class were codified and formed the roots of modern sports like kendo, judo and aikido. Samurai were still furnished with a stipend by their lord, determined by rank, although over time, the value of the stipend was devalued so much by inflation that many samurai families were forced to find other ways to make ends meet. A few, such as the Mitsui family, founders of the modern-day multinational conglomerate, gave up their samurai swords and lowered themselves to merchant status. For the overwhelming majority, this was a step too far, and they starved or lived in abject poverty rather than "lower" themselves.

Ronin: The most famous story about masterless samurai is perhaps that of the forty-seven *ronin*, the subject of a recent movie starring Keanu Reeves, as well as numerous other books, plays and media ever since the event took place. Their lord, Asano Naganori, had been ordered to perform seppuku, his lands were confiscated and his retainers made masterless samurai, *ronin*, a fate nearly as bad as death in their eyes. The sentence was handed down because Asano had dared to draw his sword within the precincts of the shogun's castle, a heinous offence. To defend his honor and legacy, forty-seven of his retainers, now *ronin*, carefully plotted revenge on Lord Kira Yoshinaka, the man who'd intentionally besmirched the honor of their master and caused his death. They pretended to live debauched lives so their intended target would lower his guard. This took fourteen months, but when the time was ripe, they gathered and attacked the culprit and his men. The attack, however, took place only after they'd sworn not to harm helpless members of the household and had informed the neighbors of their mission, to prevent their being thought of as simple robbers. The forty-seven *ronin* took Kira's head and laid it on their old master's tomb. They then presented themselves to the authorities for punishment. They were duly sentenced to cut their own bellies, and forty-six of them did on February 4, 1703. The forty-seventh man had been sent home to report the mission's success. He died in 1747 and was later commemorated beside his comrades.

Selected Bibliography

Berry, Mary Elizabeth. *The Culture of Civil War in Kyoto.* Berkeley: University of California Press, 1994.

Brown, Delmer M. "The Impact of Firearms on Japanese Warfare, 1543–98." *The Far Eastern Quarterly*, no. 7, 3 (1948): 236–253.

Cooper, Michael. *They Came to Japan: An Anthology of European Reports on Japan, 1543–1640.* Berkeley, CA: University of California Press, 1965.

Farris, William Wayne. *Japan to 1600: A Social and Economic History.* Honolulu: University of Hawaii, 2009.

Jansen, Marius. *The Making of Modern Japan*. Cambridge, MA: Belknap Press, 2000.

Kim, Young Gwan and Hahn, Sook Ja. "Homosexuality in ancient and modern Korea." *Culture, Health & Sexuality*, no. 8, 1 (2006): 59–65.

Kure, Mitsuo. *Samurai Arms, Armor, Costume*. Edison, NJ: Chartwell Books, 2007.

Morillo, Stephen. "Guns and Government: A Comparative Study of Europe and Japan." *Journal of World History*, no. 6, 1 (1995): 75–106.

Ōta, Gyūichi (J. S. A. Elisonas & J. P. Lamers, Trs. and Eds.). *The Chronicle of Lord Nobunaga*. Leiden, NL: Brill, 2011.

Screech, Timon. "The Black in Japanese Art: From the beginnings to 1850." In *The Image of the Black in African and Asian Art*, edited by David Bindman and Suzanne Preston Blier, 325–340. Cambridge, MA: Harvard University Press, 2017.

Shapinsky, Peter. *Lords of the Sea: Pirates, Violence, and Commerce in Late Medieval Japan*. Ann Arbor, MI: University of Michigan Press, 2014.

Society of Jesus. *Cartas que os padres e irmãos da Companhia de Jesus escreverão dos reynos de Japão e China II (Letters written by the fathers and brothers of the Society of Jesus from the kingdoms of Japan and China—Volume II)*. Evora, Portugal: Manoel de Lyra, 1598.

Tsang, Carol Richmond. *War and Faith: Ikko Ikki in Late Muromachi Japan*. Cambridge, MA: Harvard University Press, 2007.

Turnbull, Stephen. *The Samurai: A Military History*. London: Routledge, 1977.

Turnbull, Stephen. *The Samurai Sourcebook*. London: Cassell, 2000.

Turnbull, Stephen. *Samurai Women 1184–1877*. Oxford: Osprey, 2010.

Turnbull, Stephen. *Ninja: Unmasking the Myth*. Barnsley: Frontline Books, 2017.

Chapter 14

Ranmaru: The relationship between Nobunaga and Ranmaru counts as one of the greatest love stories in Japanese history, and their relationship remains revered and sacrosanct to this day. Ranmaru's huge *odachi* sword remains as a viewable artifact in the modern-day Honnō-ji Temple.

Sex with Nobunaga: Even if such a thing was public knowledge, no Jesuit would have written of it and our key Japanese sources, Ōta Gyūichi and Matsudaira Ietada, did not mention any personal details about Nobunaga's relationship with Yasuke other than the fact of Yasuke's first audience and warrior service with Nobunaga. [Special thanks to Cliff Pereira for his expert advice on the section about sexual practices in Africa and to Professor Timon Screech for his expert personal opinion on the depths of the Yasuke/Nobunaga relationship.]

Sex in Japan: Japan, at this time, did not take a particularly restrictive view to any sexual relationships, although different types of partnership, marriage, concubinage, casual, paid or unpaid sex, and kept mistresses or boys were often highly codified and sometimes had strict laws pertaining to them, particularly among the upper classes. Men who could afford it kept numerous concubines—Nobunaga's favorite, Kitsuno, was the mother of his first two sons—but polygamy in any formalized sense was not practiced. Multiple sexual partners for both sexes was common, as was divorce and remarriage. Traditionally, children were often held in common in the countryside, being brought up by the community rather than in exclusive nuclear families. This declined in the early modern age due in large part to the increased rule of law and hence the need to formalize inheritance and property rights. Among the lower classes, there was a lot more leeway and many families never had their relations formalized in any civil or religious fashion. Senior wives of the upper class, often the result of political marriages, were normally expected to employ courtesans and sex workers, temporarily or permanently, to entertain their husbands; and the senior wife often adopted the offspring of such liaisons, especially if no official heir was forthcoming, or a new one was needed. In Nobunaga's case, his senior wife, Nōhime, was unable to conceive, so she adopted his children by other women and is believed to have been cared for by his second son, Nobukatsu, after her husband's death. In Hideyoshi's case the opposite is rumored. His concubine, Lady Yodo, a formidable woman and daughter of Nobunaga's sister Oichi, is supposed to have conceived her son Hideyori with someone else as Hideyoshi could not do it. Not surprisingly, the Jesuits took a dim view of Japanese sexuality, especially homosexual relationships. The non-Jesuit Europeans and Africans though, thought they were in wonderland, and seemed often to be happy enough to follow the saying "when in Rome..."

Valignano leaves: Valignano eventually left from Nagasaki on board the Portuguese ship of Captain Ignacio de Lima on February 20, 1582. Accompanying him were the four young Japanese ambassadors to Rome, kinsmen of the Christian lords, Ōtomo Sōrin, Arima Harunobu and Ōmura Sumitada.

The gifted screens depicting Azuchi: After reaching Rome (which Valignano never did), the Pope duly expressed his wonder at Nobunaga's gift, and no doubt others did afterward. Europeans have always had a fascination for Asian art, and Japanese art in particular, partly set off by these early wonders which were carefully chosen to astonish. The fate of the screens thereafter is unknown. It's thought that, in the absence of expertise to preserve these magnificent works of art, they rotted away over time.

Selected Bibliography

Chatterjee, Indrani, and Eaton, Richard. *Slavery and South Asian History*. Bloomington: Indiana University Press, 2006.

Cieslik, Hubert. *Soldo Organtino: The Architect of the Japanese Mission*. Tokyo: Sophia University, 2005.

Evans-Pritchard, E. "Some Notes on Zande Sex Habits." *American Anthropologist, New Series*, No. 75, 1 (1973): 171–175.

Ihara, Saikaku, Powy Mather, E. (Trans.). *Comrade Loves of the Samurai*. Tokyo: Tuttle, 1972.

Jameson, E. W. *The Hawking of Japan: The history and development of Japanese Falconry*. Davis, CA: 1962.

Keay, John. *India: a History, From the Earliest Civilisations to the Book of the Twenty-first Century*. London: Harper Press, 2010.

Kure, Mitsuo. *Samurai Arms, Armor, Costume*. Edison, NJ: Chartwell Books, 2007.

Madut-Kuendit, Lewis Anei. *The Dinka History the Ancients of Sudan*. Perth, Australia: Africa World Books, 2015.

Murakami, Naojiro. "The Jesuit Seminary of Azuchi." *Monumenta Nipponica* No. 6, 1/2 (1943): 370–374.

Ōta, Gyūichi (J. S. A. Elisonas & J. P. Lamers, Trs. and Eds.). *The Chronicle of Lord Nobunaga*. Leiden, NL: Brill, 2011.

Russell-Wood, A. J. R. *The Portuguese Empire, 1415–1808. A World on the Move*. Baltimore and London: The Johns Hopkins University Press, 1992.

Sarkar, Jagadish Narayan. *The Art of War in Medieval India*. New Delhi: Munshiram Manoharlal Publishers, 1984.

Souza, George. *The Survival of Empire: Portuguese Trade and Society in China and the South China Sea 1630–1754*. Cambridge: Cambridge University Press, 1986.

Chapters 15 & 16

Hideyoshi's rise: Not much is known for certain about Hideyoshi's youth apart from the fact that he came from a very humble background, and his father was an *ashigaru* (common foot soldier) named Yaemon. Before 1570, his story is mostly legend, but he appears in documents from this time as one of Nobunaga's officers. In 1573, Nobunaga made him a lord, and his star never stopped rising.

Tottori castle ghost story: In the mid-1980s, seven high school students were joking around and having a late-night coffee at a café in the shadow of Tottori Castle. The conversation turned to the story of ghostly samurai on the castle mountain. Eventually, they dared each other, *kimodameshi*, or test of courage in the face of the supernatural, to enter the mountain castle grounds and brave the samurai ghosts within. Tottori Castle is on several levels; the modern entrance is at road level and steep stone stairs take one up the lower mountainside defenses. The braver students surged ahead to the second-highest level, maybe two hundred-plus feet above the moat. To go higher really would have been foolhardy in the dark; it is not only ghosts that are a threat, there are bears, and other dangerous animals living wild there too. It was 2:00 a.m., the witching hour. The streetlights below glowed, but the mountainside was pitch black and ominous. The young people, fueled by their own stories and imagination, looked around them, hoping the ghosts were nothing but stories. One student, Minoru, who was especially sensitive to supernatural entities, felt more fearful than the others. He'd seen ghosts before and climbed the slope more slowly, knowing this was not a joke. Suddenly, a huge samurai burst out of the ether, six feet tall. He was armored, wore his swords sheathed and his eyes were bloodred. There was no time to take in more details. The samurai grabbed for the students, trying to pull one away. Minoru saw it all, but the prospective victim was utterly oblivious to the ghostly presence. Minoru immediately turned and ran. In panic at the sudden flight, his friends followed him down the steep and dangerous stairs and nearly brained themselves tripping on the stone, but managed to reach the main gate and the streetlights that beckoned safety beyond. The ghost did not follow them. Panting from the run and fear, Minoru's friend asked "Why did you run?" and Minoru, his heart racing, told him of the armored warrior who'd tried to take him. Decades later, Minoru still visits Tottori Castle every April to view the cherry blossoms, but he has never returned to that one spot where years ago his twentieth-century friend was attacked by a samurai from Yasuke's day.

Ishikawa Goemon's death: In 1594, Ishikawa Goemon attempted to assassinate Hideyoshi, but his luck had run out. He was sentenced to death by boiling. His infant son was also sentenced to die with him, but the ninja managed to save the boy by holding him *above* the boiling cauldron. Hideyoshi is supposed to have generously spared the boy's life to honor the father's brave action. The type of metal bath that he died in is now known as a "Goemon bath" in his honor.

The name "ninja": It's likely the word *ninja* was hardly ever used in Yasuke's time. In fact, it probably only entered common usage in Japanese in

The execution by boiling of the ninja Ishikawa Goemon. This type of bath is called after him even today.

the post-WWII period when the concept was popularized overseas. This is supposedly because the word *ninja* is easier to render into English that many of the other terms used to refer to special operative troops in ancient Japan. Yasuke would have known them by names like *shinobi*, *Iga no mono* (a person from Iga), *rappa* (thief/ruffian), *kusa* (grass, because they hid in the grass), or *nokizaru* (roof-monkey, as they moved around on roofs rather than using the road).

Yasuke's spiritual life: Yasuke's spiritual life remains only in historic speculation. None of those who saw fit to record his life mentioned anything related to his faith. As seems most likely—due to his path through India, his skin color, and extreme height—he was born of the Dinka *(Jaang)* people in what is now South Sudan and brought up believing in a supreme creator god, Nhialic. A divine force present in all of creation, Nhialic controls the destiny of every human, plant and animal in the Dinka world. If Yasuke still held any of the spirituality he'd been born with, he would have been thanking Nhialic for his blessings. The spiritual leader of a Dinka community was the Spear Master, who was both political chief *and* holy man. Spear Masters mediated between the gods and their communities, and were the spiritual representatives on earth of Nhialic himself. Each village, as Yasuke would have seen it, had its own "Pope." The Spear Master's religious invocations ensured the giving and preservation of life, the general well-being of his people, and success against their enemies, elements and wild animals. He could also purportedly smite his enemies. Prayers were often supplemented by the sacrifice of cattle, a holy animal in the Dinka world: the source of all wealth, most nourishment and the giver of life. A cow's sacrifice was no small matter, but an action filled with significance and gravity. Enslaved during childhood, Yasuke hadn't had the chance to complete any of the ceremonies which accompanied manhood—including scarification of the face and the removal of six teeth from the lower jaw. Other Africans in Asia had been described so, but not Yasuke.

Yasuke had most likely fallen into the hands of Muslim slavers, and it is virtually certain he'd been converted to Islam. The conversion of *kaffir*, "infidel," slaves took place either willingly or forcibly, and typically happened during the slave's journey to the marketplace. Conversion to Islam is, at least on the surface, a straightforward act; all a person has to do is pronounce *Shahada*, the testimony of faith, in Arabic: "There is no true god but God (Allah), and Mohammed is his messenger." When Yasuke recited these words, he'd have been a Muslim. Whether the vast majority of newly enslaved non-Muslims who recited these words—enchained, torn from their homes, in a language they did not understand, for reasons they probably did not comprehend—actually understood what they were signing up for is unlikely. Most, in fact, were simply—as with the Jesuits and Japanese—following their new master's instructions. Had Yasuke remained a Muslim slave in a Muslim world (or in service to a Muslim master in India), however, this had various advantages. He would have had social status as a member of the faithful that was unavailable to non-Muslims, a route to emancipation *and* the legal protection that any children he fathered would have automatic freedom—as it was illegal for Muslims to enslave anybody born a Muslim. A French Jesuit reporting on Yasuke's exploits some fifty years *after* Nobunaga's death wrote that Yasuke was a Moor. A *possible* reference to Islam, although more likely to have meant only that he was a "black man" as such parlance was common at the time to describe *every* African, Arab or Indian. Prior to Valignano's employment, Yasuke no doubt took on his *third* belief system. Valignano wouldn't have engaged the services of a man like Yasuke who did not, outwardly at least, profess the Catholic faith. Yasuke was therefore, technically required to be able to recite the

Paternoster, Ave Maria, the Commandments and the Articles of Faith, as well as having been baptized—the minimum official qualifications necessary to deem a "heathen" as Catholic. In practice, many enslaved Africans were baptized with little knowledge of what was happening to them, due to lack of priests available. At various points in time, the Portuguese Crown even allowed slaving ship captains to perform mass baptisms. In Yasuke's time, this often took place in Africa *before* they boarded the ships for the destination where they were to be put to work. Fortunately for the enslaved—or so the Christian slavers told themselves—this meant that if they died from murder, despair, starvation or disease, in the terrible conditions on the slave ships, their souls had at least been saved. In Yasuke's case, he'd arrived in India via the Muslim world rather than through a Portuguese settlement in Africa. Therefore, he may well have been baptized *after* arriving in Goa, Cochin or any of the other Portuguese-controlled enclaves along the subcontinental coastline. Legally speaking, in the Portuguese world, slave children of ten or under were not given a choice about baptism; those older than ten, however, could in theory refuse. As Yasuke was employed by Valignano around the age of twenty, and perhaps not yet Catholic, he technically could have refused baptism. However, Valignano was a strict and rather demanding character, and would have insisted that any conversion be properly observed, rather than perfunctory. Living with Valignano and among the Jesuits, Yasuke had been exposed to strong and zealous professions of Catholicism on a daily basis, thereby becoming familiar with even niche areas of doctrine and practice. Yasuke outwardly showed many signs of Catholicism and attended mass and prayers, either voluntarily or as part of his duties, on a daily basis. He may well, perhaps, have truly believed.

Selected Bibliography

Berry, Mary Elizabeth. *The Culture of Civil War in Kyoto*. Berkeley: University of California Press, 1994.

Brown, Delmer M. "The Impact of Firearms on Japanese Warfare, 1543–98." *The Far Eastern Quarterly*, no. 7, 3 (1948): 236–253.

Cooper, Michael. *They Came to Japan: An Anthology of European Reports on Japan, 1543–1640*. Berkeley, CA: University of California Press, 1965.

Evans-Pritchard, E. "Some Notes on Zande Sex Habits." *American Anthropologist, New Series*, No. 75, 1 (1973): 171–175.

Farris, William Wayne. *Japan to 1600: A Social and Economic History*. Honolulu: University of Hawaii, 2009.

Ihara, Saikaku, Powy Mather, E. (Trans.). *Comrade Loves of the Samurai*. Tokyo: Tuttle, 1972.

Jansen, Marius. *The Making of Modern Japan*. Cambridge, MA: Belknap Press, 2000.

Kim, Young Gwan and Hahn, Sook Ja. "Homosexuality in ancient and modern Korea." *Culture, Health & Sexuality*, no. 8, 1 (2006): 59–65.

Kure, Mitsuo. *Samurai Arms, Armor, Costume*. Edison, NJ: Chartwell Books, 2007.

Morillo, Stephen. "Guns and Government: A Comparative Study of Europe and Japan." *Journal of World History*, no. 6, 1 (1995): 75–106.

Ōta, Gyūichi (J. S. A. Elisonas & J. P. Lamers, Trs. and Eds.). *The Chronicle of Lord Nobunaga*. Leiden, NL: Brill, 2011.

Screech, Timon. "The Black in Japanese Art: From the beginnings to 1850." In *The Image of the Black in African and Asian Art*, edited by David Bindman and Suzanne Preston Blier, 325–340. Cambridge, MA: Harvard University Press, 2017.

Shapinsky, Peter. *Lords of the Sea: Pirates, Violence, and Commerce in Late Medieval Japan*. Ann Arbor, MI: University of Michigan Press, 2014.

Society of Jesus. *Cartas que os padres e irmãos da Companhia de Jesus escreverão dos reynos de Japão e China II (Letters written by the fathers and brothers of the Society of Jesus from the kingdoms of Japan and China—Volume II)*. Evora, Portugal: Manoel de Lyra, 1598.

Tsang, Carol Richmond. *War and Faith: Ikko Ikki in Late Muromachi Japan.* Cambridge, MA: Harvard University Press, 2007.

Turnbull, Stephen. *The Samurai: A Military History.* London: Routledge, 1977.

Turnbull, Stephen. *The Samurai Sourcebook.* London: Cassell, 2000.

Turnbull, Stephen. *Ninja: Unmasking the Myth.* Barnsley: Frontline Books, 2017.

Chapter 17

The practice of collecting heads: Head collecting has a long history in Japan, and in its most basic form can be seen as the proof of a job well done. Proof that the kill or kills have been made. For the most important of heads, those of senior samurai or lords, protocol was strict and always observed if there was time. Of course in the heat of battle, that was not always possible. In Korea, during the invasions of the 1590s, the samurai tried their best to collect heads, but the sheer number of kills and the distance to courier them back to Japan, not to mention logistical problems of getting them through hostile country, meant that a nose or ear had to suffice. Many, approximately thirty-eight thousand, were interred in a mound in Kyoto called Mimizuka (which actually means "ear mound"). You can still see this gruesome monument today.

The *Shigeshoshi*: The makeup artists (almost always women) who accompanied samurai armies were a key part of the rituals attending victory in battle. To respect the fallen, heads would be cleaned and made-up, their hair dressed properly and only then displayed to the victorious general. Sometimes victors were known to have conversations with their fallen foe's head, praying to the soul or asking forgiveness. Nobunaga's treatment of Takeda Katsuyori's head was not the only time a victor crowed over his dead enemy, but it is the most infamous.

Selected Bibliography

Berry, Mary Elizabeth. *Hideyoshi.* Cambridge, MA: Council on East Asian Studies, Harvard University, 1989.

Murakami, Naojiro. "The Jesuit Seminary of Azuchi." *Monumenta Nipponica* No. 6, 1/2 (1943): 370–374.

Ōta, Gyūichi (J. S. A. Elisonas & J. P. Lamers, Trs. and Eds.). *The Chronicle of Lord Nobunaga.* Leiden, NL: Brill, 2011.

Turnbull, Stephen. *The Samurai: A Military History.* London: Routledge, 1977.

Turnbull, Stephen. *The Samurai Sourcebook.* London: Cassell, 2000.

Turnbull, Stephen. *Ninja: Unmasking the Myth.* Barnsley: Frontline Books, 2017.

Chapter 18

Tokugawa Ieyasu, the sycophant: Ieyasu was from a minor family of nobles and grew up a hostage in a rival *daimyō*'s court. He spent much of his life kowtowing to more powerful men, and was very good at it, gradually rising in power as an ally to Nobunaga and then Hideyoshi. In the end, his patience and respectful demeanor paid off when he became supreme master of Japan and shogun in 1603. His sycophancy should not be viewed negatively; he was bowing to the reality of his world, and working hard to ensure the peace of his small (but growing) realm and its people. Ultimately, the peace he founded at the end of The Age of the Country at War was one of the longest periods of sustained peace in human history, anywhere, a massive boon for the Japanese people. His early respect for the power of others and later careful and wise lawmaking made this possible.

Mount Fuji's name: Mount Fuji is probably one of the most famous mountains in the world and perhaps the most recognizable. While iconic mountain names such as the Matterhorn are known universally, few mountains other than Fuji have attained the instant recognition of its profile all over the world. The name itself is clouded in mystery, but it is thought to originate with the original settlers of the Japanese islands tens of thousands of years ago, and can therefore be said to have been a holy site for humans far into prehistory. Its name likely comes from the indigenous people of Japan, the Ainu, and is derived from an Ainu term meaning "fire," coupled with *san*, the Japanese word for "mountain."

Houchonin today: *Houchonin* were cookery masters, highly respected elite chefs to the rich and powerful. The position was often hereditary, and they were sometimes even descended from imperial branch families. As with tea masters, they sometimes taught their art to paying students, and started "schools" with ranks, and allowed students to wear ceremonial clothes of different colors as they progressed in the art (somewhat similar to colored belts in the martial arts). One warlord who studied to be a *houchonin* was Hosokawa Yusai, Akechi Mitsuhide's brother-in-law. Today there are still a few men who identify as descendants of *houchonin*, but the art has declined. One author, Eric Rath, puts this down to the fact that modern sensibilities are too squeamish to appreciate the art of carving and "bringing back to life" dead animals.

Selected Bibliography

Ōta, Gyūichi (J. S. A. Elisonas & J. P. Lamers, Trs. and Eds.). *The Chronicle of Lord Nobunaga*. Leiden, NL: Brill, 2011.

Rath, Eric, C. *Food and Fantasy in Early Modern Japan*. Berkeley: University of California Press, 2010.

Sadler, A. L. *Shogun: The Life of Tokugawa Ieyasu*. Tokyo and Vermont: Tuttle Publishing, 2009.

Turnbull, Stephen. *The Samurai: A Military History*. London: Routledge, 1977.

Turnbull, Stephen. *The Samurai Sourcebook*. London: Cassell, 2000.

Chapter 19

Akechi's betrayal: The reason for Akechi's treachery has been debated for four hundred years, and will likely be for *another* four hundred. Essentially, we will never know, but the following are some of the theories that have been proposed. (1) Ambition. Akechi felt it was his time. This is unlikely as he did not have sufficient numbers in his army, and the coup was always a very long shot. (2) Nobunaga directly caused the death of Akechi's mother by reneging on a hostage deal. This is quite likely, if not true, but the story is not certain and if so, Akechi waited a long time to take revenge as the event happened in 1578 or 1579, more than *three years* earlier. (3) Nobunaga complained about the feast Akechi'd prepared for Tokugawa Ieyasu and threw it into the garden, stamping on it, and humiliating the warlord. If this was the reason for the coup, it may have been the culmination of many slights (Nobunaga was not overly respectful to his subordinates) rather than for this act alone. (4) Betrayal by Hosokawa Yusai, Akechi's brother-in-law. Hosokawa was said to have promised to support Akechi, but actually reported the plot *to* Hideyoshi. If this is true, why didn't Hosokawa tell Nobunaga, who was much closer than Hideyoshi? In any event his family suffered greatly due to their association with Akechi. Akechi's daughter, Tama (more often known by her baptismal name of Gracia), who was married to Hosokawa's son, spent the rest of her life confined to her house or country estate, unable to show her face in public. This theory is unlikely. (5) Protecting the emperor and imperial court. Emperor Ōgimachi did not have an entirely smooth relationship with Nobunaga, despite the extensive funding he received and the other honors Nobunaga paid him. Some have suggested Nobunaga intended to cap a lifetime of surprising deeds with the abolishment of the imperial line. After all, he'd already done the same with the shogunal family, the Ashikaga, whom Akechi had served in the past. The theory goes that Akechi acted as he did to protect the emperor. This, also, is unlikely. Nobunaga had invested great resources in obtaining imperial recognition but it *is* possible that he believed that he no longer needed such support. The most likely cause is a lifetime of slights and massive frustration.

Selected Bibliography

Ōta, Gyūichi (J. S. A. Elisonas & J. P. Lamers, Trs. and Eds.). *The Chronicle of Lord Nobunaga*. Leiden, NL: Brill, 2011.

Sadler, A. L. *Shogun: The Life of Tokugawa Ieyasu*. Tokyo and Vermont: Tuttle Publishing, 2009.

Turnbull, Stephen. *The Samurai: A Military History*. London: Routledge, 1977.

Turnbull, Stephen. *The Samurai Sourcebook*. London: Cassell, 2000.

Uhlenbeck, Chris and Molenaar, Merel. *Mount Fuji, Sacred Mountain of Japan*. Leiden: Hotei Publishing, 2000.

Chapter 20

Nobunaga's head: There is no evidence for Yasuke having taken Nobunaga's head; however, Oda family lore has it so. Yasuke's memory does live on in the families that were most closely associated with him. Another story has him associated with a supposed death mask made for Nobunaga, but this is highly unlikely.

Nobunaga's legend: Oda Nobunaga is one of the most popular figures in Japanese history. Enter any bookshop in Japan—on many street corners and in most train stations of any size—and you will find a book about him, either fact or fiction. Probably both. It's open to question why he enjoys such popularity, but his decisiveness, ruthlessness and charisma are cited by the Japanese public as the prime reasons. Anybody associated with him, for example Yasuke, also basks in his glory and popularity.

Selected Bibliography

Ōta, Gyūichi (J. S. A. Elisonas & J. P. Lamers, Trs. and Eds.). *The Chronicle of Lord Nobunaga*. Leiden, NL: Brill, 2011.

Schindewolf, Brandon C. *Toki wa ima*. Unpublished undergraduate thesis, senior honors. The Ohio State University, 2010.

Society of Jesus. *Cartas que os padres e irmãos da Companhia de Jesus escreverão dos reynos de Japão e China II (Letters written by the fathers and brothers of the Society of Jesus from the kingdoms of Japan and China—Volume II)*. Evora, Portugal: Manoel de Lyra, 1598.

Turnbull, Stephen. *The Samurai: A Military History*. London: Routledge, 1977.

Turnbull, Stephen. *The Samurai Sourcebook*. London: Cassell, 2000.

Watsky, Andrew M. "Politics, and Tea. The Career of Imai Sokyu." *Monumenta Nipponica* 50, no. 1 (1995): 47–65.

PART 3

Chapters 21 & 22

The Shimazu clan's domination of Kyushu: Beginning in 1550, the Shimazu family of Satsuma Province fought to expand their territory and by 1574, they had secured their home province and neighboring Ōsumi Province. In 1577, they took parts of Hyūga Province on the eastern side of Kyushu, and fought Ōtomo Sōrin the following year. This was the occasion that a party of Jesuits had to flee for their lives. In 1584, in alliance with Arima Harunobu, they won the Battle of Okitanawate to take control of western Kyushu and had nearly achieved domination of the island, when Hideyoshi led his all-conquering armies against them.

The *Fusta*: The Jesuits' galley known as a *fusta* in Portuguese and a *galliot* in English was the first ship of its kind seen in Japan and was probably constructed in 1581/1582 by Valignano's command in Nagasaki. The Portuguese used such ships, descended from the ancient Mediterranean trireme, extensively throughout their empire, although this example, with a crew of up to three hundred must have been on the larger end. Galliots typically had between ten to twenty oars on each side, pulled by two men on each oar. Due to the energy needed to row, replacement crews were needed, and would have taken the oars in turns to keep up speed. Galliots also had sails to make use of wind when available, but were superior warships to galleons due to their ability to move even when there was no wind.

Selected Bibliography

Berry, Mary Elizabeth. *Hideyoshi*. Cambridge, MA: Council on East Asian Studies, Harvard University, 1989.

Berry, Mary Elizabeth. *The Culture of Civil War in Kyoto*. Berkeley: University of California Press, 1994.

Cieslik, Hubert. *Soldo Organtino: The Architect of the Japanese Mission.* Tokyo: Sophia University, 2005.

Cooper, Michael. *They Came to Japan: An Anthology of European Reports on Japan, 1543–1640.* Berkeley, CA: University of California Press, 1965.

Hesselink, Reinier. *The Dream of Christian Nagasaki: World Trade and the Clash of Cultures, 1560–1640.* Jefferson, NC: McFarland & Company, 2016.

Ōta, Gyūichi (J. S. A. Elisonas & J. P. Lamers, Trs. and Eds.). *The Chronicle of Lord Nobunaga.* Leiden, NL: Brill, 2011.

Sadler, A. L. *Shogun: The Life of Tokugawa Ieyasu.* Tokyo and Vermont: Tuttle Publishing, 2009.

Society of Jesus. *Cartas que os padres e irmãos da Companhia de Jesus escreverão dos reynos de Japão e China II (Letters written by the fathers and brothers of the Society of Jesus from the kingdoms of Japan and China—Volume II).* Evora, Portugal: Manoel de Lyra, 1598.

Turnbull, Stephen. *The Samurai: A Military History.* London: Routledge, 1977.

Turnbull, Stephen. *Ninja: Unmasking the Myth.* Barnsley: Frontline Books, 2017.

Chapter 23

Catholic persecution: When Catholicism became illegal, its practitioners became criminals. It could be compared to the banning of certain sects or extremist political groups like Nazis in the modern world. That is certainly how it was seen at the time and it should not be forgotten that Catholics made up a very small proportion of the overall population and were generally confined to peripheral regions. However, when people refused to recant, the puzzled Tokugawa government got serious. Experience told them that to allow rebellious elements the freedom to defy the law brought chaos and war, something that it had taken nearly one hundred thirty years to exterminate from Japan. From the Catholic perspective of course, things looked a little different. No one knows for sure how many Japanese people had converted, and by the early seventeenth century some Christian families were into their third generation of believers, but the faith was virtually pervasive in the parts of Kyushu where the Jesuits had had their greatest success. Some estimates put the number of Catholics at over three hundred thousand, around 2 percent of the estimated total Japanese population of the time. Some Jesuits, particularly Japanese ones, had gone underground to minister to their flock. The hunt started for them and then leading Catholic citizens came under pressure to set a good example. Anyone found to be assisting the missionaries was also sentenced to death, but could avoid this by apostatizing. This included foreigners as well as Japanese people. To their credit, many resisted and were subsequently tortured and many executed by fire or sword, but large numbers also recanted and stamped on the religious image, the *fumi-e,* or "stamping picture" which was used to test the faith of potential Catholics between the 1620s and 1856. Those who gave up their faith were treated well, to try to persuade the remaining Christians to follow their example. The tortures used were particularly horrific. The most common one was hanging upside down in a pit, sometimes filled with excrement or even snakes. One hand was left free to indicate readiness to recant. The victim was literally left there until they made the hand sign that they had had enough or until they had perished. The patron saint of the Philippines, Saint Lorenzo Ruiz, a Chinese Filipino, did not recant and died in the pit. One Portuguese priest, Cristóvão Ferreira, lasted five hours before making the hand signal in 1633. The worst torture was perhaps being suspended over volcanic sulfur pools until nearly dead, then being revived with cold water before being put through the whole ordeal again. It was no wonder most people were not prepared to suffer such horrors for their faith and stamped on the *fumi-e.*

Africans in China with the Jesuits: As elsewhere, slaves were essential to Portuguese activities in Macao, and a considerable number of these, perhaps around five thousand, were African males. Aside from performing manual labor, Africans in Macao participated in festivals dedicated to them and performed in orchestras and dance troupes in which "they all appeared resplendent in scarlet and other delights." They also performed for fascinated Chinese audiences as part of Jesuit missions into the Chinese interior, playing "fanfares on their trumpets and shawms (a woodwind instrument a bit like an oboe)." One man, Antonio, a Portuguese-Cantonese interpreter, was described as a "Capher Eathiopian Abissen." His ability in the local language suggests that he had been in Southern China for some time. The Catholic priest Domingo Navarette also reported Africans working with and for Chinese employers in Guangzhou, and believed them to have escaped slavery in Macao.

Selected Bibliography

Andrade, Tonio. *Lost Colony: The Untold Story of China's First Great Victory over the West.* Princeton, NJ: Princeton University Press, 2013.

Berry, Mary Elizabeth. *Hideyoshi.* Cambridge, MA: Council on East Asian Studies, Harvard University, 1989.

Boxer, C. R. *Fidalgos in the Far East, 1550–1770.* The Hague: Martinus Nijhoff, 1948.

Cocks, Richard. *Diary of Richard Cocks, Cape-Merchant in the English Factory in Japan, 1615–1622. Volume 1.* Edited by E. M. Thompson. London: The Hakluyt Society, 1883.

Curvelo, Alexandra. *Nanban Folding Screen Masterpieces: Japan-Portugal XVIIth Century.* Paris: Editions Chandeigne, 2015.

De Sousa, Lúcio. *The Early European Presence in China, Japan, The Philippines and Southeast Asia (1555–1590)—The Life of Bartolomeu Landeiro.* Macao: Macao Foundation, 2010.

De Sousa, Lúcio. *Daikokaijidai no nihonjin dorei (Japanese slaves of the Maritime Age).* Tokyo: Chuko Sosho, 2017.

Elison, George. *Deus Destroyed: The Image of Christianity in Early Modern Japan.* Cambridge, MA: Harvard University Press, 1973.

Fujita, Niel. *Japan's Encounter with Christianity: The Catholic Mission in Pre-Modern Japan.* New York: Paulist Press, 1991.

Hawley, Samuel. *The Imjin War: Japan's Sixteenth-Century Invasion of Korea and Attempt to Conquer China.* Conquistador Press, 2014.

Hesselink, Reinier. *The Dream of Christian Nagasaki: World Trade and the Clash of Cultures, 1560–1640.* Jefferson, NC: McFarland & Company, 2016.

Iwai, Miyoji. *Kamei Ryukyunokami (Kamei Lord of Ryukyu)*. Tokyo: Kadokawa, 2006.

Kaufmann, Miranda. *Black Tudors: The Untold Story*. London: Oneworld, 2017.

Nakajima, Gakusho. "Invasion of Korea and Trade with Luzon: Katō Kiyosama's scheme of Luzon trade in the late 16th century." In *The East Asian Mediterranean-Maritime Crossroads of Culture, Commerce, and Human Migration*, edited by Angela Schottenhammer. Wiesbaden: Otto Harrassowitz Verlag, 2008.

Onyeya, Nubia. *Blackamoores: Africans in Tudor England, their Presence, Status and Origins*. London: Narrative Eye, 2013.

Ōta, Gyūichi (J. S. A. Elisonas & J. P. Lamers, Trs. and Eds.). *The Chronicle of Lord Nobunaga*. Leiden, NL: Brill, 2011.

Otsuki, Fumihiko. *Daigenkai*. Tokyo: Fusanbou, 1935.

Rodrigues, João. *Vocabulário da Lingoa de Iapam*. Nagasaki: The Society of Jesus, 1603.

Screech, Timon. "The Black in Japanese Art: From the beginnings to 1850." In *The Image of the Black in African and Asian Art*, edited by David Bindman and Suzanne Preston Blier, 325–340. Cambridge, MA: Harvard University Press, 2017.

Suzuki, Akira. *Ishinzengou—sufinkusu to 34 nin no samurai (Before and After the Meiji Restoration—the Sphinx and the 34 Samurai.)* Tokyo: Shogakukan, 1988.

Turnbull, Stephen. *The Samurai: A Military History*. London: Routledge, 1977.

Chapter 24

Manga and anime: Manga as an art form can trace its history back to the twelfth century and in its premodern form to the eighteenth or early nineteenth century. Manga may represent the first graphic novels of the type we would find recognizable today. By 1995, the manga market in Japan was valued at $6 to $7 billion annually, but it had also managed to make significant strides overseas, with the French market alone valued at around $569 million in 2005. In the rest of Europe and the Middle East, the market was valued at $250 million in 2012. And in 2008, the US and Canada market was valued at $175 million. As such it can be considered one of Japan's most successful cultural exports in the modern age, and this success means that it is not now only a Japanese art form, but also a global one, with artists from every continent taking part. The word *anime* derives from the French word for animation, and refers to the moving picture variety of the comic form. The iconic movies of Miyazaki have traditionally been held to be part of this art form, but Miyazaki himself derided *anime*, describing it as "animators lacking motivation and with mass-produced, overly expressionistic products relying upon a fixed iconography of facial expressions and protracted and exaggerated action scenes but lacking depth and sophistication in that they do not attempt to convey emotion or thought." Anime, however, has produced many worldwide cultural phenomenons such as One Piece, Naruto, Sailor Moon, the Power Rangers and Pokémon.

Daigenkai: *Daigenkai*, the poetically named "Great Word Sea," was probably the first scholarly attempt to scientifically approach the Japanese language in the same manner as was becomingly increasingly common for European languages in the 1880s. *The Oxford English Dictionary* was in development at the same time. It was a project of its time and place in history, when the Japanese world was putting a massive amount of energy into both understanding new disciplines, and also reconciling them with prior traditions to form a balance in society and preserve Japan as Japan and not some mere superficial realm copied from outside influences. This effort was no mere lip service—a whole massive and proud nation of tens of millions of people threw themselves wholeheartedly into seeking this balance, from farmers seeking more nutritious cash crops and meat production in foreign methodology, to artists who created globally iconic and phenomenal movements, to constitutional lawyers creating entirely new and revolutionary legal codes. *Daigenkai*, and its shorter predecessor *Genkai*, was part of this brave new world. It was written by Ōtsuki Fumihiko, who descended from a long line of scholars of "Western" learning and advisors to the Tokugawa shoguns. His father was an expert in gunnery and in his teens, he himself was involved in the cataclysmic Boshin War of 1867/1868 which ushered in the Meiji Imperial Restoration and the end of the Tokugawa shogunate. Ōtsuki was a lexogographer, linguist and grammarian whose modern linguistic studies revolutionised understanding of the Japanese language, turning it from an art to a scientific discipline. Although Ōtsuki paid for the original publication expenses himself, it was soon republished and expanded in commercial editions that went through over a thousand printings. Modeled in part on Western monolingual dictionaries, *Genkai* gave not only basic information about words—their representations in Japanese characters and their definitions in Japanese—but also pronunciations and etymologies and historical citations of their use. Its successor, the four-volume *Daigenkai*, though published under Ōtsuki's name and based in part on his work, appeared some years after his death and was completed by other lexicographers. It is from these epic works that we know of *Kurobo* as an epithet and its origins. And furthermore, the example which gives useful information on Yasuke's story about Shikano in Inaba Province.

Selected Bibliography

Amano, Sumiki. *Momoyama biito toraibu (Momoyama Beat Tribe)*. Tokyo: Shueisha, 2010.

Cardim, Antonio Francisco. *Elogios, e ramalhete de flores borrifado com o sangue dos religiosos da Companhia de Iesu, a quem os tyrannos do imperio do Iappão tiraraõ as vidas por odio da fé catholica* (Praises, and Bouquets of Flowers Sprinkled with Jesuit Blood, to Those Whom the Japanese Tyrants Kill Through Hatred of the Catholic Faith). Lisbon: Manoel da Sylua, 1650.

Crasset, Jean. *Historie de l'eglise du Japon (History of the Japanese Church)*. Paris: Francois Montalent, 1669.

Das, Nandini. "Encounter as Process: England and Japan in the Late Sixteenth Century." *Renaissance Quarterly* 69, (2016): 1343–68.

De Lange, William. *Pars Japonica: The First Dutch Expedition to Reach the Shores of Japan*. Warren, CT: Floating World Editions, 2006.

Endō, Shūsaku. *Koronbo (The Black Man)*. Tokyo: Mainichishinbunsha, 1971.

Fróis Luís. *"Tratado em que se contêm muito sucinta e abreviadamente algumas contradições e diferenças de costumes entre a gente de Europa e esta província de Japão"* (Treatise in which are contained very succinctly and briefly some contradictions and differences of customs between the people of Europe and this province of Japan). In *Kulturgegensaetze Europa–Japan, 1585* edited by Schuette, Josef Franz, 94–266. Tokyo, 1955.

Gill, Robin. *Topsy-Turvey 1585*. Key Biscayne, FL: Paraverse Press, 2005.

Hesselink, Reinier. *The Dream of Christian Nagasaki: World Trade and the Clash of Cultures, 1560–1640.* Jefferson, NC: McFarland & Company, 2016.

Kurusu, Yoshio. *Kurosuke.* Tokyo: Iwasaki Shoten, 1968.

Massarella, Derek. *A World Elsewhere: Europe's Encounter with Japan in the 16th and 17th Centuries.* New Haven, CT: Yale University Press, 1990.

Morimoto, Masahiro. *Matsudaira Ietada niki (Matsudaira Ietada's Diary).* Tokyo: Kadokawa Sensho, 1999.

Murakami Naojiro. *Yasokai shi Nihon tsushin* [Japanese correspondence of Jesuit missionaries] (2 vols.; Tokyo, 1907), vol. 2, pp. 86–87.

Murakami, Naojiro and Yanigaya, Takeo. *Ieszusukai Nihonnenpo jou (Jesuit Reports from Japan, Volume I).* Tokyo: Omatsudo, 2002.

Okazaki, Takashi. *Afro Samurai.* Los Angeles: Seven Seas Entertainment, 2008.

Ōta, Gyūichi (J. S. A. Elisonas & J. P. Lamers, Trs. and Eds.). *The Chronicle of Lord Nobunaga.* Leiden, NL: Brill, 2011.

Otsuki, Fumihiko. *Daigenkai.* Tokyo: Fusanbou, 1935.

Russel, John, G. "The other other: The black presence in the Japanese experience." In *Japan's Minorities: The Illusion of Homogeneity.* Edited by Weiner, Michael, 84–115. Abingdon: Routledge, 2009.

Screech, Timon. "The English and the Control of Christianity in the Early Edo Period." *Japan Review,* no. 24 (2012): 3–40.

The Society of Jesus. *Cartas que os padres e irmãos da Companhia de Jesus escreverão dos reynos de Japão e China II (Letters written by the fathers and brothers of the Society of Jesus from the kingdoms of Japan and China—Volume II)*. Evora, Portugal: Manoel de Lyra, 1598.

Solier, François. *Histoire ecclésiastique des îles et royaume du Japon (The Ecclesiastical History of the Islands and Kingdoms of Japan)*. Paris: Sebastian Cramoisy, 1627.

Tremml-Werner, B. *Spain, China, and Japan in Manila, 1571–1644: Local Comparisons and Global Connections*. Amsterdam: Amsterdam University Press, 2015.

INDEX

Page numbers in *italics* indicate photographs.

A

Adams, William, 355, 360–361, 374

African Americans in Japan, 356, 421–422

Africans, *118, 119*
    acceptance of homosexuality, 208–209
    attire of, 119, 120
    as bodyguards, 350, 351–352
    Japanese attitude toward, 117, 121, 369–370
    Japanese images of, 416
    in Nagasaki, 84
    population of in Japan, 345, 421–422, 430

    represented in comic books, 368–369
    as ship crewmembers, 119–120
    slavery and, 55–59, 79.
    See also slavery
    as soldiers, 79
    Valignano's opinions on, 77–78

Africans in China, 450

African slavery. *See* slavery

Afro Samurai, 375

The Age of the Country at War, 53, 95–96, 140, 141, 196–197, 245, 413

Akbar the Great, 174–175, 177–178, *179*, 180

Akechi, Mitsuhide
  about, 162–163, 251
  aftermath of Nobunaga
    overthrow, 307–308
  and ceremonial preparation,
    269
  death of, 312, 314
  final battle of, 312–314
  mother's execution, 275
  and siege of Tottori Castle, 223
  treason and, 277–278, 279–281,
    283–284, 445–446
  victory of, 301–303
  warring against Mori clan,
    269–271
ama, the, 102, 419–420
anime and manga, 373, 375, *377*,
  377–379, 452–453
animist gods, 72
architecture, early Japanese, 43,
  127, 138, 186–189, 235, 336
Arima, Harunobu
  about, 29, 33–34, 47, 48, 316,
    386–387
  baptism of, 84
  battle against Ryūzōji, 325–330
  control of Shimabara
    peninsula, 333
  and Jesuit cannon, 319–320
  Jesuit conversion of, 48
  marriage of, 49, 51
  mistress and, 49
  and Satsuma clan, 85, 316–317,
    322, 333
  and siege of Hinoe Castle,
    46–48

warring with Ryūzōji clan, 36
Arima, Naozumi, 353
Arima-Satsuma alliance, 316–317,
  322, 333
ashigaru, 196, 197, 437
Ashikaga Dynasty, 428
Ashikaga, Yoshiaki, 140–141, 164,
  428
Asia, slavery in. *See* slavery
assassination attempts on
  Nobunaga, 224–225
attire
  Africans in Japan, *118*, *119*,
    119–120, *121*, 416
  European, 135
  of Hideyoshi, 338
  Japanese, 103, 445
  of Jesuits, 72
  of Nobunaga, 169
  of sailors, 100–101
  of samurai, 198–199, *199*
Azande warriors, 209
Azuchi
  about, 133, 186–189, 391–392
  destruction of, 308–309
  mission in, 185–186
  and New Year celebration,
    235–239
Azuchi Castle, *187*
  about, 182–184
  landslide of, 238

B

Balthazar, 68, *68*, 69, 415
Battle of Adwa, 370

Battle of Dangpo, 345

Battle of Nagashino, *144*, 145, 240

Battle of Okitanawate, 322–329, 331–333, 335, 448

biblical dramas. *See* drama and theatre

bigotry. *See* racism

blackened teeth, 416

black history, contemporary study of, 377, 378–379

black manga, 377–378

black ships, *21*, 27–28, 408

Bowen, Keville A., 376–379

bowing, as cultural practice, 425

Bracey, Marty, 375

breech loaders, 319–320, 321

buccaneers, Portuguese, 350–351

Buddhism, 72, 126, 417

Buddhist temples, destruction of, 48, 114, 135–136, 317

Buddhist warrior monks, 53, 54, 106, 135, 163, 167, 198, 210

buildings. *See* architecture, early Japanese

Bungo (Oita Prefecture), 89, 336–337

burial at sea, 404–405

Bushidō (way of warriors), 195–196

C

Cabral, Antonio, 179–181

Cabral, Francisco, 35–36, 42, 43, 71, 317

Cambridge, Asuka, 373

cannon, 319–320, 323–324, 327, 328

Cardim, Antonio Francisco, 361

Cartas, 358, 361, 364–365

Catholicism/Catholics.

  See also Christianity; Jesuits; missionaries

  Asian missions and, about, 359–360

  attitude toward homosexuality, 209

  conflict with Satsuma clan, 334–335

  expulsion by the Mori clan, 418

  literacy and, 67

  mass execution of, 347–348, *349*

  persecution of, 449

  recanting, 349, 449–450

Cavalcade of Horses, 116, 160, 163, 165–171

Charlevoix, Pierre-François-Xavier de, 362

Chinese language, 41, 62, 74

Chinese lunar calendar, 404

Christendom, turmoil in, 359–360

Christianity.

  See also Catholicism/Catholics

  Africans, understanding of, 77

  ban of, 387

  and blacks, 76–78

  Japanese conversion to, 31

  Ōtomo and, 89

  spread of, 39, 40, 65

  toleration of, 316, 407

Christmas divinity play, 68–69

The Chronicle of Nobunaga (Ōta), 362

civil wars in Japan, 33, 53, 124, 140, 316, 405, 432

clothing. *See* attire

Cocks, Richard, 353

Coelho, Gaspar, 317, 332–333, 334, 336–337, 338, 339, 384

colonialism, end of, 366, 371

comic books, 368, 373, 375, 378, 452–453

commerce. *See* trade

computer gaming, 374–376

Confucian ethics, 66

contemporary media, 373–375

conversions, Jesuit, 42–43

Crasset, Jean, 361

cultural practices
   bowing, 425
   death beauticians, 250, 273, 443
   death poems, 248–249
   divorce and marriage, 417–418
   gift giving, 134, 236–237
   hawking as pastime, 214–216
   head collecting, 242–243, *243,* 252–253, 443

currency, 130, 335–336

D

Daigenkai, 369, 453–454

Daikoku, 168

Daikokuten, 149, 426

daimyō (provincial rulers), 53, 164–165, 197

dairi (divine sovereign), 53, 163, 164, 165

death beauticians, 250, 273, 443

death poems, 248–249

de Brito, Leonel, 28

diet
   Japanese, 86, 264–265, 282
   of Jesuits in Japan, 39, 102
   seafaring, 22–23

Dinka, 55–59, 208–209, 412, 426, 439–440

discrimination. *See* racism

disease, 21, 22, 27, 44, 342, 441

divinity plays, 67, 414–415

divorce and marriage, 417–418

drama and theatre, 67, 414–415

E

Easter Sunday gathering, 110–113

education, 62, 64, 92–93

Erin-ji, destruction of, 254–255

Ethiopia, rise in esteem of, 370–371

Ethiopian royal party, *371*

European weapons, 323

F

Fernandes, Ambrosius, 24, 322

Festival of the Dead *(Obon),* 217–218

fiefs, bestowal of, 257–259

firearms. *See* weapons

Flower Knights, 208

folding screen pictures, 117–121, *118, 119*

food. *See* diet

Fróis, Luís
   about, 381–382
   after overthrow of Nobunaga, 303

at Easter Sunday gathering, 112
greeting Valignano, 35
fusta (Jesuit galley), 322, 328, 332–333, 448

G
gaming, computer, 374
Gamō, Katahide, 309, 310
gift giving, 134, 236–237
Gifu, 174
Goa-Africa slave trade, 430.
See also slavery
government, Japanese methods of, 139–141
Gozen, Dota, 188, 237
Great Saint Nichiren, 136
Great Shining Deity, 255
guns, 48, 280, 300, 319–320, 404, 425–425.
See also weapons
Gyōyu, 273–274
Gyūichi, Ōta, 147, 382

H
Habshi slave soldiers, 79, 175–181, 429, 430
Hachiman, 91
hairstyle of samurai, 199
hakama trousers, 198
Hamzaban, 178–181
Hanzō, Hattori, 311, 387–388
hari-kiri. See seppuku
Harunobu Arima. See Arima, Harunobu
hatamoto (personal attendants), 197, 360

hawking, as pastime, 214–216
head collecting, 242–243, *243*, 252–253, 443
hereditary military dictatorship, 140
Hideyoshi
about, 221–222, 384–385
attack on Hōjō clan, 340
attire of, 338
and ban of slave trade, 339
battle against Akechi, 312–315
claiming Kyushu, 339–340
expansion of power, 315–316, 437
gifts to Nobunaga, 236–237
and Imjin War of 1592–1598, 341–342
and island of Kyushu, 334–335
and the Koreans, 340–341
as Nobunaga's successor, 336
as Oda clan head, 314
plans to invade China, 340–341
siege of Mori castle, 312, *313*
siege of Tottori Castle, 227–229
Tokugawa as vassal of, 334
and Toyotomi clan name, 336
warring against Mori clan, 269
warring against Satsuma clan, 337–338
Hideyoshi (drama), 373
Hijiyama Castle, destruction of, 226
Hino Castle, 310
Hinoe Castle
about, 46, 410–411

under siege, 46–48

Yasuke and Valignano to, 44–46

Hirate, Masahide, 143

Historie de l'eglise du Japon (Crasset), 361

hitatare kimono, 198

Hitoana, 265–266

Hōjō clan, 340

Hōjō, Lady, 248–249

homosexuality, 206–209

Honnō-ji Temple, 95, 132, 136, *137,* 161–162, 271, 276, 424

houchonin (master chefs), 264–265, 445

Hughes, Langston, 371

human trafficking. *See* slavery

Hwarang (Flower Knights), 208

Hyouge Mono, 373

I

Iapam, 407

Ietada, Matsudaira, 261, 355, 364, 382, 435

Iga

ambush of Nobunaga, 232–234, 263

war on, 223–231

warriors of, 223–224

Imai, Sōkyū, 281

Imjin War of 1592–1598, 341, 342

imperialism, 369, 371

India, during Yasuke's time, 174–182

Indians, as Jesuits, 77

Indian slavery, 80.

See also slavery

Indies, about, 41

Ishikawa, Goemon, 224–225

death of, 438, *439*

Islam, 76–77, 209

J

Jaang (the Dinka). *See* the Dinka

Japan.

about, 388–390

civil wars, 33, 53, 124, 140, 316, 405, 432

communication between Korea and China, 74

economic issues, 389–390

Jesuits in, about, 409–410

name of, 407

political rank in, 163–164

sexual practices in, 66, 435–436

in sixteenth century, 52–54

slave trade and. *See* slavery

trade in. *See* trade

and WWII, 372

Japanese world concept, 423

Japão, 407

Jesuit Church in Kyoto, *128*

Jesuits

about, 39–41

and Arima, 318

to Arima camp, 320–321

attire of, 72

cannon to Arima, 319–320

commerce and, 61

control of Nagasaki, 332–333, 338

conversion of Japan, 42–43, 65

drama and theatre, 414–415

expulsion from Japan, 339, 348

Indians as, 77

influence on Takayama, 310

military force of, 61

as missionaries in Japan. *See* missionaries

and Nobunaga overthrow, 307–312

and Ōmura, 25, 333

plots of, 413

and Satsuma clan, 333–334

schools of, 410

secret evangelism of, 348–349

sexual relationships and, 49

Shimazu clan, conflicts with, 70

slavery and. *See* slavery

Yasuke leaving service of, 335

Yasuke speculated as retainer to, 346–350

Jesuits in Japan, 383–384, 388–389, 409–410

"Jezebel the Witch," 89–92, 100, 336–337, 387

K

kaishakunin, 295

Kamei, Korenori, 345

Kameyama Castle, 273

kami (animist gods), 72

kamikaze (divine winds), 245

Kanō School, 117, 422

Kashiwara, 231

Katō, Kiyomasa, 339, 341–345, 342, 387

Kawada, Lord of Sagami, 354

Kawatamachi, 86

Kazusa, 63

Kiso, Lord, 240

Kiyomizu Temple, 426

Knights (Murao), 378

Koei Tecmo, 374

Kuchinotsu Harbor, *27*, 29, 63

kuge (imperial nobility of Kyoto), 163, 165, 428

kunizukushi (cannon), 319–320

Kuro (man from), 344–345

Kurobo, definition of, 342, 369

Kurobo, Yasuke as, 342–345, 387

Kuronbo (Endō), 366–369, 454

Kurosuke (Kurusu), 365–367, 368, 378

Kyoto

about, 124–127, 392, 420–421

ancient nobility of, 428

collapse of centralized power, 197

under Oda rule, 219

Kyoto, Jesuit Church in, *128*

L

Lake Biwa, 173

lascars (Indian workers), 119–120

literacy, 67–69.

See also education

literature

contemporary stories of Yasuke, 365–369

early stories of Yasuke, 358–365

Lodensteyn, Jan Joosten van, 361

M

Macao, 390–391

Manabe, Rokuro, 224–225

manga and anime, 368, 373, 375, 377–379, 452–453

marriage and divorce, 417–418

Martinho, 321, 323–324

marujuji, 323

Masahito, Crown Prince, 281

massed musket volley fire, *144*, 145, 286, 425

mass executions of Catholics, 349

master chefs *(houchonin)*, 264–265, 445

meals. *See* diet

media, contemporary, 373–376

Mexia, Lourenço, 204, 257, 358, 382–383

military dictatorship, hereditary, 140

military slaves, 79.

See also slavery

military supplies. *See* weapons

Minamoto clan, 140, 196

Mirza, Ibrahim Husain, 174–175

missionaries, 359–360.

See also Catholicism/Catholics; Jesuits

arrival in Japan, 31

Jesuits as, 409–410

life of in Japan, 64–65

missionary work

about, 38, 40, 43, 65

and literacy, 67–69

racism and, 76–78

Miyamoto, Ariana, 373

Momoyama Beat Tribe, 374

money. *See* currency

Mongol invasions, 245

monogamy, 49

Mori, Bōmaru, 205–206

Mori, Ranmaru, 205, *206*

about, 435

death of, 295–296

defense of Honnō-ji Temple, 283–288, *287*

lands given to, 259

retreat from Honnō-ji Temple, 289–293

Mori, Rikimaru, 205–206

Mori castle, siege of, 269

Mori clan

about, 97, 191, 392–393

expulsion of Catholics, 418

warring against Akechi, 269–271

warring against Hideyoshi, 269

Mount Fuji

about, 444–445

goddess of, 265–266

movies, 375–376

Mughal conquest, 175, 177–178

munitions. *See* weapons

Murai, Sadakatsu, 281

Murakami, Lord, 97–98

Murakami, Naojiro, 364

Murakami pirates, 100–102

Muromachi period, 428

muzzle-loading cannon, 320

N
Nagasaki
about, 316, 391
growth of, 51, 84–87
Jesuit control of, 84–87, 332–333, 338
and mass execution of Catholics, *349*
Nagasaki Harbor, *85*
naginata, 66–67, 201, 276, 286, 287.
See also weapons
nanban byobu ("screens of southern barbarians"), 118
nanshoku ("way of adolescent boys"), 206–209
Nanzen-ji temple, 72
Nata, the, 90
New Year celebration, 235–237, 239
Nichiren Buddhism, 72, 126
Nijō Palace, 293–294
ninja
about, 83–84, 92, 438–439
Iga warriors as, 223–224
women as, 231–232
Nioh (game), 374–375
Niwa, Nagahide, 166, 186, 277
Nobunaga Concerto, 375
Nobunaga: King of Zipangu, 373
Nobunaga's Ambition, 374
Nōhime (Nobinaga's wife), 188, 219–220, 237, 435

Norquist, Ken, 376
Northeast Africa, 46, *57*, 412
Noshima Island, 103

O
Obon (Festival of the Dead), 217–218
Oda, Hidenobu, 314
Oda, Nobukatsu, 151, 167, 214, 220, 223, 230, 233, 314
Oda, Nobunaga, *141, 366*
about, 53–54, 139, 141–146
to aid Hideyoshi in battle, 276–277
attire of, 168–169
attitude toward Buddhism, 135
Azuchi horseback spectacle, 213
to Azuchi with Yasuke, 172–174
bestowal of fiefs, 257–259
Cavalcade of Horses, 112–113, 116, 168–169
competition of, 191
contemporary popularity of, 447
as *daimyō*, 164–165
death of, 294
defeat of Saika bandits, 244
deposing Yoshiaki, 140–141
execution of Iga residents, 225–226
execution of servants, 210–211
expansion of rule, 219–229
family of, 219–220
governing style, 256–257

as guest of Tokugawa Ieyasu, 259–268

hawking, pastime of, 214–216

Honnō-ji Temple and, 278–279, 283–288, *287,* 289–293

Iga ambush on, 232–234, 263

intelligence of, 133, 142, 154

Jesuits and, 134–135

meeting with Valignano, 112

meeting Yasuke, 138–139, 148–155

and New Year celebration, 235–239

ninja assassination attempts on, 224–225

and Ranmaru, about, 435

seminary visit of, 234–235

severed head of, 447

sexual practices of, 207, 435–436

and siege of Tottori Castle, 222, 225, 226–229

sumo tournament, 212–213

in Suwa, 255–256

Takeda as competitor, 191

tea ceremony and, 281–282

territorial government of, 256

Tokugawa, expression of loyalty by, 268–271

traditions of, 203

treason of Akechi. *See* Akechi, Mitsuhide, treason and

warring of, 219–230

wife of. *See* Nōhime

Yasuke and. *See* Yasuke

Oda, Nobutada

about, 151

attack on, 298–300

and attack on Honnō-ji Temple, 293–295

death of, 300

invasion of Takeda lands, 240–241

against Katsuyori, 243, 244, 246–248

Oda, Nobutaka, 151, 220, 314, 315–316

Oda mokkou crest, 130, 192

Ōgimachi, 165, 281, 427–428, 446

Ogura, Jingorou, 205–206

Ogura, Matsuju, 205–206

Ohaguro (blackened teeth), 416

Oichi, 188, 220, 237, 464

Ōmura, Sumitada

about, 50–51, 333, 387

conflicts with Arima, 87

and the Jesuits, 25

Nagasaki gifted to Jesuits, 51–52

and Valignano, 52

Organtino, Gnecchi Soldo, 129, 133

about, 384

at Azuchi Castle, 185–186

and Nobunaga overthrow, 308, 309–311

seminary visit by Nobunaga, 234–235

and Yasuke meeting Nobunaga, 138–139, 148, 153, 155

Oromo, male-to-male relationships of, 209

Osaka Castle, 336–337
Ōtomo, Sōrin, 336–337
    about, 89–90, 387
    castle of, 92–93
    and Murakami pirates, 100
    and Nobunaga, 100
    wife of, 90–92
Otsuka, Mataichiro, 205–206

P
pearl divers, 102, 419–420
Pennsylvania Humanities Council, 376
pictures, folding screen, 117–121, *118, 119*
pirates, 95–97, 100–102, 351–353, 419
poems, death, 248–249
political rank in Japan, 163–164
polygamy, 344, 417, 435
Portuguese, about, 181–182
Portuguese buccaneers, 350–351
prejudice, 43, 75–78, 77, 368, 369–370, 372–373
printing press, 413
prostitution, 415–416.
    See also sexual relationships
provincial rulers, 53, 164–165, 197

R
racism, 43, 75–78, 77, 368, 369–370, 372–373
rakuichi rakuza ("free markets and open guilds"), 256–257
Ranmaru. *See* Mori, Ranmaru
Reformation, the, 359–360

religion, 29, 67, 72, 126, 198, 308, 316, 391, 409
renga, 274–275, 277
Rin School, 120, 344
ritual suicid. *See* seppuku
rocket launchers, 419
Rokkaku, Yoshikata, 254–255
role-playing games, 374
ronin, 86, 200, 433
Russo-Japanese War, 370
Ryusa, Konishi, 105
Ryūzōji, Takanobu
    about, 29, 316, 325
    battle against Arima, 36, 85, 326–330
    warring against Arima-Satsuma alliance, 322

S
Saika bandits, defeat of, 244
sailors
    about, 350
    attire of, 100–101
    as short-term hires, 405
Sakai, *105, 107*
    about, 104, 392, 432
    under Oda rule, 219
sampan boats, 44–45, *45*
samurai, *199*
    about, 32, 47, 194–197, 432
    as hereditary caste, 356
    loyalty of, 200
    myths of, 266
    role of, 204
    sexual practices of, 206–209

training of, 211

women as, 200–202, 247

Satomura, Jōha, 273–274

Satsuma clan

about, 316, 392–393

and Arima, 316–317

conflict with Catholics, 85, 92, 334–335

control of Jesuits, 333–334

and Kyushu, 332

warring against Hideyoshi, 337–338

schools and seminaries, 64, 72, 92–93, 185–186

sea, burial at, 404–405

sea divers, women, 102, 419–420

Sea Lords

about, 95–97, 100–102, 197, 418

attacks by, 419

sea travel, 20–22, 69

segregation, 373

seiza (kneeling position), 159

seminaries and schools, 64, 72, 92–93, 185–186

seppuku, 143, 144, 200, 220, 254, 293, 294–295, 312, 315, 387, 423–424

severed heads. *See* head collecting

sexual relationships, 49, 66, 206–209, 344, 417, 435–436

shigeshoshi (death beauticians), 250, 273, 443

Shimabara garrison, attempted siege of, 317

Shimazu, Iehisa, 322, 327

Shimazu, Yoshihisa, 322

Shimazu clan, 70, 448

Shingon Buddhism, 126

shinobi, 83–84

Shinpu, destruction of, 247–248

Shinto shrines, destruction of, 48, 114

shogunate, about, 163–164

shoguns, about, 52–53, 140

shudo ("way of adolescent boys"), 206–209

shugo (governors), 164

Sidotti, Giovanni Battista, 362

silk trade, 181

slavery, 54, 55–59, 62, 78, 79–82, 175–181, 339, 351–352, 379, 429–430

social activism, 372

sohei (warrior monks), 53, 54, 106, 135, 145, 163, 167, 198, 210

Spheres of Influence, 374

stereotypes. *See* racism

sumo wrestlers, 212, 213, 285, 363, *363*

Surat, 175

Suwa, Lady, 247

Suwa, town of, 255

T

Taiga dramas, 373

taiko (war drums), 218, 326

Taira-Minamoto War (Genpei War), 195

Taitei no Ken, 375

Takahashi, Toramatsu, 205–206, 293

Takamatsu, siege of, 269, *313*

Takatō Castle, taking of, 246–247

Takatsuki, 110–111

Takayama, Ukon
  about, 104–105, 109–110, 150,
    251, 348, 385–386
  battle against Akechi, 312–313
  and Easter Sunday gathering,
    111–112
  Jesuit influence on, 310
  and Nobunaga, 112
  and overthrow of Nobunaga,
    277, 280–281
  and siege of Tottori Castle,
    223, 226

Takeda, Katsuyori, 145, 240, 245
  competitor of Oda, Nobunaga,
    191
  defeat of, 252
  suicide of, 250
  warring with Nobutada, 246–
    248

Takeda clan
  last stand of, 249, 249–250
  as threat to Nobunaga, 240
  war against, 242–250

Tamaki, Mitsuya, 354, 355, 356, 357

tax code in Japan, 195

tea ceremony, 237, 276, 281–282

Teen Reading Lounge, 376, 377

teeth, blackened, 416

temples, destruction of. See
  Buddhist temples, destruction of

Tenjiku, 34

tenka fubu, 145, 270

Tenkyū, 253–254

theatre and drama, 67, 414–415

Tokugawa, Ieyasu
  about, 334, 385, 444
  aftermath of Nobunaga
    overthrow, 311–312
  loyalty to Nobunaga, 258, 267,
    268–271
  Nobunaga guest of, 260–268

Tomé, 354

tono (lord), 47

torture of Catholics, 349

Tottori Castle, 391
  ghost story, 437–438
  siege of, 222–223, 227–228

Toyama, Lady, 248

Toyotomi (Hideyoshi), 336

trade
  Japanese, 50, 86, 102, 104–105,
    256–257, 405–407
  Portuguese, 181

trafficking, human. See slavery

travel by sea, 20–22, 69

U

Uesugi, Kenshin, 83, 163, 355

umazoroe (Cavalcade of Horses),
  116, 160, 163, 165–171

Uwai, Kakken, 333, 334–335

V

Valignano, Alessandro, 40
  about, 23, 33, 39, 42, 383–384
  arrival in Japan, 19–20, 23–24,
    29–37
  in Azuchi, 185, 216–217
  and Catholic seminary in
    Kazusa, 63–64

and Easter Sunday gathering, 110–113

education and, 62

establishment of mission, 71

gunrunning and, 61

to Hinoe Castle, 44–46

on homosexuality, 207

on Japanese language, 72–74

to Kyoto, 99–103

meeting with Nobunaga, 112, 156

in Nagasaki, 84–88

on non-European languages and customs, 62

and Ōmura Sumitada, 52

opinions on Africans, 75–78

relationship to Rome, 61–63

role of in Japan, 41

in Sakai, 104–109

slavery and, 61, 78

success of, 88

in Utsuki, 91–92

as Visitor to the Indies, 41

Yasuke as bodyguard to, 20–21, 30, 60–61, 69–70, 78, 82

Ventura, 349–350

video games, 374, 375

the Visitor. *See* Valignano, Alessandro

volley fire, musket, *144,* 145, 286, 425

W

wabisabi (and tea culture), 237, 276

war drums, 218, 326

warfare

guns. *See* guns

in India, 175, 177–179

massed musket volley fire, *144,* 145, 286, 425

rocket launchers, 419

Yasuke's experience with. *See* Yasuke

warlords, about, 53

warlords, individual, 384–388

warrior monks, 53, 54, 106, 135, 145, 163, 167, 197–198, 210

warriors, Azande, 209

warriors, Japanese, 32

warriors, of Iga, 223–224

warriors, way of *(Bushidō),* 195–196

"way of adolescent boys," 206–209

"the way of tea," 237, 276, 281–282

weapons, 48, 50, 52, 66–67, 101, 178, 190–191, 280, 404

breech loaders, 319–320, 321

cannon, 319–320, 323–324, 327

European, 216, 323

guns, 48, 280, 300, 319–320, 404, 425

Japanese, 323

of Jesuits, 61

manufacture of, 104, 216–217

musket volley fire, *144,* 145, 286, 425

naginata, 66–67, 201, 276, 286, 287

rocket launchers, 419

trident, 120

The Witch of Bungo. *See* "Jezebel the Witch"

women

ninja as, 231–232

pearl divers, 102, 419–420

as samurai, 200–202, 247

women in Japan, attitude toward, 66, 417–417

writing box, 120, *121*, 344

WWII and the Japanese, 372

X

Xavier, Francis, 76, 409

Y

yakata (Nobunaga's palace), 188–189

Yasuke, *363, 367, 378*

about, 25–26, 34

as African spectacle in Kyoto, 122–124, 127–131

after Nobunaga overthrow, 311, 314–315, 316

to aid Hideyoshi in battle, 276–277

arrival in Japan, 23–24, 29–38

to Azuchi, 172–174

at Azuchi Castle, 184–185

in battle against Ryūzōji, 325

becomes samurai, 192–193

birthplace of, *57*, 153

as bodyguard to Nobunaga, 189–191

as bodyguard to Valignano, 20–21, 30, 60–61, 69–70, 78, 82

and cannon expertise, 319–320, 321, 323–325

in Cavalcade of Horses, 167–168

in contemporary comics, 373, 375, 377

in contemporary media, 365–369, 373–375

and death of Nobunaga, 294

defense of Honnō-ji Temple, 283–288, 289–303

derivation of name, 403–404

and divinity plays, 67–69

early stories of, 361–365

and Easter Sunday gathering, 110–113

enslavement of, 55–59, 79–80, 174–182, 227

evidence of, 358–359

experience with warfare, 175, 177–179, 190–191

fleeing to Nobutada, 296–299

gifted to Arima, 318

gifted to Nobunaga, 156, 160, 426

in Goa, 181–182

as *hatamoto*, 197

hawking, participant in, 214–216

to Hinoe Castle, 44–46

at Honnō-ji Temple, 278–279

Iga ambush on, 232–234

and Japanese language, 63, 74

joining Nobutada's forces with Nobunaga, 250–251

in Kasuza, 63

to Kyoto, 99–103

leisure time of, 65–66, 86

meeting Nobunaga, 131, 138–139, 148–155

and mission life, 64–65

in Nagasaki, 84–88

and New Year's horsemanship, 239

observations of the Japanese, 32

origins of, 411–413

personality of, 204

as Portuguese oarsman, 181

and possibility of being a ruler, 257–258

retreat from Honnō-ji Temple, 289–293

return to Jesuits of, 301–303

in Sakai, 104–109

as samurai, 202–203, 204–205, 356–357

spiritual life of, 439–441

stories of, 358–369

summoned by Nobunaga, 113–114

in sumo tournament, 212–213

surrender of, 300–301

in Suwa, 255–256

training Japanese militia, 85–86

training of, 66–67

travel to Japan, 19–20

in Utsuki, 91–92

Yasuke, speculations about after Ryūzōji battle

about, 331–332

as ancestor of Tamaki, 354, 357

as Chinese pirate, 351–353

descendants of, 355–356

as employed by a Tokugawa follower, 355

employment value of, 335

as Jesuit muscle, 346–350

as Kurobo, 342–345

leaving service of Jesuits, 335

as part of Coelho's military force, 332–333

as Portuguese buccaneer, 350–351

as retainer to Katō, 342

as retainer to Naozumi, 353

as sailor, 350

wealth of, 335–336

working for Kamei, 345

Yellow Peril, 370

Yohoken, Paulo, 64–65

Yorozu, Genkuro, 376

Young, 375

Z

Zen Buddhism, 72, 126

Zheng, Zhilong, 352–353